Trainee Guide

Electrical

Level 1

National Center for Construction Education and Research

Wheels of Learning
Standardized Craft Training

Prentice Hall
Upper Saddle River, New Jersey   Columbus, Ohio

Prentice-Hall, Inc.
A Simon & Schuster Company
Upper Saddle River, New Jersey 07458

This information is general in nature and intended for training purposes only. Actual performance of activities described in this manual requires compliance with all applicable operations procedures under the direction of qualified personnel. References in this manual to patented or proprietary devices do not constitute a recommendation for their use.

Printed in the United States of America

10 9 8 7 6 5 4 3 2 1

ISBN: 0-13-085875-7

Prentice-Hall International (UK) Limited, *London*
Prentice-Hall of Australia Pty. Limited, *Sydney*
Prentice-Hall of Canada, Inc., *Toronto*
Prentice-Hall Hispanoamericana, S. A., *Mexico*
Prentice-Hall of India Private Limited, *New Delhi*
Prentice-Hall of Japan, Inc., *Tokyo*
Simon & Schuster Asia Pte. Ltd., *Singapore*
Editora Prentice-Hall do Brasil, Ltda., *Rio de Janeiro*

This volume is one of many in the *Wheels of Learning* craft training program. This program, covering more than 20 standardized craft areas, including all major construction skills, was developed over a period of years by industry and education specialists. Sixteen of the largest construction and maintenance firms in the U.S. committed financial and human resources to the teams that wrote the curricula and planned the national accredited training process. These materials are industry-proven and consist of competency-based textbooks and instructor guides.

The *Wheels of Learning* was developed by the National Center for Construction Education and Research in response to the training needs of the construction and maintenance industries. The NCCER is a nonprofit educational entity affiliated with the University of Florida and supported by the following industry and craft associations:

Partnering Associations

- ABC Texas Gulf Coast Chapter
- American Fire Sprinkler Association
- American Society for Training and Development
- American Vocational Association
- American Welding Society
- Army Corps of Engineers
- Associated Builders and Contractors, Inc.
- Associated General Contractors of America
- Association for Career and Technical Education
- The Business Roundtable
- Carolinas AGC, Inc.
- Carolinas Electrical Contractors Association
- Construction Industry Institute
- Design-Build Institute of America
- Merit Contractors Association of Canada
- Metal Building Manufacturers Association
- National Association of Minority Contractors
- National Association of Women in Construction
- National Insulation Association
- National Ready Mixed Concrete Association
- National Utility Contractors Association
- National Vocational Technical Honor Society
- North American Crane Bureau
- Painting and Decorating Contractors of America
- Portland Cement Association
- Steel Erectors Association of America
- University of Florida
- Vocational Industrial Clubs of America
- Women Construction Owners and Executives, USA

Some of the features of the *Wheels of Learning* program include:

- A proven record of success over many years of use by industry companies.
- National standardization providing "portability" of learned job skills and educational credits that will be of tremendous value to trainees.
- Recognition: upon successful completion of training with an accredited sponsor, trainees receive an industry-recognized certificate and transcript from the NCCER.
- Each level meets or exceeds Bureau of Apprenticeship and Training (BAT) requirements for related classroom training (CFR 29:29).
- Well illustrated, up-to-date, and practical information. All standardized manuals are reviewed annually in a continuous improvement process.

Acknowledgments

This manual would not exist were it not for the dedication and unselfish energy of those volunteers who served on the Technical Review Committees. A sincere thanks is extended to the:

1996 – 1999 Technical Review Committee

Bane Allman
Mike Basham
Jerry Bass
Michael E. Dahl
Gary I. Edgington
Timothy A. Ely
Al Hamilton
Don Hostetler
E. L. Jarrell
L.J. LeBlanc
Jim Mitchem
Robert Mueller
Christine Porter
Mike Powers
Rob Soileau
Joe Sullivan
Sherry Yarbrough

1992 – 1996 Technical Review Committee

Bane Allman
S.A. ("Barney") Barnette
John Brayshaw
Gary I. Edgington
Timothy A. Ely
Al Hamilton
Allen A. Hill
Don Hostetler
L.J. LeBlanc
Daniel C. Miller
Jim Mitchem
Bob Mueller
Burton Smith

Contents

Electrical Safety

Module 26101

NATIONAL
CENTER FOR
CONSTRUCTION
EDUCATION AND
RESEARCH

ELECTRICAL SAFETY

OBJECTIVES

Upon completion of this module, the trainee will be able to:

1. Demonstrate safe working procedures in a construction environment.
2. Explain the purpose of OSHA and how it promotes safety on the job.
3. Identify electrical hazards and how to avoid or minimize them in the workplace.
4. Explain safety issues concerning lockout/tagout procedures, personal protection using assured grounding and isolation programs, confined space entry, respiratory protection, and fall protection systems.

Prerequisites

Successful completion of the following Task Modules is required before beginning study of this Task Module: Core Curricula.

Required Trainee Materials

1. Trainee Task Module
2. Copy of the latest edition of the *National Electrical Code*
3. *OSHA Electrical Safety Guidelines* (pocket guide)
4. Appropriate Personal Protective Equipment

Note: The designations "National Electrical Code," "NE Code," and "NEC," where used in this document, refer to the *National Electrical Code®*, which is a registered trademark of the National Fire Protection Association, Quincy, MA. *All National Electrical Code (NEC) references in this module refer to the 1999 edition of the NEC.*

Course Map

This course map shows all of the task modules in the first level of the Electrical curricula. The suggested training order begins at the bottom and proceeds up. Skill levels increase as a trainee advances on the course map. The training order may be adjusted by the local Training Program Sponsor.

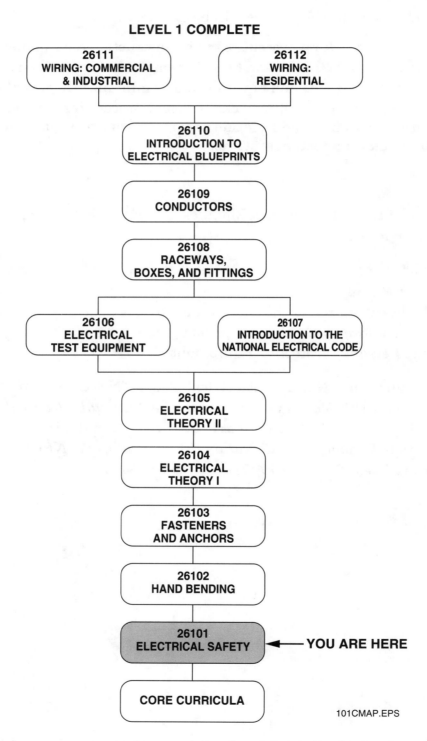

LEVEL 1 COMPLETE

26111
WIRING: COMMERCIAL & INDUSTRIAL

26112
WIRING: RESIDENTIAL

26110
INTRODUCTION TO ELECTRICAL BLUEPRINTS

26109
CONDUCTORS

26108
RACEWAYS, BOXES, AND FITTINGS

26106
ELECTRICAL TEST EQUIPMENT

26107
INTRODUCTION TO THE NATIONAL ELECTRICAL CODE

26105
ELECTRICAL THEORY II

26104
ELECTRICAL THEORY I

26103
FASTENERS AND ANCHORS

26102
HAND BENDING

26101
ELECTRICAL SAFETY ◄— **YOU ARE HERE**

CORE CURRICULA

101CMAP.EPS

TABLE OF CONTENTS

TABLE OF CONTENTS (Continued)

Trade Terms Introduced In This Module

Double-insulated/ungrounded tool: An electrical tool that is constructed so that the case is insulated from electrical energy. The case is made of a nonconductive material.

Fibrillation: Very rapid irregular contractions of the muscle fibers of the heart that result in the heartbeat and pulse going out of rhythm with each other.

Grounded tool: An electrical tool with a three-prong plug at the end of its power cord or some other means to ensure that stray current travels to ground without passing through the body of the user. The ground plug is bonded to the conductive frame of the tool.

Ground fault circuit interrupter (GFCI): A protective device that functions to deenergize a circuit or portion thereof within an established period of time when a current to ground exceeds some predetermined value that is less than that required to operate the overcurrent protective device of the supply circuit.

Polychlorinated biphenyls (PCBs): Toxic chemicals that may be contained in liquids used to cool certain types of large transformers and capacitors.

1.0.0 INTRODUCTION

As an electrician, you will be exposed to many potentially hazardous conditions which will exist on the job site. No training manual, set of rules and regulations, or listing of hazards can make working conditions completely safe. However, it is possible for an electrician to work a full career without serious accident or injury. To reach this goal, you need to be aware of potential hazards and stay constantly alert to these hazards. You must take the proper precautions and practice the basic rules of safety. You must be safety-conscious at all times. Safety should become a habit. Keeping a safe attitude on the job will go a long way in reducing the number and severity of accidents. Remember that your safety is up to you.

As an apprentice electrician, you need to be especially careful. You should only work under the direction of experienced personnel who are familiar with the various job site hazards and the means of avoiding them.

The most life-threatening hazards on a construction site are:

- Falls when you are working in high places
- Electrocution caused by coming into contact with live electrical circuits
- The possibility of being crushed by falling materials or equipment
- The possibility of being struck by flying objects or moving equipment/vehicles such as trucks, forklifts, and construction equipment

Other hazards include cuts, burns, back sprains, and getting chemicals or objects in your eyes. Most injuries, both those that are life-threatening and those that are less severe, are preventable if the proper precautions are taken.

2.0.0 ELECTRICAL SHOCK

Electricity can be described as a potential that results in the movement of electrons in a conductor. This movement of electrons is called *electrical current*. Some substances, such as silver, copper, steel, and aluminum, are excellent conductors. The human body is also a conductor. The conductivity of the human body greatly increases when the skin is wet or moistened with perspiration.

Electrical current flows along the path of least resistance to return to its source. The source return point is called the *neutral* or *ground* of a circuit. If the human body contacts an electrically energized point and is also in contact with the ground or another point in the circuit, the human body becomes a path for the current to return to its source. *Table 1* shows the effects of current passing through the human body. One mA is one milliamp, or one one-thousandth of an ampere.

Current Value	Typical Effects
Less than 1mA	No sensation.
1 to 20mA	Sensation of shock, possibly painful. May lose some muscular control between 10 and 20mA.
20 to 50mA	Painful shock, severe muscular contractions, breathing difficulties.
50 to 200mA	Same symptoms as above, only more severe, up to 100mA. Between 100 and 200mA, ventricular fibrillation may occur. Typically results in almost immediate death unless special medical equipment and treatment are available.
Over 200mA	Severe burns and muscular contractions. The chest muscles contract and stop the heart for the duration of the shock.

Table 1. Current Level Effects On The Human Body

A primary cause of death from electrical shock is when the heart's rhythm is overcome by an electrical current. Normally, the heart's operation uses a very low level electrical signal to cause the heart to contract and pump blood. When an abnormal electrical signal, such as current from an electrical shock, reaches the heart, the low level heartbeat signals are overcome. The heart begins twitching in an irregular manner and goes out of rhythm with the pulse. This twitching is called **fibrillation**. Unless the normal heartbeat rhythm is restored using special defibrillation equipment (paddles), the individual will die. No known case of heart fibrillation has ever been corrected without the use of defibrillation equipment by a qualified medical practitioner. Other effects of electrical shock may include immediate heart stoppage and burns. In addition, the body's reaction to the shock can cause a fall or other accident. Delayed internal problems can also result.

Note: Electric shocks or burns are a major cause of accidents in our industry. According to the Bureau of Labor Statistics, electrical shock is the leading cause of death in the electrical industry.

2.1.0   THE EFFECT OF CURRENT

The amount of current measured in amperes that passes through a body determines the outcome of an electrical shock. The higher the voltage, the greater the chance for a fatal shock. In a one-year study in California, the following results were observed by the State Division of Industry Safety:

- Thirty percent of all electrical accidents were caused by contact with conductors. Of these accidents, 66% involved low-voltage conductors [those carrying 600 volts (V) or less].
- Portable, electrically-operated hand tools made up the second largest number of injuries (15%). Almost 70% of these injuries happened when the frame or case of the tool became energized. These injuries could have been prevented by following proper safety practices, using grounded or double-insulated tools, and using **ground fault circuit interrupter (GFCI)** protection.

In one ten-year study, investigators found 9,765 electrical injuries that occurred in accidents. Over 18% of these injuries involved contact with voltage levels of over 600 volts. A little more than 13% of these high-voltage injuries resulted in death. These high-voltage totals included limited-amperage contacts, which are often found on electronic equipment. When tools or equipment touch high-voltage overhead lines, the chance that a resulting injury will be fatal climbs to 28%. Of the low-voltage injuries, 1.4% were fatal.

CAUTION: High voltage, defined as 600 volts or more, is almost ten times as likely to kill as low voltage. However, on the job you spend most of your time working on or near lower voltages. Due to the frequency of contact, most electrocution deaths actually occur at low voltages. Attitude about the harmlessness of lower voltages undoubtedly contributes to this statistic.

These statistics have been included to help you gain respect for the environment where you work and to stress how important safe working habits really are.

2.1.1   Body Resistance

Electricity travels in closed circuits, and its normal route is through a conductor. Shock occurs when the body becomes part of the electric circuit (*Figure 1*). The current must enter the body at one point and leave at another. Shock normally occurs in one of three ways: the person must come in contact with both wires of the electric circuit; one wire of the electric circuit and the ground; or a metallic part that has become *hot* by being in contact with an energized wire while the person is also in contact with the ground.

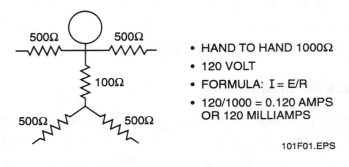

- HAND TO HAND 1000Ω
- 120 VOLT
- FORMULA: $I = E/R$
- 120/1000 = 0.120 AMPS OR 120 MILLIAMPS

101F01.EPS

Figure 1. Body Resistance

To fully understand the harm done by electrical shock, we need to understand something about the physiology of certain body parts: the skin, the heart, and muscles.

- Skin covers the body and is made up of three layers. The most important layer, as far as electric shock is concerned, is the outer layer of dead cells referred to as the *horny layer*. This layer is composed mostly of a protein called *keratin*, and it is the keratin that provides the largest percentage of the body's electrical resistance. When it is dry, the outer layer of skin may have a resistance of several thousand ohms, but when it is moist, there is a radical drop in resistance, as is also the case if there is a cut or abrasion that pierces the horny layer. The amount of resistance provided by the skin will vary widely from individual to individual. A workman with a thick horny layer will have a much higher resistance than a child. The resistance will also vary widely at different parts of the body. For instance, the workman with high-resistance hands may have low-resistance skin on the back of his calf. The skin, like any insulator, has a breakdown voltage at which it ceases to act as a resistor and is simply *punctured*, leaving only the lower-resistance body tissue to impede the flow of current in the body. The breakdown voltage will vary with the individual, but is in the area of 600V. Since most industrial power distribution systems operate at 480V or higher, technicians working at these levels need to have special awareness of the shock potential.

- The heart is the pump that sends life-sustaining blood to all parts of the body. The blood flow is caused by the contractions of the heart muscle, which is controlled by electrical impulses. The electrical impulses are delivered by an intricate system of nerve tissue with built-in timing mechanisms which make the chambers of the heart contract at exactly the right time. An outside electric current of as little as 75 milliamperes can upset the rhythmic, coordinated beating of the heart by disturbing the nerve impulses. When this happens, the heart is said to be in *fibrillation,* and the pumping action stops. Death will occur quickly if the normal beat is not restored. Remarkable as it may seem, what is needed to defibrillate the heart is a shock of an even higher intensity.

- The other muscles of the body are also controlled by electrical impulses delivered by nerves. Electric shock can cause loss of muscular control, resulting in the inability to let go of an electrical conductor. Electric shock can also cause injuries of an indirect nature in which involuntary muscle reaction from the electric shock can cause bruises, fractures, and even deaths resulting from collisions or falls.

Severity of shock – The severity of shock received when a person becomes a part of an electric circuit is affected by three primary factors: the amount of current flowing through the body (measured in amperes), the path of the current through the body, and the length of time the body is in the circuit. Other factors which may affect the severity of the shock are the frequency of the current, the phase of the heart cycle when shock occurs, and the general health of the person prior to the shock. Effects can range from a barely perceptible tingle to immediate cardiac arrest. Although there are no absolute limits, or even known values that show the exact injury at any given amperage range, *Table 1* lists the general effects of electric current on the body for different current levels. As this table illustrates, a difference of only 100 milliamperes exists between a current that is barely perceptible and one that can kill.

A severe shock can cause considerably more damage to the body than is visible. For example, a person may suffer internal hemorrhages and destruction of tissues, nerves, and muscle. In addition, shock is often only the beginning in a chain of events. The final injury may well be from a fall, cuts, burns, or broken bones.

2.1.2 Burns

The most common shock-related injury is a burn. Burns suffered in electrical accidents may be of three types: electrical burns, arc burns, and thermal contact burns.

- Electrical burns are the result of electric current flowing through the tissues or bones. Tissue damage is caused by the heat generated by the current flow through the body. An electrical burn is one of the most serious injuries you can receive, and should be given immediate attention. Since the most severe burning is likely to be internal, what may appear at first to be a small surface wound could, in fact, be an indication of severe internal burns.

- Arc burns make up a substantial portion of the injuries from electrical malfunctions. The electric arc between metals can be up to 35,000°F, which is about four times hotter than the surface of the sun. Workers several feet from the source of the arc can receive severe or fatal burns. Since most electrical safety guidelines recommend safe working distances based on shock considerations, workers can be following these guidelines and still be at risk from arc. Electric arcs can occur due to poor electrical contact or failed insulation. Electrical arcing is caused by the passage of substantial amounts of current through the vaporized terminal material (usually metal or carbon).

CAUTION: Since the heat of the arc is dependent on the short circuit current available at the arcing point, arcs generated on 480V systems can be just as dangerous as those generated at 13,000V.

- The third type of burn is a thermal contact burn. It is caused by contact with objects thrown during the blast associated with an electric arc. This blast comes from the pressure developed by the near-instantaneous heating of the air surrounding the arc, and from the expansion of the metal as it is vaporized. (Copper expands by a factor in excess of 65,000 times in boiling.) These pressures can be great enough to hurl people, switchgear, and cabinets considerable distances. Another hazard associated with the blast is the hurling of molten metal droplets, which can also cause thermal contact burns and associated damage. A possible beneficial side effect of the blast is that it could hurl a nearby person away from the arc, thereby reducing the effect of arc burns.

3.0.0   REDUCING YOUR RISK

There are many things that can be done to greatly reduce the chance of receiving an electrical shock. Always comply with your company's safety policy and all applicable rules and regulations, including job site rules. In addition, the Occupational Safety and Health Administration (OSHA) publishes the *Code of Federal Regulations (CFR)*. CFR Part 1910 covers the OSHA standards for general industry and CFR Part 1926 covers the OSHA standards for the construction industry.

Do not approach any electrical conductors closer than indicated in *Table 2* unless you are sure they are deenergized and your company has designated you as a qualified individual. Also, the values given in the table are *minimum* safe clearance distances; if you already have standard distances established, these are provided only as supplemental information. These distances are listed in CFR 1910.333/1926.416.

Voltage Range (Phase-to-Phase)	Minimum Approach Distance
300V and less	Avoid contact
Over 300V, not over 750V	1 ft. 0 in. (30.5 cm)
Over 750V, not over 2kV	1 ft. 6 in. (46 cm)
Over 2kV, not over 15kV	2 ft. 0 in. (61 cm)
Over 15kV, not over 37kV	3 ft. 0 in. (91 cm)
Over 37kV, not over 87.5kV	3 ft. 6 in. (107 cm)
Over 87.5kV, not over 121kV	4 ft. 0 in. (122 cm)
Over 121kV, not over 140kV	4 ft. 6 in. (137 cm)

Table 2. Approach Distances For Qualified Employees – Alternating Current

3.1.0   PROTECTIVE EQUIPMENT

You should also become familiar with common personal protective equipment. In particular, know the voltage rating of each piece of equipment. Rubber gloves are used to prevent the skin from coming into contact with energized circuits. A separate leather cover protects the rubber glove from punctures and other damage (see *Figure 2*). OSHA addresses the use of protective equipment, apparel, and tools in CFR 1910.335(a). This article is divided into two sections: *Personal Protective Equipment* and *General Protective Equipment and Tools*.

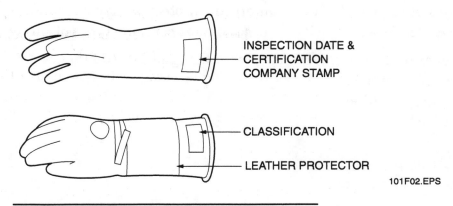

Figure 2. Rubber Gloves And Leather Protector

The first section, *Personal Protective Equipment*, includes the following requirements:

- Employees working in areas where there are potential electrical hazards shall be provided with, and shall use, electrical protective equipment that is appropriate for the specific parts of the body to be protected and for the work to be performed.

- Protective equipment shall be maintained in a safe, reliable condition and shall be periodically inspected or tested, as required by CFR 1910.137/1926.95.

- If the insulating capability of protective equipment may be subject to damage during use, the insulating material shall be protected.

- Employees shall wear nonconductive head protection wherever there is a danger of head injury from electric shock or burns due to contact with exposed energized parts.

- Employees shall wear protective equipment for the eyes and face wherever there is danger of injury to the eyes or face from electric arcs or flashes or from flying objects resulting from electrical explosion.

The second section, *General Protective Equipment and Tools*, includes the following requirements:

- When working near exposed energized conductors or circuit parts, each employee shall use insulated tools or handling equipment if the tools or handling equipment might make contact with such conductors or parts. If the insulating capability of insulated tools or handling equipment is subject to damage, the insulating material shall be protected.

- Fuse handling equipment, insulated for the circuit voltage, shall be used to remove or install fuses when the fuse terminals are energized.

- Ropes and handlines used near exposed energized parts shall be nonconductive.

- Protective shields, protective barriers, or insulating materials shall be used to protect each employee from shock, burns, or other electrically-related injuries while that employee is working near exposed energized parts that might be accidentally contacted or where dangerous electric heating or arcing might occur. When normally enclosed live parts are exposed for maintenance or repair, they shall be guarded to protect unqualified persons from contact with the live parts.

The types of electrical safety equipment, protective apparel, and protective tools available for use are quite varied. We will discuss the most common types of safety equipment. These include:

- Rubber protective equipment, including gloves and blankets
- Protective apparel
- Personal clothing
- Hot sticks
- Fuse pullers
- Shorting probes
- Eye and face protection

3.1.1   1910.335(a)(1)/1926.951(a) Rubber Protective Equipment

At some point during the performance of their duties, all electrical workers will be exposed to energized circuits or equipment. Two of the most important articles of protection for electrical workers are insulated rubber gloves and rubber blankets, which must be matched to the voltage rating for the circuit or equipment. Rubber protective equipment is designed for the protection of the user. If it fails during use, a serious injury could occur.

Rubber protective equipment is available in two types. Type 1 designates rubber protective equipment that is manufactured of natural or synthetic rubber that is properly vulcanized, and Type 2 designates equipment that is ozone resistant, made from any elastomer or combination of elastomeric compounds. Ozone is a form of oxygen that is produced from electricity and is present in the air surrounding a conductor under high voltages. Normally, ozone is found at voltages of 10kV and higher, such as those found in electric utility transmission and distribution systems. Type 1 protective equipment can be damaged by *corona cutting*, which is the cutting action of ozone on natural rubber when it is under mechanical stress. Type 1 rubber protective equipment can also be damaged by ultraviolet rays. However, it is very important that the rubber protective equipment in use today be made of natural rubber or Type 1 equipment. Type 2 rubber protective equipment is very stiff and is not as easily worn as Type 1 equipment.

Various classes – The American National Standards Institute (ANSI) and the American Society for Testing of Materials (ASTM) have designated a specific classification system for rubber protective equipment. The voltage ratings are as follows:

- Class 0   1,000V
- Class 1   7,500V
- Class 2   17,500V
- Class 3   26,500V
- Class 4   36,000V

Inspection of protective equipment – Before rubber protective equipment can be worn by personnel in the field, all equipment must have a current test date stenciled on the equipment, and it must be inspected by the user. Insulating gloves must be tested each day by the user before they can be used. They must also be tested during the day if their insulating value is ever in question. Because rubber protective equipment is going to be used for personal protection and serious injury could result from its misuse or failure, it is important that an adequate safety factor be provided between the voltage on which it is to be used and the voltage at which it was tested.

All rubber protective equipment must be marked with the appropriate voltage rating and last inspection date. The markings that are required to be on rubber protective equipment must be applied in a manner that will not interfere with the protection that is provided by the equipment.

WARNING!   Never work on anything energized without direct instruction from your employer!

Gloves – Both high- and low-voltage rubber gloves are of the gauntlet type and are available in various sizes. To get the best possible protection and service life, here are a few general rules that apply whenever they are used in electrical work:

- Always wear leather protectors over your gloves. Any direct contact with sharp or pointed objects may cut, snag, or puncture the gloves and take away the protection you are depending on. Leather protectors are required by the National Fire Protection Association's Standard 70-E if the insulating capabilities of the gloves are subject to damage. [The standards of the National Fire Protection Association (NFPA) are incorporated into the OSHA standards.]
- Always wear rubber gloves right side out (serial number and size to the outside). Turning gloves inside out places a stress on the preformed rubber.
- Always keep the gauntlets up. Rolling them down sacrifices a valuable area of protection.
- Always inspect and field check gloves before using them. Always check the inside for any debris. The inspection of gloves is covered in more detail later in this section.
- Use light amounts of talcum powder or cotton liners with the rubber gloves. This gives the user more comfort, and it also helps to absorb some of the perspiration that can damage the gloves over years of use.
- Wash the rubber gloves in lukewarm, clean, fresh water after each use. Dry the gloves inside and out prior to returning to storage. Never use any type of cleaning solution on the gloves.
- Once the gloves have been properly cleaned, inspected, and tested, they must be properly stored. They should be stored in a cool, dry, dark place that is free from ozone, chemicals, oils, solvents, or other materials that could damage the gloves. Such storage should not be

in the vicinity of hot pipes or direct sunlight. Both gloves and sleeves should be stored in their natural shape and kept in a bag or box inside of their protectors. They should be stored undistorted, right side out, and unfolded.

- Gloves can be damaged by many different chemicals, especially petroleum-based products such as oils, gasoline, hydraulic fluid inhibitors, hand creams, pastes, and salves. If contact is made with these or other petroleum-based products, the contaminant should be wiped off immediately. If any signs of physical damage or chemical deterioration are found (e.g., swelling, softness, hardening, stickiness, ozone deterioration, or sun checking), the protective equipment must not be used.
- Never wear watches or rings while wearing rubber gloves; this can cause damage from the inside out and defeats the purpose of using rubber gloves. Never wear anything conductive.
- Rubber gloves must be tested every six months by a certified testing laboratory. Always check the inspection date before using gloves.
- Use rubber gloves only for their intended purpose, not for handling of chemicals or other work. This also applies to the leather protectors.

Before rubber gloves are used, a visual inspection and an air test should be made. This should be done prior to use and as many times during the day as you feel necessary. To perform a visual inspection, stretch a small area of the glove, checking to see that no defects exist, such as:

- Embedded foreign material
- Deep scratches
- Pin holes or punctures
- Snags or cuts

Gloves and sleeves can be inspected by rolling the outside and inside of the protective equipment between the hands. This can be done by squeezing together the inside of the gloves or sleeves to bend the outside area and create enough stress to the inside surface to expose any cracks, cuts, or other defects. When the entire surface has been checked in this manner, the equipment is then turned inside out, and the procedure is repeated. It is very important not to leave the rubber protective equipment inside out.

Remember, any damage at all reduces the insulating ability of the rubber glove. Look for signs of deterioration from age, such as hardening and slight cracking. Also, if the glove has been exposed to petroleum products, it should be considered suspect because deterioration can be caused by such exposure. Gloves that are found to be defective must be turned in for disposal. Never leave a damaged glove lying around; someone may think it is a good glove and not perform an inspection prior to using it.

After visually inspecting the glove, other defects may be observed by applying the air test.

Step 1   Stretch the glove and look for any defects as shown in *Figure 3*.

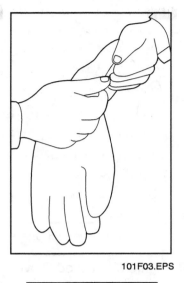

101F03.EPS

Figure 3. Inspection

Step 2   Twirl the glove around quickly or roll it down from the glove gauntlet to trap air inside as shown in *Figure 4*.

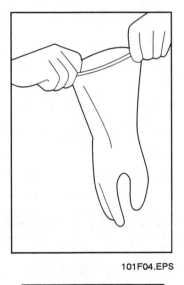

101F04.EPS

Figure 4. Trapping Air

Step 3   Trap the air by squeezing the gauntlet with one hand. Use the other hand to squeeze the palm, fingers, and thumb to check for weaknesses and defects. See *Figures 5* and *6*.

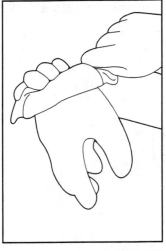

101F05.EPS

Figure 5. Inflated Glove

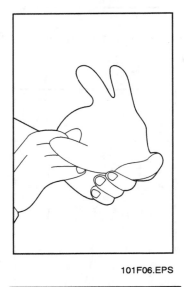

101F06.EPS

Figure 6. Inspecting Glove

Step 4   Hold the glove up to your ear to try and detect any escaping air.

Step 5   If the glove does not pass this inspection, it must be turned in for disposal.

CAUTION: Never use compressed gas for the air test as this can damage
the glove.

Insulating blankets – An insulating blanket is a versatile cover-up device best suited for the protection of maintenance technicians against accidental contact with energized electrical equipment.

These blankets are designed and manufactured to provide insulating quality and flexibility for use in covering. Insulating blankets are designed only for covering equipment and should not be used on the floor. (Special rubber mats are available for floor use.) Use caution when installing on sharp edges or covering pointed objects.

Blankets must be inspected yearly and should be checked before each use. To check rubber blankets, place the blanket on a flat surface and roll the blanket from one corner to the opposite corner. If there are any irregularities in the rubber, this method will expose them. After the blanket has been rolled from each corner, it should then be turned over and the procedure repeated.

Insulating blankets are cleaned in the same manner as rubber gloves. Once the protective equipment has been properly cleaned, inspected, and tested, it must be properly stored. It should be stored in a cool, dry, dark place that is free from ozone, chemicals, oils, solvents, or other materials that could damage the equipment. Such storage should not be in the vicinity of hot pipes or direct sunlight. Blankets may be stored rolled in containers that are designed for this use; the inside diameter of the roll should be at least two inches.

3.1.2 Protective Apparel

Besides rubber gloves, there are other types of special application protective apparel, such as fire suits, face shields, and rubber sleeves.

Manufacturing plants should have other types of special application protective equipment available for use, such as high-voltage sleeves, high-voltage boots, nonconductive protective helmets, nonconductive eyewear and face protection, and switchboard blankets.

All equipment should be inspected before use and during use, as necessary. The equipment used and the extent of the precautions taken depend on each individual situation; however, it is better to be overprotected than underprotected when you are trying to prevent electrocution.

When working with high voltages, flash suits may be required in some applications. Some plants require them to be worn for all switching and rack-in or rack-out operations.

Face shields should also be worn during all switching operations where arcs are a possibility. The thin plastic type of face shield should be avoided because it will melt when exposed to the extremely high temperatures of an electrical arc.

Rubber sleeves are another type of protective apparel that should be worn during switching operations and breaker racking. Sleeves must be inspected yearly.

3.1.3 Personal Clothing

Any individual who will perform work in an electrical environment or in plant substations should dress accordingly. Avoid wearing synthetic fiber clothing; these types of materials will melt when exposed to high temperatures and will actually increase the severity of a burn. Wear cotton clothing, fiberglass toe boots or shoes, and hard hats. Use hearing protection where needed.

3.1.4 Hot Sticks

Hot sticks are insulated tools designed for the manual operation of disconnecting switches, fuse removal and insertion, and the application and removal of temporary grounds.

A hot stick is made up of two parts, the head or *hood* and the insulating rod. The head can be made of metal or hardened plastic, while the insulating section may be wood, plastic, laminated wood, or other effective insulating materials. There are also telescoping sticks available.

Most plants have hot sticks available for different purposes. Select a stick of the correct type and size for the application.

Storage of hot sticks is important. They should be hung up vertically on a wall to prevent any damage. They should also be stored away from direct sunlight and prevented from being exposed to petroleum products. The preferred method of storage is to place the stick in a long section of capped pipe.

3.1.5 Fuse Pullers

Use the plastic or fiberglass style of fuse puller for removing and installing low-voltage cartridge fuses. All fuse pulling and replacement operations must be done using fuse pullers.

The best type of fuse puller is one that has a spread guard installed. This prevents the puller from opening if resistance is met when installing fuses.

3.1.6 Shorting Probes

Before working on deenergized circuits that have capacitors installed, you must discharge the capacitors using a safety shorting probe. When using a shorting probe, first connect the test clip to a good ground to make contact. If necessary, scrape the paint from the metal surface. Then, hold the shorting probe by the handle and touch the probe end of the shorting rod to the points to be shorted. The probe end can be hooked over the part or terminal to provide a constant connection to ground. Never touch any metal part of the shorting probe while grounding circuits or components. Whenever possible, especially when working on or near any deenergized high-voltage circuits, shorting probes should be connected and then left attached to the deenergized portion of any circuit for the duration of the work. This action serves an extra safety precaution against any accidental application of voltage to the circuit.

3.1.7 Eye And Face Protection

NFPA 70-E requires that protective equipment for the eyes and face shall be used whenever there is danger of injury to the eyes or face from electrical arcs or flashes, or from flying or falling objects resulting from electrical explosion.

3.2.0 VERIFY CIRCUITS ARE DEENERGIZED

You should always assume that all the circuits are energized until you have verified that the circuit is deenergized. Follow these steps to verify that a circuit is deenergized:

Step 1   Ensure the circuit is properly tagged and locked out (OSHA 1910.333/1926.417).

Step 2   Verify the test instrument operation on a known source.

Step 3   Using the test instrument, check the circuit to be deenergized. The voltage should be zero.

Step 4   Verify the test instrument operation once again on a known power source.

3.3.0 OTHER PRECAUTIONS

There are several other precautions you can take to help make your job safer. For example:

- Always remove all jewelry (e.g., rings, watches, bracelets, and necklaces) before working on electrical equipment. Most jewelry is made of conductive material and wearing it can result in a shock, as well as other injuries if the jewelry gets caught in moving components.
- When working on energized equipment, it is safer to work in pairs. In doing so, if one of the workers experiences a harmful electrical shock, the other worker can quickly deenergize the circuit and call for help.
- Plan each job before you begin it. Make sure you understand exactly what it is you are going to do. If you are not sure, ask your supervisor.
- You will need to look over the appropriate prints and drawings to locate isolation devices and potential hazards. Never defeat safety interlocks. Remember to plan your escape route before starting work. Know where the nearest phone is and the emergency number to dial for assistance.
- If you realize that the work will go beyond the scope of what was planned, stop and get instructions from your supervisor before continuing. Do not attempt to plan as you go.
- It is critical that you stay alert. Work places are dynamic, and situations relative to safety are always changing. If you leave the work area to pick up material, take a break, or have lunch, reevaluate your surroundings when you return. Remember, plan ahead.

4.0.0 OSHA

The purpose of the Occupational Safety and Health Administration (OSHA) is "to assure safe and healthful working conditions for working men and women." OSHA is authorized to enforce standards and assist and encourage the states in their efforts to ensure safe and healthful working conditions. OSHA assists states by providing for research, information, education, and training in the field of occupational safety and health.

The law that established OSHA specifies the duties of both the employer and employee with respect to safety. Some of the key requirements are outlined below. This list does not include everything, nor does it override the procedures called for by your employer.

- Employers shall provide a place of employment free from recognized hazards likely to cause death or serious injury.
- Employers shall comply with the standards of the act.
- Employers shall be subject to fines and other penalties for violation of those standards.

WARNING! OSHA states that employees have a duty to follow the safety rules laid down by the employer. Additionally, some states can reduce the amount of benefits paid to an injured employee if that employee was not following known, established safety rules. Your company may also terminate you if you violate an established safety rule.

4.1.0 SAFETY STANDARDS

The OSHA standards are split into several sections. As discussed earlier, the two which affect you the most are CFR 1926, construction specific, and CFR 1910, which is the standard for general industry. Either or both may apply depending on where you are working and what you are doing. If a job site condition is covered in the 1926 book, then that standard takes precedent. However, if a more stringent requirement is listed in the 1910 standard, it should also be met. An excellent example is the current difference in the two standards on confined spaces; if someone gets hurt or killed, the decision to use the less stringent 1926 standard could be called into question. OSHA's *General Duty Clause* states that an employer should have known all recognized hazards and removed the hazard or protected the employee.

To protect workers from the occupational injuries and fatalities caused by electrical hazards, OSHA has issued a set of design safety standards for electrical utilization systems. These standards are 1926.400-449 and 1910.302-308. OSHA also currently recognizes the 1984 edition of the *National Electrical Code (NEC)* for installations with exceptions for 1926.404(b)(1) and 1926.405(a)(2)(ii)(E)(F)(G) and (J). The OSHA electrical standards for construction are included in *Appendix A* in the back of this module.

Note: OSHA does *not* recognize the current edition of the NEC; it generally takes several years for that to occur.

The CFR 1910 standard must be followed whenever the construction standard CFR 1926 does not address an issue which is covered by CFR 1910, or for a pre-existing installation. If the CFR 1910 standard is more stringent then CFR 1926, then the more stringent standard should be followed. OSHA does not update their standards in a timely manner, and as such, there are often differences in similar sections of the two standards. Safety should always be first, and the more protective work rules should always be chosen.

4.1.1   1910.302-308/1926.402-408 Design Safety Standards For Electrical Systems

This section contains design safety regulations for all the electrical equipment and installations used to provide power and light to employee workplaces. The articles listed are outlined in the following sections.

4.1.2   1910.302/1926.402 Electric Utilization Systems

This article identifies the scope of the standard. Listings are included to show which electrical installations and equipment are covered under the standard, and which installations and equipment are not covered under the standard. Furthermore, certain sections of the standard apply only to utilization equipment installed after March 15, 1972, and some apply only to equipment installed after April 16, 1981. Article 1910.302 (1926.402) addresses these oddities and provides guidance to clarify them.

4.1.3   1910.303/1926.403 General Requirements

This article covers topics that mostly concern equipment installation clearances, identification, and examination. Some of the major subjects addressed in this article are:

- Equipment installation examinations
- Splicing
- Marking
- Identification of disconnecting means
- Workspace around electrical equipment

4.1.4   1910.304/1926.404 Wiring Design And Protection

This article covers the application, identification, and protection requirements of grounding conductors, outside conductors, service conductors, and equipment enclosures. Some of the major topics discussed are:

- Grounded conductors
- Outside conductors
- Service conductors
- Overcurrent protection
- System grounding requirements

4.1.5   1910.305/1926.405 Wiring Methods, Components, And Equipment For General Use

In general, this article addresses the wiring method requirements of raceways, cable trays, pull and junction boxes, switches, and switchboards; the application requirements of temporary wiring installations; the equipment and conductor requirements for general wiring; and the protection requirements of motors, transformers, capacitors, and storage batteries. Some of the major topics are:

*   Wiring methods
*   Cabinets, boxes, and fittings
*   Switches
*   Switchboards and panelboards
*   Enclosures for damp or wet locations
*   Conductors for general wiring
*   Flexible cords and cables
*   Portable cables
*   Equipment for general use

4.1.6   1910.306/1926.406 Specific Purpose Equipment And Installations

This article addresses the requirements of special equipment and installations not covered in other articles. Some of the major types of equipment and installations found in this article are:

*   Electric signs and outline lighting
*   Cranes and hoists
*   Elevators, dumbwaiters, escalators, and moving walks
*   Electric welders
*   Data processing systems
*   X-ray equipment
*   Induction and dielectric heating equipment
*   Electrolytic cells
*   Electrically-driven or controlled irrigation machines
*   Swimming pools, fountains, and similar installations

4.1.7   1910.307/1926.407 Hazardous (Classified) Locations

This article covers the requirements for electric equipment and wiring in locations that are classified because they contain: (1) flammable vapors, liquids, and/or gases, or combustible dust or fibers; and (2) the likelihood that a flammable or combustible concentration or quantity is present. Some of the major topics covered in this article are:

*   Scope
*   Electrical installations in hazardous locations
*   Conduit
*   Equipment in Division 2 locations

The OSHA standard for hazardous (classified) locations is included in *Appendix A* at the back of this module.

4.1.8   1910.308/1926.408 Special Systems

This article covers the wiring methods, grounding, protection, identification, and other general requirements of special systems not covered in other articles. Some of the major subtopics found in this article are:

- Systems over 600 volts nominal
- Emergency power systems
- Class 1, 2, and 3 remote control, signaling, and power-limited circuits
- Fire-protective signaling systems
- Communications systems

4.1.9   1910.331/1926.416 Scope

This article serves as an overview of the following articles and also provides a summary of the installations that this standard allows qualified and unqualified persons to work on or near, as well as the installations that this standard does *not* cover.

4.1.10   1910.332 Training

The training requirements contained in this article apply to employees who face a risk of electric shock. Some of the topics that appear in this article are:

- Content of training
- Additional requirements for unqualified persons
- Additional requirements for qualified persons
- Type of training

4.1.11   1910.333/1926.416-417 Selection And Use Of Work Practices

This article covers the implementation of safety-related work practices necessary to prevent electrical shock and other related injuries to the employee. Some of the major topics addressed in this article are listed below:

- General
- Working on or near exposed deenergized parts
- Working on or near exposed energized parts

4.1.12 1910.334/1926.431 Use Of Equipment

This article was added to reinforce the regulations pertaining to portable electrical equipment, test equipment, and load break switches. Major topics include:

- Portable electric equipment
- Electric power and lighting circuits
- Test instruments and equipment

4.1.13 1910.335/1926.416 Safeguards For Personnel Protection

This article covers the personnel protection requirements for employees in the vicinity of electrical hazards. It addresses regulations that protect personnel working on equipment as well as personnel working nearby. Some of the major topics are:

- Use of protective equipment
- Alerting techniques

Now that background topics have been covered and an overview of the OSHA electrical safety standards has been provided, it is time to move on to topics related directly to safety. As we discuss these topics, we will continually refer to the OSHA standards to identify the requirements that govern them.

OSHA 1926 Subpart K also addresses electrical safety requirements that are necessary for the practical safeguarding of employees involved in construction work.

4.2.0 SAFETY PHILOSOPHY AND GENERAL SAFETY PRECAUTIONS

The most important piece of safety equipment in performing work in an electrical environment is common sense. All areas of electrical safety precautions and practices draw upon common sense and attention to detail. One of the most dangerous conditions in an electrical work area is a poor attitude toward safety.

WARNING!   Only qualified individuals may work on electrical equipment. Your employer will determine who is qualified. Remember, your employer's safety rules must always be followed.

As stated in CFR 1910.333(a)/1926.403, safety-related work practices shall be employed to prevent electric shock or other injuries resulting from either direct or indirect electrical contact when work is performed near or on equipment or circuits that are or may be energized. The specific safety-related work practices shall be consistent with the nature and extent of the associated electrical hazards. The following are considered some of the basic and necessary attitudes and electrical safety precautions that lay the groundwork for a proper safety program. Before going on any electrical work assignment, these safety precautions should be reviewed and adhered to.

- *All work on electrical equipment should be done with circuits deenergized and cleared or grounded* – It is obvious that working on energized equipment is much more dangerous than working on equipment that is deenergized. Work on energized electrical equipment should be avoided if at all possible. CFR 1910.333(a)(1)/1926.403 states that live parts to which an employee may be exposed shall be deenergized before the employee works on or near them, unless the employer can demonstrate that deenergizing introduces additional or increased hazards or is not possible because of equipment design or operational limitations. Live parts that operate at less than 50 volts to ground need not be deenergized if there will be no increased exposure to electrical burns or to explosion due to electric arcs.
- *All conductors, buses, and connections should be considered energized until proven otherwise* – As stated in 1910.333(b)(1)/1926.417, conductors and parts of electrical equipment that have not been locked out or tagged out in accordance with this section should be considered energized. Routine operation of the circuit breakers and disconnect switches contained in a power distribution system can be hazardous if not approached in the right manner. Several basic precautions that can be observed in switchgear operations are:
  - Wear proper clothing made of 100% cotton or fire-resistant fabric.
  - Eye, face, and head protection should be worn. Turn your head away whenever closing devices.
  - Whenever operating circuit breakers in low-voltage or medium-voltage systems, always stand off to the side of the unit.
  - Always try to operate disconnect switches and circuit breakers under a no-load condition.
  - Never intentionally force an interlock on a system or circuit breaker.
  - Always verify what you are closing a device into; you could violate a lockout or close into a hard fault.

Often, a circuit breaker or disconnect switch is used for providing lockout on an electrical system. To ensure that a lockout is not violated, perform the following procedures when using the device as a lockout point:

- Breakers must always be locked out and tagged as discussed previously whenever you are working on a circuit that is tied to an energized breaker. Breakers capable of being opened and racked out to the disconnected position should have this done. Afterward, approved safety locks must be installed. The breaker may be removed from its cubicle completely to prevent unexpected mishaps. Always follow the standard rack-out and removal procedures that were supplied with the switchgear. Once removed, a sign must be hung on the breaker identifying its use as a lockout point, and approved safety locks must be installed when the breaker is used for isolation. In addition, the closing springs should be discharged.

- Some of the circuit breakers used are equipped with keyed interlocks for protection during operation. These locks are generally called *kirklocks* and are relied upon to ensure proper sequence of operation only. These are not to be used for the purpose of locking out a circuit or system. Where disconnects are installed for use in isolation, they should never be opened under load. When opening a disconnect manually, it should be done quickly with a positive force. Again, lockouts should be used when the disconnects are open.

- Whenever performing switching or fuse replacements, always use the protective equipment necessary to ensure personnel safety. *Never* make the assumption that because things have gone fine the last 999 times, they will not go wrong this time. Always prepare yourself for the worst case accident when performing switching.

- Whenever reenergizing circuits following maintenance or removal of a faulted component, extreme care should be used. Always verify that the equipment is in a condition to be reenergized safely. All connections should be insulated and all covers should be installed. Have all personnel stand clear of the area for the initial reenergization. *Never* assume everything is in perfect condition. Verify the conditions!

The following procedure is provided as a guideline for ensuring that equipment and systems will not be damaged by reclosing low-voltage circuit breakers into faults. If a low-voltage circuit breaker has opened for no apparent reason, perform the following:

Step 1 Verify that the equipment being supplied is not physically damaged and shows no obvious signs of overheating or fire.

Step 2 Make all appropriate tests to locate any faults.

Step 3 Reclose the feeder breaker. Stand off to the side when closing the breaker.

Step 4 If the circuit breaker trips again, do not attempt to reclose the breaker. In a plant environment, Electrical Engineering should be notified, and the cause of the trip should be isolated and repaired.

The same general procedure should be followed for fuse replacement, with the exception of transformer fuses. If a transformer fuse blows, the transformer and feeder cabling should be inspected and tested before reenergizing. A blown fuse to a transformer is very significant because it normally indicates an internal fault. Transformer failures are catastrophic in nature and can be extremely dangerous. If applicable, contact the in-plant Electrical Engineering Department prior to commencing any effort to reenergize a transformer.

Power must always be removed from a circuit when removing and installing fuses. The air break disconnects (or quick disconnects) provided on the upstream side of a large transformer must be opened prior to removing the transformer's fuses. Otherwise, severe arcing will occur as the fuse is removed. This arcing can result in personnel injury and equipment damage.

To replace fuses servicing circuits below 600 volts:

• Secure power to fuses or ensure all downstream loads have been disconnected.
• Always use a positive force to remove and install fuses.

When replacing fuses servicing systems above 600 volts:

• Open and lock out the disconnect switches.
• Unlock the fuse compartment.
• Verify that the fuses are deenergized.
• Attach the fuse removal hot stick to the fuse and remove it.

4.3.0 ELECTRICAL REGULATIONS

OSHA has certain regulations that apply to job site electrical safety. These regulations include the following:

• All electrical work shall be in compliance with the latest NEC and OSHA standards.

Note: OSHA may not recognize the current edition of the NEC, which can sometimes cause problems; however, OSHA typically will *not* cite for any differences.

• The noncurrent-carrying metal parts of fixed, portable, and plug-connected equipment shall be grounded. It is best to choose **grounded tools**. However, portable tools and appliances protected by an approved system of double insulation need not be grounded. *Figure 7* shows an example of a **double-insulated/ungrounded tool**.

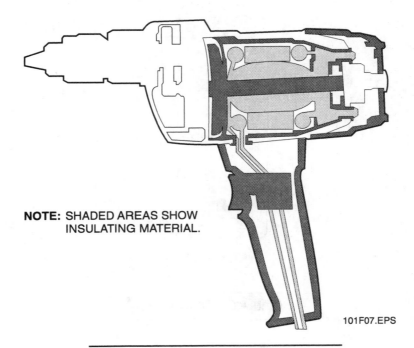

NOTE: SHADED AREAS SHOW INSULATING MATERIAL.

101F07.EPS

Figure 7. Double-Insulated Electric Drill

- Extension cords shall be the three-wire type, shall be protected from damage, and shall not be fastened with staples or hung in a manner that could cause damage to the outer jacket or insulation. Never run an extension cord through a doorway or window that can pinch the cord. Also, never allow vehicles or equipment to drive over cords.

- Exposed lamps in temporary lights shall be guarded to prevent accidental contact, except where lamps are deeply recessed in the reflector. Temporary lights shall not be suspended, except in accordance with their listed labeling.

- Receptacles for attachment plugs shall be of an approved type and properly installed. Installation of the receptacle will be in accordance with the listing and labeling for each receptacle and shall be GFCI-protected if the setting is a temporarily-wired construction site. If permanent receptacles are used with extension cords, then you must use GFCI protection.

- Each disconnecting means for motors and appliances and each service feeder or branch circuit at the point where it originates shall be legibly marked to indicate its purpose and voltage.

- Flexible cords shall be used in continuous lengths (no splices) and shall be of a type listed in *NEC Table 400-4*.

- Personnel safety ground fault protection for temporary construction is required. There are two methods for accomplishing this: an assured grounding program (this method is still used but should be used in conjunction with a GFCI program) or ground fault protection receptacles or breakers. Each employer will set the standard and method to be used. *Figure 8* shows a typical ground-fault circuit interrupter. The OSHA standard for ground fault protection on construction sites is included in *Appendix A* at the back of this module.

101F08.EPS

Figure 8. Typical GFCI

4.3.1 OSHA Lockout/Tagout Rule

OSHA released the 29 CFR 1926 lockout/tagout rule in December 1991. This rule covers the specific procedure to be followed for the "servicing and maintenance of machines and equipment in which the unexpected energization or startup of the machines or equipment, or releases of stored energy, could cause injury to employees. This standard establishes minimum performance requirements for the control of such hazardous energy."

WARNING!   This procedure is provided for your information only. The OSHA procedure provides only the minimum requirements for lockouts/tagouts. Consult the lockout/tagout procedure for your company and the plant or job site at which you are working. Remember that your life could depend on the lockout/tagout procedure. It is critical that you use the correct procedure for your site.

The purpose of the OSHA procedure is to ensure that equipment is isolated from all potentially hazardous energy (e.g., electrical, mechanical, hydraulic, chemical, or thermal), and tagged and locked out before employees perform any servicing or maintenance activities where the unexpected energization, startup, or release of stored energy could cause injury. All employees shall be instructed in the lockout/tagout procedure.

CAUTION: Although 99% of your work may be electrical, be aware that you may also need to lock out mechanical equipment.

The following is an example of a lockout/tagout procedure. Make sure to use the procedure that is specific to your employer or job site.

I. *Introduction*

A. This lockout/tagout procedure has been established for the protection of personnel from potential exposure to hazardous energy sources during construction, installation, service, and maintenance of electrical energy systems.

B. This procedure applies to and must be followed by all personnel who may be potentially exposed to the unexpected startup or release of hazardous energy (e.g., electrical, mechanical, pneumatic, hydraulic, chemical, or thermal).

Exception: This procedure does not apply to process and/or utility equipment or systems with cord and plug power supply systems when the cord and plug are the only source of hazardous energy, are removed from the source, and remain under the exclusive control of the authorized employee.

Exception: This procedure does not apply to troubleshooting (diagnostic) procedures and installation of electrical equipment and systems when the energy source cannot be deenergized because continuity of service is essential or shutdown of the system is impractical. Additional personal protective equipment for such work is required and the safe work practices identified for this work must be followed.

II. Definitions

- *Affected employee* – Any person working on or near equipment or machinery when maintenance or installation tasks are being performed by others during lockout/tagout conditions.
- *Appointed authorized employee* – Any person appointed by the job site supervisor to coordinate and maintain the security of a group lockout/tagout condition.
- *Authorized employee* – Any person authorized by the job site supervisor to use lockout/tagout procedures while working on electrical equipment.
- *Authorized supervisor* – The assigned job site supervisor who is in charge of coordination or procedures and maintenance of security of all lockout/tagout operations at the job site.
- *Energy isolation device* – An approved electrical disconnect switch capable of accepting approved lockout/tagout hardware for the purpose of isolating and securing a hazardous electrical source in an open or safe position.
- *Lockout / tagout hardware* – A combination of padlocks, danger tags, and other devices designed to attach to and secure electrical isolation devices.

IH. Training

A. Each authorized supervisor, authorized employee, and appointed authorized employee shall receive initial and as-needed user-level training in lockout/tagout procedures.
B. Training is to include recognition of hazardous energy sources, the type and magnitude of energy sources in the workplace, and the procedures for energy isolation and control.
C. Retraining will be conducted on an as-needed basis whenever lockout/tagout procedures are changed or there is evidence that procedures are not being followed properly.

IV. Protective Equipment and Hardware

A. Lockout/tagout devices shall be used exclusively for controlling hazardous electrical energy sources.
B. All padlocks must be numbered and assigned to one employee only.
C. No duplicate or master keys will be made available to anyone except the site supervisor.
D. A current list with the lock number and authorized employee's name must be maintained by the site supervisor.
E. Danger tags must be of the standard white, red, and black *DANGER—DO NOT OPERATE* design and shall include the authorized employee's name, the date, and the appropriate network company (use permanent markers).
F. Danger tags must be used in conjunction with padlocks as shown in *Figure 9*.

Figure 9. Lockout/Tagout Device

V. *Procedures*

A. Preparation for lockout/tagout:
   1. Check the procedures to ensure that no changes have been made since you last used a lockout/tagout.
   2. Identify all authorized and affected employees involved with the pending lockout/tagout.

B. Sequence for lockout/tagout:
   1. Notify all authorized and affected personnel that a lockout/tagout is to be used and explain the reason why.
   2. Shut down the equipment or system using the normal OFF or STOP procedures.
   3. Lock out energy sources and test disconnects to be sure they cannot be moved to the ON position and open the control cutout switch. If there is not a cutout switch, block the magnet in the switch open position before working on electrically-operated equipment/apparatus such as motors, relays, etc. Remove the control wire.
   4. Lock and tag the required switches in the open position. Each authorized employee must affix a separate lock and tag. An example is shown in *Figure 10*.
   5. Dissipate any stored energy by attaching the equipment or system to ground.
   6. Verify that the test equipment is functional via a known power source.
   7. Confirm that all switches are in the open position and use test equipment to verify that all parts are deenergized.
   8. If it is necessary to temporarily leave the area, upon returning, retest to ensure that the equipment or system is still deenergized.

C. Restoration of energy:
   1. Confirm that all personnel and tools, including shorting probes, are accounted for and removed from the equipment or system.
   2. Completely reassemble and secure the equipment or system.

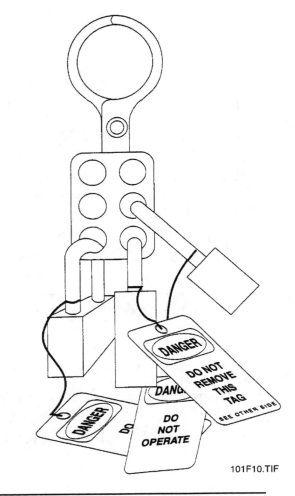

Figure 10. Multiple Lockout/Tagout Device

3. Replace and/or reactivate all safety controls.
4. Remove locks and tags from isolation switches. Authorized employees must remove their own locks and tags.
5. Notify all affected personnel that the lockout/tagout has ended and the equipment or system is energized.
6. Operate or close isolation switches to restore energy.

VI. *Emergency Removal Authorization*

A. In the event a lockout/tagout device is left secured, and the authorized employee is absent, or the key is lost, the authorized supervisor can remove the lockout/tagout device.
B. The authorized employee must be informed that the lockout/tagout device has been removed.
C. Written verification of the action taken, including informing the authorized employee of the removal, must be recorded in the job journal.

4.4.0 OTHER OSHA REGULATIONS

There are other OSHA regulations which you need to be aware of on the job site. For example:

- OSHA requires the posting of hard hat areas. Be alert to those areas and always wear your hard hat properly, with the bill in front. Hard hats should be worn whenever overhead hazards exist, or there is the risk of exposure to electric shock or burns.
- You should wear safety shoes on all job sites. Keep them in good condition.
- Do not wear clothing with exposed metal zippers, buttons, or other metal fasteners. Avoid wearing loose-fitting or torn clothing.
- Protect your eyes. Your eyesight is threatened by many activities on the job site. Always wear safety glasses with full side shields. In addition, the job may also require protective equipment such as face shields or goggles.

4.4.1 Testing For Voltage

OSHA also requires that you inspect or test existing conditions before beginning work on electrical equipment or lines. Usually, you will use a voltmeter/sensor or voltage tester to do this. You should assume that all electrical equipment and lines are energized until you have determined that they are not. Do not proceed to work on or near energized parts until the operating voltage is determined.

After the electrical equipment to be worked on has been locked and tagged out, the equipment must be verified as deenergized before work can proceed. This section sets the requirements that must be met before any circuits or equipment can be considered deenergized. First, and most importantly, only qualified persons may verify that a circuit or piece of equipment is deenergized. Before approaching the equipment to be worked on, the qualified person shall operate the equipment's normal operating controls to check that the proper energy sources have been disconnected.

Upon opening a control enclosure, the qualified person shall note the presence of any components that may store electrical energy. Initially, these components should be avoided.

To verify that the lockout was adequate and the equipment is indeed deenergized, a qualified person must use appropriate test equipment to check for power, paying particular attention to induced voltages and unrelated feedback voltage.

Ensure that your testing equipment is working properly by performing the *live-dead-live* check before each use. To perform this test, first check your voltmeter on a known live voltage source. This known source must be in the same range as the electrical equipment you will be working on. Next, without changing scales on your voltmeter, check for the presence of power in the equipment you have locked out. Finally, to ensure that your voltmeter did not malfunction, check it again on the known live source. Performing this test will assure you that your voltage testing equipment is reliable.

In accordance with OSHA section 1910.333(b)(2)(iv)/1926.417(d)(4)(ii), if the circuit to be tested normally operates at more than 600 volts, the live-dead-live check must be performed.

Once it has been verified that power is not present, stored electrical energy that might endanger personnel must be released. A qualified person must use the proper devices to release the stored energy, such as using a shorting probe to discharge a capacitor.

5.0.0 LADDERS AND SCAFFOLDS

Ladders and scaffolds account for about half of the injuries from workplace electrocutions. The involuntary recoil which can occur when a person is shocked can cause them to be thrown from a ladder or high place.

5.1.0 LADDERS

Many job site accidents involve the misuse of ladders. Make sure to follow these general rules every time you use any ladder. Following these rules can prevent serious injury or even death.

- Before using any ladder, inspect it. Look for loose or missing rungs, cleats, bolts, or screws, and check for cracked, broken, or badly worn rungs, cleats, or side rails.
- If you find a ladder in poor condition, do not use it. Report it and tag it for repair or disposal.
- Never modify a ladder by cutting it or weakening its parts.
- Do not set up ladders where they may be run into by others, such as in doorways or walkways. If it is absolutely necessary to set up a ladder in such a location, protect the ladder with barriers.
- Do not increase a ladder's reach by standing it on boxes, barrels, or anything other than a flat surface.
- Check your shoes for grease, oil, or mud before climbing a ladder. These materials could make you slip.
- Always face the ladder and hold on with both hands when climbing up or down.
- Never lean out from the ladder. Keep your belt buckle centered between the rails. If something is out of reach, get down and move the ladder.

WARNING!   When performing electrical work, always use ladders made of nonconductive material.

5.1.1 Straight And Extension Ladders

There are some specific rules to follow when working with straight and extension ladders:

- Always place a straight ladder at the proper angle. The distance from the ladder feet to the base of the wall or support should be about one-fourth the working height of the ladder (see *Figure 11*).

ELECTRICAL — TRAINEE TASK MODULE 26101

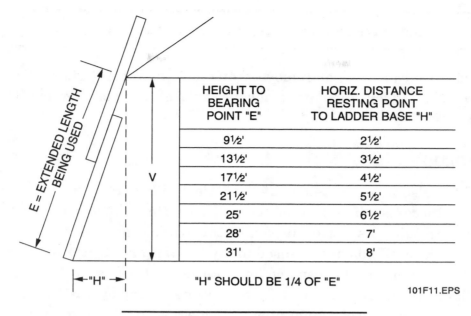

HEIGHT TO BEARING POINT "E"	HORIZ. DISTANCE RESTING POINT TO LADDER BASE "H"
9½'	2½'
13½'	3½'
17½'	4½'
21½'	5½'
25'	6½'
28'	7'
31'	8'

"H" SHOULD BE 1/4 OF "E"

101F11.EPS

Figure 11. Straight Ladder Positioning

- Secure straight ladders to prevent slipping. Use ladder shoes or hooks at the top and bottom. Another method is to secure a board to the floor against the ladder feet. For brief jobs, someone can hold the straight ladder.
- Side rails should extend above the top support point by at least 36 inches (see *Figure 11*).
- It takes two people to safely extend and raise an extension ladder. Extend the ladder only after it has been raised to an upright position.
- Never carry an extended ladder.
- Never use two ladders spliced together.
- Ladders should not be painted because paint can hide defects.

5.1.2   Step Ladders

There are also a few specific rules to use with a step ladder:

- Always open the step ladder all the way and lock the spreaders to avoid collapsing the ladder accidentally.
- Use a step ladder that is high enough for the job so that you do not have to reach. Get someone to hold the ladder if it is more than 10 feet high.
- Never use a step ladder as a straight ladder.
- Never stand on or straddle the top two rungs of a step ladder.
- Ladders are not shelves.

WARNING!	Do not leave tools or materials on a step ladder.

Sometimes you will need to move or remove protective equipment, guards, or guardrails to complete a task using a ladder. Remember, always replace what you moved or removed before leaving the area.

5.2.0 SCAFFOLDS

Working on scaffolds also involves being safe and alert to hazards. In general, keep scaffold platforms clear of unnecessary material or scrap. These can become deadly tripping hazards or falling objects. Carefully inspect each part of the scaffold as it is erected. Your life may depend on it! Makeshift scaffolds have caused many injuries and deaths on job sites. Use only scaffolding and planking materials designed and marked for their specific use. When working on a scaffold, follow the established specific requirements set by OSHA for the use of fall protection. When appropriate, wear an approved harness with a lanyard properly anchored to the structure.

Note: The following requirements represent a compilation of the more stringent requirements of both CFR 1910 and CFR 1926.

The following are some of the basic OSHA rules for working safely on scaffolds:

- Scaffolds must be erected on sound, rigid footing that can carry the maximum intended load.
- Guardrails and toe boards must be installed on the open sides and ends of platforms which are higher than six feet above the ground or floor.
- There must be a screen of ½-inch maximum openings between the toe board and the midrail where persons are required to work or pass under the scaffold.
- Scaffold planks must extend over their end supports not less than six inches nor more than 12 inches and must be properly blocked.
- If the scaffold does not have built-in ladders which meet the standard, then it must have an attached ladder access.
- All employees must be trained to erect, dismantle, and use scaffold(s).
- Unless it is impossible, fall protection must be worn while building or dismantling all scaffolding.
- Work platforms must be completely decked for use by employees.
- Your hard hat is the first line of protection from falling objects. Your hard hat, however, cannot protect your shoulders, arms, back, or feet from the danger of falling objects. The person working below depends on those working above. When you are working above the ground, be careful so that material, including your tools, cannot fall from your work site. Use trash containers or other similar means to keep debris from falling and never throw or sweep material from above.

6.0.0 LIFTS, HOISTS, AND CRANES

On the job, you may be working in the operating area of lifts, hoists, or cranes. The following safety rules are for those who are working in the area with overhead equipment but are not directly involved in its operation.

- Stay alert and pay attention to the warning signals from operators.
- Never stand or walk under a load, regardless of whether it is moving or stationary.
- Always warn others of moving or approaching overhead loads.
- Never attempt to distract signal persons or operators of overhead equipment.
- Obey warning signs.
- Do not use equipment that you are not qualified to operate.

7.0.0 LIFTING

Back injuries cause many lost working hours every year. That is in addition to the misery felt by the person with the hurt back! Learn how to lift properly and size up the load. To lift, first stand close to the load. Then, squat down and keep your back straight. Get a firm grip on the load and keep the load close to your body. Lift by straightening your legs. Make sure that you lift with your legs and not your back. Do not be afraid to ask for help if you feel the load is too heavy. See *Figure 12* for an example of proper lifting.

101F12.TIF

Figure 12. Proper Lifting

8.0.0 BASIC TOOL SAFETY

When using any tools for the first time, read the operator's manual to learn the recommended safety precautions. If you are not certain about the operation of any tool, ask the advice of a more experienced worker. Before using a tool, you should know its function and how it works.

Always use the right tool for the job. Incorrectly using tools is one of the leading causes of job site injury. Using a hammer as a pry bar or a screwdriver as a chisel can cause damage to the tool and injure you in the process.

9.0.0 CONFINED SPACE ENTRY PROCEDURES

Occasionally, you may be required to do your work in a manhole or vault. If this is the case, there are some special safety considerations that you need to be aware of. For details on the subject of working in manholes and vaults, refer to 1910.146/1926.21(a)(6)(i) and (ii) and the *National Electrical Safety Code*. The general precautions are listed in the following paragraphs.

9.1.0 GENERAL GUIDELINES

A confined space includes (but is not limited to) any of the following: a manhole, boiler, tank, trench (four feet or deeper), tunnel, hopper, bin, sewer, vat, pipeline, vault, pit, air duct, or vessel. A confined space is identified as follows:

- It has limited entry and exit.
- It is not intended for continued human occupancy.
- It has poor ventilation.
- It has the potential for entrapment/engulfment.
- It has the potential for accumulating a dangerous atmosphere.
- Entry into a confined space occurs when any part of the body crosses the plane of entry. No employee shall enter a confined space unless the employee has been trained in confined space entry procedures.
- All hazards must be eliminated or controlled before a confined space entry is made.
- All appropriate personal protective equipment shall be worn at all times during confined space entry and work. The minimum required equipment includes a hard hat, safety glasses, full body harness, and life line.
- Ladders used for entry must be secured.
- A rescue retrieval system must be in use when entering confined spaces and while working in permit-required confined spaces (discussed later). Each employee must be capable of being rescued by the retrieval system.
- Only no-entry rescues will be performed by company personnel. Entry rescues will be performed by trained rescue personnel identified on the entry permit.
- The area outside the confined space must be properly barricaded, and appropriate warning signs must be posted.

- Entry permits can be issued and signed by job site supervisors only. Permits must be kept at the confined space while work is being conducted. At the end of the shift, the entry permits must be made part of the job journal and retained for one year.

9.2.0 CONFINED SPACE HAZARD REVIEW

Before determining the proper procedure for confined space entry, a hazard review shall be performed. The hazard review shall include, but not be limited to, the following conditions:

- The past and current uses of the confined space
- The physical characteristics of the space including size, shape, air circulation, etc.
- Proximity of the space to other hazards
- Existing or potential hazards in the confined space, such as:
  - Atmospheric conditions (oxygen levels, flammable/explosive levels, and/or toxic levels)
  - Presence/potential for liquids
  - Presence/potential for particulates
- Potential for mechanical/electrical hazards in the confined space (including work to be done)

Once the hazard review is completed, the supervisor, in consultation with the project managers and/or safety manager, shall classify the confined space as one of the following:

- A nonpermit confined space
- A permit-required confined space controlled by ventilation
- A permit-required confined space

Once the confined space has been properly classified, the appropriate entry and work procedures must be followed.

9.3.0 ENTRY AND WORK PROCEDURES

Nonpermit spaces – A hazard review checklist must be completed before a confined space is designated as a *nonpermit space*. The checklist must be made part of the job journal, and a copy of the checklist must be sent to the safety office. A nonpermit confined space must meet the following criteria:

- There is no actual or potential atmospheric hazard.

Note: Using ventilation to clear the atmosphere does not meet this criteria.

- There are no actual or potential physical, electrical, or mechanical hazards capable of causing harm or death.

Documentation using the hazards checklist and entry permit forms, and verifying that the confined space is hazard-free, must be made available to employees and maintained at the confined space while work is conducted. If it is necessary to enter the space to verify it is hazard-free or to eliminate hazards, entry must be made under the requirements of a permit-required space.

An employee may enter the confined space using the minimum fall protection of harness and anchored life line. Once in the space, the employee may disconnect the life line and reconnect it before exiting.

If the work being done creates a hazard, the space must be reclassified as a permit-required space. If any other atmospheric, physical, electrical, or mechanical hazards arise, the space is to be evacuated immediately and reclassified as a permit-required entry space.

Permit-required spaces controlled by ventilation – A hazard review checklist must be completed before a confined space is designated as a *permit-required space controlled by ventilation*. The checklist must be made part of the job journal and a copy of the checklist must be sent to the safety office. A permit-required confined space controlled by ventilation must meet the following criteria:

- The only hazard in the confined space is an actual/potential atmospheric hazard.
- Continuous forced-air ventilation maintains a safe atmosphere (i.e., within the limits designated on the entry permit).
- Inspection and monitoring data are documented.
- No other physical, electrical, or mechanical hazard exists.

An entry permit must be issued and signed by the job site supervisor and be kept at the confined space while work is being conducted.

Atmospheric testing must be conducted before entry into the confined space and in the following order:

- Oxygen content
- Flammable gases and vapors
- Toxic contaminants

Unacceptable atmospheric conditions must be eliminated with forced air ventilation. If continuous forced air ventilation is required to maintain an acceptable atmosphere, employees may not enter until forced air ventilation has eliminated any hazardous atmosphere. Periodic atmospheric testing must be conducted during the work shift to ensure that the atmosphere remains clear. Periodic monitoring must be documented on the entry permit. If atmospheric conditions change, employees must exit the confined space immediately, and atmospheric conditions must be re-evaluated. Continuous communication must be maintained with the employees working in the confined space.

If *hot work* is to be performed, a hot work permit is required, and the hazard analysis must document that the hot work does not create additional hazards that are not controlled by ventilation only. Hot work is defined as any work that produces arcs, sparks, flames, heat, or other sources of ignition.

A rescue plan using trained rescue personnel must be in place prior to the start of work in the confined space. All employees should be aware of the rescue plan and how to activate it.

Permit-required confined spaces – A hazard review checklist must be completed before a confined space is designated as a *permit-required confined space*. The checklist must be made part of the job journal and a copy must be sent to the safety office. A permit-required space meets the following criteria:

- There are actual/potential hazards, other than a hazardous atmosphere.
- Ventilation alone does not eliminate atmospheric hazards.
- Conditions in and around the confined space must be continually monitored.

An entry permit must be issued and signed by the job site supervisor. The permit is to be kept at the confined space while work is being performed in the space.

Atmospheric testing must be conducted before entry into the confined space and in the following order:

- Oxygen content
- Flammable gases and contaminants
- Toxic contaminants

Unacceptable atmospheric conditions must be eliminated/controlled prior to employee entry. Methods of elimination may include isolation, purging, flushing, or ventilating. Continuous atmospheric monitoring must be conducted while employees are in the confined space. Triggering of a monitoring alarm means employees should evacuate the confined space immediately. Any other physical hazards must be eliminated or controlled by engineering and work practice controls before entry. Additional personal protective equipment should be used as a follow-up to the above methods. An attendant, whose job it is to monitor conditions in and around the confined space and to maintain contact with the employees in the space, must be stationed outside the confined space for the duration of entry operations.

If hot work is to be performed, a hot work permit is required, and the hazard analysis must document the additional hazards and precautions to be considered.

A rescue plan using trained rescue personnel must be in place before confined space entry. The attendant should be aware of the rescue plan and have the means to activate it.

10.0.0 FIRST AID

You should be prepared in case an accident does occur on the job site or anywhere else. First aid training that includes certification classes in CPR and artificial respiration could be the best insurance you and your fellow workers ever receive. Make sure that you know where first aid is available at your job site. Also, make sure you know the accident reporting procedure. Each job site should also have a first aid manual or booklet giving easy-to-find emergency treatment procedures for various types of injuries. Emergency first aid telephone numbers should be readily available to everyone on the job site. Refer to CFR 1910.151/1926.23 and 1926.50 for specific requirements.

11.0.0 SOLVENTS AND TOXIC VAPORS

The solvents that are used by electricians may give off vapors that are toxic enough to make people temporarily ill or even cause permanent injury. Many solvents are skin and eye irritants. Solvents can also be systemic poisons when they are swallowed or absorbed through the skin.

Solvents in spray or aerosol form are dangerous in another way. Small aerosol particles or solvent vapors mix with air to form a combustible mixture with oxygen. The slightest spark could cause an explosion in a confined area because the mix is perfect for fast ignition. There are procedures and methods for using, storing, and disposing of most solvents and chemicals. These procedures are normally found in the Material Safety Data Sheets (MSDSs) available at your facility.

An MSDS is required for all materials that could be hazardous to personnel or equipment. These sheets contain information on the material, such as the manufacturer and chemical makeup. As much information as possible is kept on the hazardous material to prevent a dangerous situation; or, in the event of a dangerous situation, the information is used to rectify the problem in as safe a manner as possible. See *Figure 13* for an example of procedures you may find on the job.

11.1.0 PRECAUTIONS WHEN USING SOLVENTS

It is always best to use a nonflammable, nontoxic solvent whenever possible. However, any time solvents are used, it is essential that your work area be adequately ventilated and that you wear the appropriate personal protective equipment:

- A chemical face shield with chemical goggles should be used to protect the eyes and skin from sprays and splashes.
- A chemical apron should be worn to protect your body from sprays and splashes. Remember that some solvents are acid-based. If they come into contact with your clothes, solvents can eat through your clothes to your skin.

Section VII — Precautions for Safe Handling and Use

Steps to Be Taken in Case Material is Released or Spilled
Isolate from oxidizers, heat, sparks, electric equipment, and open flames.

Waste Disposal Method
Recycle or incinerate observing local, state and federal health, safety
and pollution laws.

Precautions to Be Taken in Handling and Storing
Store in a cool dry area. Observe label cautions and instructions.

Other Precautions
SEE ATTACHMENT PARA #3

Section VIII — Control Measures

Respiratory Protection (Specify Type)
Suitable for use with organic solvents

Ventilation	Local Exhaust	Special
	preferable	none
	Mechanical (General)	Other
	acceptable	none

Protective Gloves	Eye Protection
recommended (must not dissolve in solvents)	goggles

Other Protective Clothing or Equipment
none

Work/Hygenic Practices
Use with adequate ventilation. Observe label cautions.

101F13.TIF

Figure 13. Portion Of An MSDS

- A paper filter mask does not stop vapors; it is used only for nuisance dust. In situations where a paper mask does not supply adequate protection, chemical cartridge respirators might be needed. These respirators can stop many vapors if the correct cartridge is selected. In areas where ventilation is a serious problem, a self-contained breathing apparatus (SCBA) must be used. Make sure that you have been given a full medical evaluation and that you are properly trained in using respirators at your site.

11.2.0 RESPIRATORY PROTECTION

Protection against high concentrations of dust, mist, fumes, vapors, gases, and/or oxygen deficiency is provided by appropriate respirators.

Appropriate respiratory protective devices should be used for the hazardous material involved and the extent and nature of the work performed.

An air-purifying respirator is, as its name implies, a respirator that removes contaminants from air inhaled by the wearer. The respirators may be divided into the following types: particulate-removing (mechanical filter), gas- and vapor-removing (chemical filter), and a combination of particulate-removing and gas- and vapor-removing.

Particulate-removing respirators are designed to protect the wearer against the inhalation of particulate matter in the ambient atmosphere. They may be designed to protect against a single type of particulate, such as pneumoconiosis-producing and nuisance dust, toxic dust, metal fumes or mist, or against various combinations of these types.

Gas and vapor-removing respirators are designed to protect the wearer against the inhalation of gases or vapors in the ambient atmosphere. They are designated as gas masks, chemical cartridge respirators (nonemergency gas respirators), and self-rescue respirators. They may be designed to protect against a single gas such as chlorine; a single type of gas, such as acid gases; or a combination of types of gases, such as acid gases and organic vapors.

If you are required to use a respiratory protective device, you must be evaluated by a physician to ensure that you are physically fit to use a respirator. You must then be fitted and thoroughly instructed in the respirator's use.

WARNING!   Do not use any respirator unless you have been fitted for it and thoroughly understand its use. As with all safety rules, follow your employer's respiratory program and policies.

Any employee whose job entails having to wear a respirator must keep their face free of facial hair in the seal area.

Respiratory protective equipment must be inspected regularly and maintained in good condition. Respiratory equipment must be properly cleaned on a regular basis and stored in a sanitary, dustproof container.

12.0.0   ASBESTOS

Asbestos is a mineral-based material that is resistant to heat and corrosive chemicals. Depending on the chemical composition, asbestos fibers may range in texture from coarse to silky. The properties that make asbestos fibers so valuable to industry are its high tensile strength, flexibility, heat and chemical resistance, and good frictional properties.

Asbestos fibers enter the body by inhalation of airborne particles or by ingestion and can become embedded in the tissues of the respiratory or digestive systems. Years of exposure to asbestos can cause numerous disabling or fatal diseases. Among these diseases are asbestosis, an emphysema-like condition; lung cancer; mesothelioma, a cancerous tumor that spreads rapidly in the cells of membranes covering the lungs and body organs; and gastrointestinal cancer.

12.1.0 MONITORING

Employers who have a workplace or work operation covered by OSHA 3096 (*Asbestos Standard for the Construction Industry*) must perform initial monitoring to determine the airborne concentrations of asbestos to which employees may be exposed. If employers can demonstrate that employee exposures are below the action level and/or excursion limit by means of objective or historical data, initial monitoring is not required. If initial monitoring indicates that employee exposures are below the action level and/or excursion limit, then periodic monitoring is not required. Within regulated areas, the employer must conduct daily monitoring unless all workers are equipped with supplied-air respirators operated in the positive-pressure mode. If daily monitoring by statistically reliable measurements indicates that employee exposures are below the action level and/or excursion limit, then no further monitoring is required for those employees whose exposures are represented by such monitoring. Employees must be given the chance to observe monitoring, and affected employees must be notified as soon as possible following the employer's receipt of the results.

12.2.0 REGULATED AREAS

The employer must establish a regulated area where airborne concentrations of asbestos exceed or can reasonably be expected to exceed the locally determined exposure limit, or when certain types of construction work are performed, such as cutting asbestos-cement sheets and removing asbestos-containing floor tiles. Only authorized personnel may enter regulated areas. All persons entering a regulated area must be supplied with an appropriate respirator. No smoking, eating, drinking, or applying cosmetics is permitted in regulated areas. Warning signs must be displayed at each regulated area and must be posted at all approaches to regulated areas. These signs must bear the following information:

DANGER

ASBESTOS

CANCER AND LUNG DISEASE HAZARD

AUTHORIZED PERSONNEL ONLY

RESPIRATORS AND PROTECTIVE CLOTHING

ARE REQUIRED IN THIS AREA

Where feasible, the employer shall establish negative-pressure enclosures before commencing asbestos removal, demolition, and renovation operations. The setup and monitoring requirements for negative-pressure enclosures are as follows:

- A competent person shall be designated to set up the enclosure and ensure its integrity and supervise employee activity within the enclosure.

- Exemptions are given for small-scale, short duration maintenance or renovation operations.
- The employer shall conduct daily monitoring of the exposure of each employee who is assigned to work within a regulated area. Short-term monitoring is required whenever asbestos concentrations will not be uniform throughout the workday and where high concentrations of asbestos may reasonably be expected to be released or created in excess of the local limit.

In addition, warning labels must be affixed on all asbestos products and to all containers of asbestos products, including waste containers, that may be in the workplace. The label must include the following information:

```
┌─────────────────────────────────────┐
│            DANGER                    │
│                                      │
│     CONTAINS ASBESTOS FIBERS         │
│                                      │
│       AVOID CREATING DUST            │
│                                      │
│   CANCER AND LUNG DISEASE HAZARD     │
└─────────────────────────────────────┘
```

12.3.0 METHODS OF COMPLIANCE

To the extent feasible, engineering and work practice controls must be used to reduce employee exposure to within the permissible exposure limit (PEL). The employer must use one or more of the following control methods to achieve compliance:

- Local exhaust ventilation equipped with high-efficiency particulate air (HEPA) filter dust collection systems
- General ventilation systems
- Vacuum cleaners equipped with HEPA filters
- Enclosure or isolation of asbestos dust-producing processes
- Use of wet methods, wetting agents, or removal encapsulants during asbestos handling, mixing, removal, cutting, application, and cleanup
- Prompt disposal of asbestos-containing wastes in leak-tight containers

Prohibited work practices include the following:

- The use of high-speed abrasive disc saws that are not equipped with appropriate engineering controls
- The use of compressed air to remove asbestos-containing materials, unless the compressed air is used in conjunction with an enclosed ventilation system

Where engineering and work practice controls have been instituted but are insufficient to reduce employee exposure to a level that is at or below the PEL, respiratory protection must be used to supplement these controls.

13.0.0 BATTERIES

Working around wet cell batteries can be dangerous if the proper precautions are not taken. Batteries often give off hydrogen gas as a byproduct. When hydrogen mixes with air, the mixture can be explosive in the proper concentration. For this reason, smoking is strictly prohibited in battery rooms, and only insulated tools should be used. Proper ventilation also reduces the chance of explosion in battery areas. Follow your company's procedures for working near batteries. Also, ensure that your company's procedures are followed for lifting heavy batteries.

13.1.0 ACIDS

Batteries also contain acid, which will eat away human skin and many other materials. Personal protective equipment for battery work typically includes chemical aprons, sleeves, gloves, face shields, and goggles to prevent acid from contacting skin and eyes. Follow your site procedures for dealing with spills of these materials. Also, know the location of first aid when working with these chemicals.

13.2.0 WASH STATIONS

Because of the chance that battery acid may contact someone's eyes or skin, wash stations are located near battery rooms. Do not connect or disconnect batteries without proper supervision. Everyone who works in the area should know where the nearest wash station is and how to use it. Battery acid should be flushed from the skin and eyes with large amounts of water or with a neutralizing solution.

CAUTION: If you come in contact with battery acid, report it immediately to your supervisor.

14.0.0 PCBs

Polychlorinated biphenyls (PCBs) are chemicals that were marketed under various trade names as a liquid insulator/cooler in older transformers. In addition to being used in older transformers, PCBs are also found in some large capacitors and in the small ballast transformers used in street lighting and ordinary fluorescent light fixtures. Disposal of these materials is regulated by the EPA and must be done through a regulated disposal company; use extreme caution and follow your facility procedures.

WARNING! Do not come into contact with PCBs. They present a variety of serious health risks, including lung damage and cancer.

15.1.0 FALL PROTECTION PROCEDURES

Fall protection must be used when employees are on a walking or working surface that is six feet or more above a lower level and has an unprotected edge or side. The areas covered include, but are not limited to:

- Finished and unfinished floors or mezzanines
- Temporary or permanent walkways/ramps
- Finished or unfinished roof areas
- Elevator shafts and hoist-ways
- Floor, roof, or walkway holes
- Working six feet or more above dangerous equipment

Exception: If the dangerous equipment is unguarded, fall protection must be used at all heights regardless of the fall distance.

Note: Walking/working surfaces do not include ladders, scaffolds, vehicles, or trailers. Also, an unprotected edge or side is an edge/side where there is no guardrail system at least 39 inches high.

Fall protection is not required during inspection, investigation, or assessment of job site conditions before or after construction work.

These fall protection guidelines do not apply to the following areas. Fall protection for these areas is located in the subparts cited in parenthesis.

- Cranes and derricks (1926 subpart N/1910 subpart N)
- Scaffolding (1926 subpart L/1910 subpart D)
- Electrical power transmission and distribution (1926 subpart V/1910 subpart R)
- Stairways and ladders (1926 subpart X/1910 subpart D)
- Excavations (1926 subpart P)

Fall protection must be selected in order of preference as listed below. Selection of a lower-level system (e.g., safety nets) must be based only on feasibility of protection. The list includes, but is not limited to, the following:

- Guardrail systems and hole covers
- Personal fall arrest systems
- Safety nets

These fall protection procedures are designed to warn, isolate, restrict, or protect workers from a potential fall hazard.

15.2.0 TYPES OF FALL PROTECTION SYSTEMS

The type of system selected shall depend on the fall hazards associated with the work to be performed. First, a hazard analysis shall be conducted by the job site supervisor prior to the start of work. Based on the hazard analysis, the job site supervisor and project manager, in consultation with the safety manager, will select the appropriate fall protection system. All employees will be instructed in the use of the fall protection system before starting work.

SUMMARY

Safety must be your concern at all times so that you do not become either the victim of an accident or the cause of one. Safety requirements and safe work practices are provided by OSHA and your employer. It is essential that you adhere to all safety requirements and follow your employer's safe work practices and procedures. Also, you must be able to identify the potential safety hazards of your job site. The consequences of unsafe job site conduct can often be expensive, painful, or even deadly. Report any unsafe act or condition immediately to your supervisor. You should also report all work-related accidents, injuries, and illnesses to your supervisor immediately. Remember, proper construction techniques, common sense, and a good safety attitude will help to prevent accidents, injuries, and fatalities.

References

For advanced study of topics covered in this task module, the following books are suggested:

29 CFR Parts 1900 – 1910, Standards for General Industry, Occupational Safety and Health Administration, US Department of Labor.

29 CFR Part 1926, Standards for the Construction Industry, Occupational Safety and Health Administration, US Department of Labor.

National Electrical Code Handbook, Latest Edition, National Fire Protection Association, Quincy, MA.

National Electrical Safety Code, Latest Edition, National Fire Protection Association, Quincy, MA.

REVIEW QUESTIONS

1. The most life-threatening hazards on a construction site are _____.
   a. falls
   b. electrocution
   c. being crushed or struck by falling or flying objects
   d. all of the above

2. If a person's heart begins to fibrillate due to an electrical shock, the solution is to _____.
   a. leave the person alone until the fibrillation stops
   b. administer heart massage
   c. use the Heimlich maneuver
   d. have a qualified person use emergency defibrillation equipment

3. The majority of injuries due to electrical shock are caused by _____.
   a. electrically-operated hand tools
   b. contact with low-voltage conductors
   c. contact with high-voltage conductors
   d. lightning

4. Class 0 rubber gloves are used when working with voltages less than _____.
   a. 1,000 volts
   b. 7,500 volts
   c. 17,500 volts
   d. 26,500 volts

5. An important use of a hot stick is to _____.
   a. keep cattle moving
   b. keep your hands warm
   c. replace fuses
   d. test circuits to see if they are live

6. Which of these statements correctly describes a double-insulated power tool?
   a. There is twice as much insulation on the power cord.
   b. It can safely be used in place of a grounded tool.
   c. It is made entirely of plastic or other non-conducting material.
   d. The entire tool is covered in rubber.

7. Which of the following applies in a lockout/tagout procedure?
   a. Only the supervisor can install lockout/tagout devices.
   b. If several employees are involved, the lockout/tagout equipment is applied only by the first employee to arrive at the disconnect.
   c. Lockout/tagout devices applied by one employee can be removed by another employee as long as it can be verified that the first employee has left for the day.
   d. Lockout/tagout devices are installed by every authorized employee involved in the work.

8. What is the proper distance from the feet of a straight ladder to the wall?
   a. One-fourth the working height of the ladder
   b. One-half the height of the ladder
   c. Three feet
   d. One-fourth of the square root of the height of the ladder

9. What are the minimum and maximum distances (in inches) that a scaffold plank can extend beyond its end support?
   a. 4, 8
   b. 6, 10
   c. 6, 12
   d. 8, 12

10. Which of these conditions applies to a permit-required confined space, but not to a permit-required space controlled by ventilation?
    a. A hazard review checklist must be completed.
    b. An attendant, whose job is to monitor the space, must be stationed outside the space.
    c. Unacceptable atmospheric conditions must be eliminated.
    d. Atmospheric testing must be conducted.

ANSWERS TO REVIEW QUESTIONS

Answer	**Section**
1. d	1.0.0
2. d	2.0.0
3. b	2.1.0
4. a	3.1.1
5. c	3.1.4
6. b	4.3.0
7. d	4.3.1
8. a	5.1.1
9. c	5.2.0
10. b	9.3.0

SUMMARY OF OSHA CONSTRUCTION STANDARDS

CONSTRUCTION SAFETY AND
HEALTH OUTREACH PROGRAM

U.S. Department of Labor
OSHA Office of Training and Education

ELECTRICAL STANDARDS FOR CONSTRUCTION

INTRODUCTION

Electricity has long been recognized as a serious workplace hazard, exposing employees to such dangers as electric shock, electrocution, fires, and explosions.

Experts in electrical safety have traditionally looked toward the widely used *National Electrical Code* (NEC) for help in the practical safeguarding of persons from these hazards. The Occupational Safety and Health Administration (OSHA) recognized the important role of the NEC in defining basic requirements for safety in electrical installations by including the entire 1971 NEC by reference in Subpart K of 29 *Code of Federal Regulations* Part 1926 (Construction Safety and Health Standards).

In a final rule dated July 11, 1986, OSHA updated, simplified, and clarified Subpart K, 29 CFR 1926. The revisions serve these objectives:

- NEC requirements that directly affect employees in construction workplaces have been placed in the text of the OSHA standard, eliminating the need for the NEC to be incorporated by reference.

- Certain requirements that supplemented the NEC have been integrated in the new format.

- Performance language is utilized and superfluous specifications omitted and changes in technology accommodated.

In addition, the standard is easier for employers and employees to use and understand. Also, the OSHA revision of the electrical standards has been made more flexible, eliminating the need for constant revision to keep pace with the NEC, which is revised every three years.

SUBPART K

The NEC provisions directly related to employee safety are included in the body of the standard itself - making it unnecessary to continue the adoption by reference of the NEC. Subpart K is divided into four major groups plus a general definitions section:

- Installation Safety Requirements
  [29 CFR 1926.402 - 1926.415]

- Safety-Related Work Practices
  [29 CFR 1926.416 - 1926.430]

- Safety-Related Maintenance and Environmental Considerations
  [29 CFR 1926.431 - 1926.440]

- Safety Requirements for Special Equipment
  [29 CFR 1926.441 - 1926.448]

- Definitions
  [29 CFR 1926.449]

101APX02.TIF

I. INSTALLATION SAFETY REQUIREMENTS

Part I of the standard is very comprehensive. Only some of the major topics and brief summaries of these requirements are included in this discussion.

Sections 29 CFR 1926.402 through 1926.408 contain installation safety requirements for electrical equipment and installations used to provide electric power and light at the jobsite. These sections apply to installations, both temporary and permanent, used on the jobsite; but they *do not* apply to existing permanent installations that were in place before the construction activity commenced.

If an installation is made in accordance with the 1984 *National Electrical Code*, it will be considered to be in compliance with Sections 1926.403 through 1926.408, except for:

1926.404(b)(1)	Ground-fault protection for employees
1926.405(a)(2)(ii)(E)	Protection of lamps on temporary wiring
1926.405(a)(2)(ii)(F)	Suspension of temporary lights by cords
1926.405(a)(2)(ii)(G)	Portable lighting used in wet or conductive locations
1926.405(a)(2)(ii)(J)	Extension cord sets and flexible cords

101APX03.TIF

Approval

The electrical conductors and equipment used by the employer must be approved.

Examination, Installation, and Use of Equipment

The employer must ensure that electrical equipment is free from recognized hazards that are likely to cause death or serious physical harm to employees. Safety of equipment must be determined by the following:

- Suitability for installation and use in conformity with the provisions of the standard. Suitability of equipment for an identified purpose may be evidenced by a listing, by labeling, or by certification for that identified purpose.

- Mechanical strength and durability. For parts designed to enclose and protect other equipment, this includes the adequacy of the protection thus provided.

- Electrical insulation.

- Heating effects under conditions of use.

- Arcing effects.

- Classification by type, size, voltage, current capacity, and specific use.

- Other factors that contribute to the practical safeguarding of employees who use or are likely to come in contact with the equipment.

Guarding

Live parts of electric equipment operating at 50 volts or more must be guarded against accidental contact. Guarding of live parts must be accomplished as follows:

- Location in a cabinet, room, vault, or similar enclosure accessible only to qualified persons.

101APX04.TIF

- Use of permanent, substantial partitions or screens to exclude unqualified persons.

- Location on a suitable balcony, gallery, or platform elevated and arranged to exclude unqualified persons.

- Elevation of eight feet or more above the floor.

Entrance to rooms and other guarded locations containing exposed live parts must be marked with conspicuous warning signs forbidding unqualified persons to enter.

Electric installations that are over 600 volts and that are open to unqualified persons must be made with metal-enclosed equipment or enclosed in a vault or area controlled by a lock. In addition, equipment must be marked with appropriate caution signs.

Overcurrent Protection

The following requirements apply to overcurrent protection of circuits rated 600 volts, nominal, or less.

- Conductors and equipment must be protected from overcurrent in accordance with their ability to safely conduct current and the conductors must have sufficient current-carrying capacity to carry the load.

- Overcurrent devices must not interrupt the continuity of the grounded conductor unless all conductors of the circuit are opened simultaneously, except for motor-running overload protection.

- Overcurrent devices must be readily accessible and not located where they could create an employee safety hazard by being exposed to physical damage or located in the vicinity of easily ignitable material.

- Fuses and circuit breakers must be so located or shielded that employees will not be burned or otherwise injured by their operation, e.g., arcing.

101APX05.TIF

Grounding of Equipment Connected by Cord and Plug

Exposed noncurrent-carrying metal parts of cord- and plug-connected equipment that may become energized must be grounded in the following situations:

- When in a hazardous (classified) location.

- When operated at over 150 volts to ground, except for guarded motors and metal frames of electrically heated appliances if the appliance frames are permanently and effectively insulated from ground.

- When one of the types of equipment listed below. But see Item 6 for exemption.

    1. Hand held motor-operated tools.

    2. Cord- and plug-connected equipment used in damp or wet locations or by employees standing on the ground or on metal floors or working inside metal tanks or boilers.

    3. Portable and mobile X-ray and associated equipment.

    4. Tools likely to be used in wet and/or conductive locations.

    5. Portable hand lamps.

    6. [Exemption] Tools likely to be used in wet and/or conductive locations need not be grounded if supplied through an isolating transformer with an ungrounded secondary of not over 50 volts. Listed or labeled portable tools and appliances protected by a system of double insulation, or its equivalent, need not be grounded. If such a system is employed, the equipment must be distinctively marked to indicate that the tool or appliance uses a system of double insulation.

101APX06.TIF

II. SAFETY-RELATED WORK PRACTICES

Protection of Employees

The employer must not permit an employee to work near any part of an electric power circuit that the employee could contact in the course of work, unless the employee is protected against shock by de-energizing the circuit and grounding it or by guarding it effectively by insulation or other means.

Where the exact location of underground electric power lines is unknown, employees using jack hammers or hand tools that may contact a line must be provided with insulated protective gloves.

Even before work is begun, the employer must determine by inquiry, observation, or instruments where any part of an exposed or concealed energized electric power circuit is located. This is necessary because a person, tool or machine could come into physical or electrical contact with the electric power circuit.

The employer is required to advise employees of the location of such lines, the hazards involved, and protective measures to be taken as well as to post and maintain proper warning signs.

Passageways and Open Spaces

The employer must provide barriers or other means of guarding to ensure that workspace for electrical equipment will not be used as a passageway during the time when energized parts of electrical equipment are exposed. Walkways and similar working spaces must be kept clear of electric cords. Other standards cover load ratings, fuses, cords, and cables.

Lockout and Tagging of Circuits

Tags must be placed on controls that are to be deactivated during the course of work on energized or de-energized equipment or circuits. Equipment or circuits that are de-energized must be rendered inoperative and have tags attached at all points where such equipment or circuits can be energized.

101APX07.TIF

III. SAFETY-RELATED MAINTENANCE AND ENVIRONMENTAL CONSIDERATIONS

Maintenance of Equipment

The employer must ensure that all wiring components and utilization equipment in hazardous locations are maintained in a dust-tight, dust-ignition-proof, or explosion-proof condition without loose or missing screws, gaskets, threaded connections, seals, or other impairments to a tight condition.

Environmental Deterioration of Equipment

Unless identified for use in the operating environment, no conductors or equipment can be located:

- In damp or wet locations.

- Where exposed to gases, fumes, vapors, liquids, or other agents having a deteriorating effect on the conductors or equipment.

- Where exposed to excessive temperatures.

Control equipment, utilization equipment, and busways approved for use in dry locations only must be protected against damage from the weather during building construction.

For protection against corrosion, metal raceways, cable armor, boxes, cable sheathing, cabinets, elbows, couplings, fittings, supports, and support hardware must be of materials appropriate for the environment in which they are installed.

101APX08.TIF

IV. SAFETY REQUIREMENTS FOR SPECIAL EQUIPMENT

Batteries

Batteries of the unsealed type must be located in enclosures with outside vents or in well-ventilated rooms arranged to prevent the escape of fumes, gases, or electrolyte spray into other areas. Other provisions include the following:

Ventilation--to ensure diffusion of the gases from the battery and to prevent the accumulation of an explosive mixture.

Racks and trays--treated to make them resistant to the electrolyte.

Floors--acid-resistant construction unless protected from acid accumulations.

Face shields, aprons, and rubber gloves--for workers handling acids or batteries.

Facilities for quick drenching of the eyes and body--within 25 feet (7.62 m) of battery handling areas.

Facilities--for flushing and neutralizing spilled electrolytes and for fire protection.

Battery Charging

Battery charging installations must be located in areas designated for that purpose. When batteries are being charged, vent caps must be maintained in functioning condition and kept in place to avoid electrolyte spray. Also, charging apparatus must be protected from damage by trucks.

101APX09.TIF

HAZARDOUS (CLASSIFIED) LOCATIONS

The *National Electrical Code* (NEC) defines hazardous locations as those areas "where fire or explosion hazards may exist due to flammable gases or vapors, flammable liquids, combustible dust, or ignitable fibers or flyings."

A substantial part of the NEC is devoted to the discussion of hazardous locations. That's because electrical equipment can become a source of ignition in these volatile areas. Articles 500 through 504, and 510 through 517 provide classification and installation standards for the use of electrical equipment in these locations. The writers of the NEC developed a short-hand method of describing areas classified as hazardous locations. One of the purposes of this discussion is to explain this classification system. Hazardous locations are classified in three ways by the *National Electrical Code*: TYPE, CONDITION, and NATURE.

Hazardous Location Types

Class I Locations
According to the NEC, there are three types of hazardous locations. The first type of hazard is one which is created by the presence of <u>flammable gases or vapors</u> in the air, such as natural gas or gasoline vapor. When these materials are found in the atmosphere, a potential for explosion exists, which could be ignited if an electrical or other source of ignition is present. The Code writers have referred to this first type of hazard as <u>Class I</u>. So, a *Class I Hazardous Location* is one in which *flammable gases or vapors* may be present in the air in sufficient quantities to be explosive or ignitable. Some typical Class I locations are:

- Petroleum refineries, and gasoline storage and dispensing areas;

- Dry cleaning plants where vapors from cleaning fluids can be present;

- Spray finishing areas;

- Aircraft hangars and fuel servicing areas; and

101APX10.TIF

- Utility gas plants, and operations involving storage and handling of liquified petroleum gas or natural gas.

All of these are Class I . . . gas or vapor . . . hazardous locations. All require special Class I hazardous location equipment.

Class II Locations

The second type of hazard listed by the *National Electrical Code* are those areas made hazardous by the presence of combustible <u>dust</u>. These are referred to in the Code as "Class II Locations." Finely pulverized material, suspended in the atmosphere, can cause as powerful an explosion as one occurring at a petroleum refinery. Some typical Class II locations are:

- Grain elevators;

- Flour and feed mills;

- Plants that manufacture, use or store magnesium or aluminum powders;

- Producers of plastics, medicines and fireworks;

- Producers of starch or candies;

- Spice-grinding plants, sugar plants and cocoa plants; and

- Coal preparation plants and other carbon handling or processing areas.

Class III Locations

Class III hazardous locations, according to the NEC, are areas where there are <u>easily-ignitable fibers or flyings</u> present, due to the types of materials being handled, stored, or processed. The fibers and flyings are not likely to be suspended in the air, but can collect around machinery or on lighting fixtures and where heat, a spark or hot metal can ignite them. Some typical Class III locations are:

101APX11.TIF

- Textile mills, cotton gins;

- Cotton seed mills, flax processing plants; and

- Plants that shape, pulverize or cut wood and create sawdust or flyings.

Hazardous Location Conditions

In addition to the types of hazardous locations, the *National Electrical Code* also concerns itself with the kinds of conditions under which these hazards are present. The Code specifies that hazardous material may exist in several different kinds of conditions which, for simplicity, can be described as, first, normal conditions, and, second, abnormal conditions.

In the normal condition, the hazard would be expected to be present in everyday production operations or during frequent repair and maintenance activity.

When the hazardous material is expected to be confined within closed containers or closed systems and will be present only through accidental rupture, breakage or unusual faulty operation, the situation could be called "abnormal."

The Code writers have designated these two kinds of conditions very simply, as Division 1 - normal and Division 2 - abnormal. Class I, Class II and Class III hazardous locations can be either Division 1 or Division 2.

Good examples of Class I, Division 1 locations would be the areas near open dome loading facilities or adjacent to relief valves in a petroleum refinery, because the hazardous material would be present during normal plant operations.

Closed storage drums containing flammable liquids in an inside storage room would not normally allow the hazardous vapors to escape into the atmosphere. But, what happens if one of the containers is leaking? You've got a Division 2 - abnormal - condition . . . a Class I, Division 2 hazardous location.

101APX12.TIF

So far we've covered the three types of hazardous locations:

Class I - gas or vapor
Class II - dust, and
Class III - fibers and flyings.

And secondly, kinds of conditions:

Division 1 - normal conditions, and
Division 2 - abnormal conditions.

Now let's move on to a discussion of the <u>nature</u> of hazardous substances.

Nature of Hazardous Substances

The gases and vapors of Class I locations are broken into four groups by the Code: A, B, C, and D. These materials are grouped according to the ignition temperature of the substance, its explosion pressure, and other flammable characteristics.

The only substance in Group A is acetylene. Acetylene makes up only a very small percentage of hazardous locations. Consequently, little equipment is available for this type of location. Acetylene is a gas with extremely high explosion pressures.

Group B is another relatively small segment of classified areas. This group includes hydrogen and other materials with similar characteristics. If you follow certain specific restrictions in the Code, some of these Group B locations, other than hydrogen, can actually be satisfied with Group C and Group D equipment.

Group C and Group D are by far the most usual Class I groups. They comprise the greatest percentage of all Class I hazardous locations. Found in Group D are many of the most common flammable substances such as butane, gasoline, natural gas and propane.

101APX13.TIF

In Class II - dust locations - we find the hazardous materials in Groups E, F, and G. These groups are classified according to the <u>ignition temperature</u> and the <u>conductivity</u> of the hazardous substance. Conductivity is an important consideration in Class II locations, especially with metal dusts.

Metal dusts are categorized in the Code as Group E. Included here are aluminum and magnesium dusts and other metal dusts of similar nature.

Group F atmospheres contain such materials as carbon black, charcoal dust, coal and coke dust.

In Group G we have grain dusts, flour, starch, cocoa, and similar types of materials.

Review

Let's quickly review. Hazardous locations are classified in three ways by the *National Electrical Code*: TYPE, CONDITION, and NATURE.

There are three <u>types</u> of hazardous conditions: Class I - gas and vapor, Class II - dust, and Class III - fibers and flyings.

There are two kinds of hazardous <u>conditions</u>: Division 1 - normal, and Division 2 - abnormal.

And finally, there is the nature of the hazardous substance . . . where we find Groups A, B, C, and D in Class I locations, and, in Class II locations: Groups E, F, and G.

Let's illustrate our Code "translation" with an example. How would we classify a storage area where LP gas is contained in closed tanks? LP gas is a Class I substance (gas or vapor). It's Division 2 because it would only be in the atmosphere if an accidental rupture or leakage occurred, and it is Group D material.

The table below summarizes the various hazardous (classified) locations.

101APX14.TIF

SUMMARY OF CLASS I, II, III HAZARDOUS LOCATIONS				
CLASSES	GROUPS	DIVISIONS		
		1	2	
I Gases, vapors, and liquids (Art. 501)	A: Acetylene B: Hydrogen, etc. C: Ether, etc. D: Hydrocarbons, fuels, solvents, etc.	Normally explosive and hazardous	Not normally present in an explosive concentration (but may accidentally exist)
II Dusts (Art. 502)	E: Metal dusts (conductive,* and explosive) F: Carbon dusts (some are conductive,* and all are explosive) G: Flour, starch, grain, combustible plastic or chemical dust (explosive)	Ignitable quantities of dust normally are or may be in suspension, or conductive dust may be present	Dust not normally suspended in an ignitable concentration (but may accidentally exist). Dust layers are present.
III Fibers and flyings (Art. 503)	Textiles, wood-working, etc. (easily ignitable, but not likely to be explosive)	Handled or used in manufacturing	Stored or handled in storage (exclusive of manufacturing)

* NOTE: Electrically conductive dusts are dusts with a resistivity less than 10^5 ohm-centimeter.

101APX15.TIF

Hazardous Location Equipment

Sources of Ignition

Now that we've completed our Code translation, we're ready to move to the next part of our discussion - hazardous location equipment. To do this, let's first take a look at the ways in which electrical equipment can become a source of ignition. There are three of them.

<u>Arcs and sparks</u> produced by the normal operation of equipment, like motor starters, contactors, and switches, can ignite a hazardous location atmosphere.

<u>The high temperatures</u> of some heat-producing equipment, such as lamps and lighting fixtures, can ignite flammable atmospheres if they exceed the ignition temperature of the hazardous material. The *National Electrical Code* requires special marking of heat - producing equipment with temperatures above 100°C (212°F).

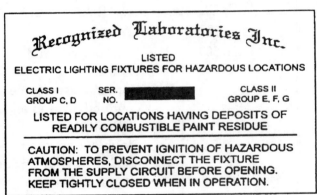

Recognized Laboratories Inc.

LISTED
ELECTRIC LIGHTING FIXTURES FOR HAZARDOUS LOCATIONS

CLASS I   SER.   CLASS II
GROUP C, D   NO.   GROUP E, F, G

LISTED FOR LOCATIONS HAVING DEPOSITS OF
READILY COMBUSTIBLE PAINT RESIDUE

CAUTION: TO PREVENT IGNITION OF HAZARDOUS
ATMOSPHERES, DISCONNECT THE FIXTURE
FROM THE SUPPLY CIRCUIT BEFORE OPENING.
KEEP TIGHTLY CLOSED WHEN IN OPERATION.

<u>Electrical equipment failure</u> is another way an explosion could be set off. A burn out of a lamp socket or shorting of a terminal could spark a real disaster in a hazardous location.

Equipment Design and Construction

Now let's get down to specific hardware and how it is designed and constructed to be suitable for hazardous locations . . . starting with those designed for Class I . . . gas or vapor . . . applications.

The first requirement for a Class I enclosure is <u>strength</u>. The enclosure must be strong enough to contain an explosion <u>within</u>. The walls must be thick enough to withstand the internal strain. It has to be explosion-proof in case gas or vapors get inside. Secondly, it must function at a temperature <u>below</u> the ignition temperature

101APX16.TIF

of the surrounding atmosphere.

The equipment must also provide a way for the burning gases to escape from the device as they expand during an internal explosion; but, only after they have been cooled off and their flames "quenched." This escape route for the exploding gases is provided through several types of flame paths.

One type is the ground surface flame path. Here the surfaces are ground, mated, and held to a tolerance of 15 ten-thousandths of an inch. This permits gases to escape, but only after they've been sufficiently cooled, so they won't ignite the volatile surrounding atmosphere.

Another kind of flame path is the threaded flame path. After an explosion, the gas travels out the threaded joint . . . but as it does, it cools off.

Exploded gases may also escape around the shafts of operators used in the enclosure. But, here again, close tolerances are used to quench the burning gas.

Examples of two flame paths are shown below.

101APX17.TIF

FLAME PATHS

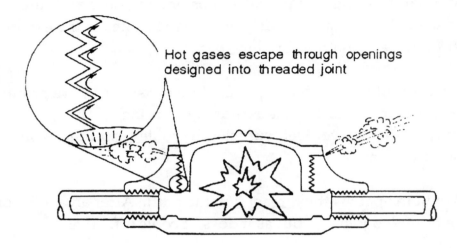

Hot gases escape through openings
designed into threaded joint

OPENINGS DESIGNED INTO THREADED JOINT

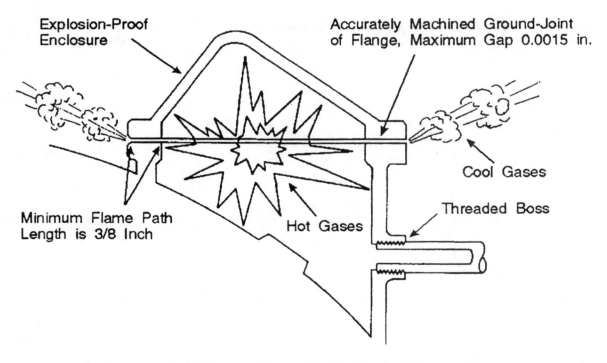

Explosion-Proof
Enclosure

Accurately Machined Ground-Joint
of Flange, Maximum Gap 0.0015 in.

Cool Gases

Threaded Boss

Minimum Flame Path
Length is 3/8 Inch

Hot Gases

OPENINGS DESIGNED INTO GROUND JOINT

101APX18.TIF

You can see how important it is to make certain that all flame paths are protected during installation and maintenance, and even during handling, shipping, and storage of explosion-proof material. Even slight damage to a flame path can permit burning gases to escape, igniting the surrounding atmosphere. Also, all cover bolts must be installed for the same reason. A single missing bolt could allow the release of flaming gases.

In designing equipment for Class I, Division 1 locations, it is assumed that the hazardous gases or vapors will be present and eventually seep into the enclosure, so there is a very real chance for an internal explosion to occur.

In the case of Class II, however, the assumptions are different and so the design is different. In Class II, the explosive dust is kept away from equipment housed within the enclosure so that no internal explosion can take place and there is no longer any need for heavy explosion-containing construction, or flame paths. This difference explains why Class I, Division 1 equipment can be called explosion-proof, and Class II equipment is called dust-ignition proof. Class II equipment has a different set of requirements:

1.   It must seal out the dust.

2.   It must operate below the ignition temperature of the hazardous substance.

3.   It must allow for a <u>dust blanket</u>. That is, the build-up of dust collecting on top of the device that can cause it to run "hot" and ignite the surrounding atmosphere.

For Class III equipment, there is very little difference in the design from Class II. Class III equipment must minimize entrance of fibers and flyings; prevent the escape of sparks, burning material or hot metal particles resulting from failure of equipment; and operate at a temperature that will prevent the ignition of fibers accumulated on the equipment.

101APX19.TIF

There are many enclosures, devices, and fixtures suitable for all three classes. This simply means that it meets the specifications for each individual type. A Class I device which could contain an explosion of a specified gas would also have to prevent dust from entering the enclosure to be suitable for Class II. The close tolerance of the flame path which cools the burning gases is also close enough to exclude explosive dust so that a gasket would not be needed.

Proper installation of hazardous location equipment calls for the use of seals. Special fittings are required to keep hot gases from traveling through the conduit system igniting other areas if an internal explosion occurs in a Class I device. They are also needed in certain situations to keep flammable dusts from entering dust-ignition-proof enclosures through the conduit. As shown in the figure below, when arcs and sparks cause ignition of flammable gases and vapors, the equipment contains the explosion and vents only cool gases into the surrounding hazardous area.

101APX20.TIF

HAZARDOUS LOCATION EQUIPMENT SEALS

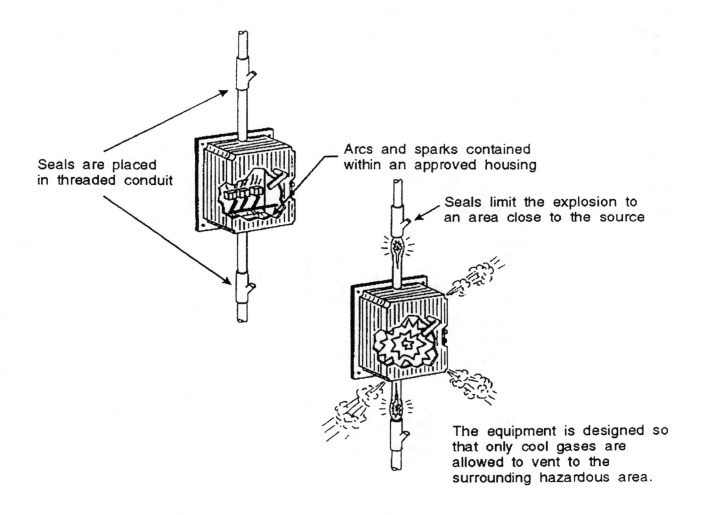

Seals are placed
in threaded conduit

Arcs and sparks contained
within an approved housing

Seals limit the explosion to
an area close to the source

The equipment is designed so
that only cool gases are
allowed to vent to the
surrounding hazardous area.

101APX21.TIF

Sealing fittings are designed to be filled with a chemical compound after the wires have been pulled. As the compound hardens, it seals passageways for dusts and gases. As shown in the figure below, in each conduit run entering an enclosure for switches, circuit breakers, fuses, relays, resistors, or other apparatus which may produce arcs, sparks, or high temperatures within Class I locations, conduit seals shall be placed as close as practicable and in no case more than 18 inches (457 mm) from such enclosures. Again, consult the Code for specific rules for the use of seals.

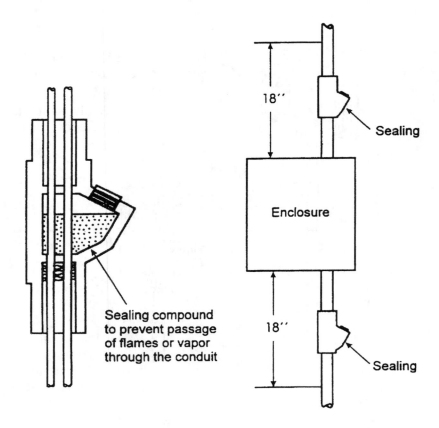

Sealing compound
to prevent passage
of flames or vapor
through the conduit

18″   Sealing

Enclosure

18″   Sealing

101APX22.TIF

Rigorous standards for hazardous location equipment have been set. Nationally Recognized Testing Laboratories conduct actual explosion tests under laboratory conditions. For each Class I enclosure they experiment with different mixtures of gas and air . . . from very lean mixtures (a small percentage of gas) to very rich mixtures (a high percentage of gas) until they find the one that creates the greatest explosion pressure. To pass inspection, the equipment must not only prevent the ignition of the surrounding atmosphere, but also be able to withstand a hydrostatic test where oil is pumped into the enclosure at high pressure to test the limits of its strength. The device will not pass unless it can resist rupture at four times the maximum pressure found in the explosion tests. For example, if explosion testing shows a maximum pressure for a junction box of 250 pounds per square inch (psi), to get approval, the box must be able to withstand 1,000 psi of hydrostatic pressure - FOUR TIMES the maximum anticipated pressure of 250 psi.

101APX23.TIF

Summary

Regardless of the cause of a hazardous location, it is necessary that every precaution be taken to guard against ignition of the atmosphere. Electrical equipment can be a potential source of ignition through one of three ways:

1. Arcs and sparks
2. High temperatures
3. Electrical equipment failure

Hazardous location equipment is designed and constructed to eliminate the potential for ignition of the atmosphere.

The *National Electrical Code* is the "Bible" of the Electrical Industry, and the primary source of reference for hazardous locations. The NEC is also the basis for OSHA standard 1926.407, Hazardous (Classified) Locations. There are several OSHA standards that require the installation of electrical wiring and equipment in hazardous (classified) locations according to the requirements of Subpart K, Electrical. The NEC should be consulted as a supplement to the OSHA standards for additional background information concerning hazardous locations.

101APX24.TIF

GROUND-FAULT PROTECTION ON CONSTRUCTION SITES

INSULATION AND GROUNDING

Insulation and grounding are two recognized means of preventing injury during electrical equipment operation. Conductor insulation may be provided by placing nonconductive material such as plastic around the conductor. Grounding may be achieved through the use of a direct connection to a known ground such as a metal cold water pipe.

Consider, for example, the metal housing or enclosure around a motor or the metal box in which electrical switches, circuit breakers, and controls are placed. Such enclosures protect the equipment from dirt and moisture and prevent accidental contact with exposed wiring. However, there is a hazard associated with housings and enclosures. A malfunction within the equipment--such as deteriorated insulation--may create an electrical shock hazard. Many metal enclosures are connected to a ground to eliminate the hazard. If a "hot" wire contacts a grounded enclosure, a ground fault results which normally will trip a circuit breaker or blow a fuse. Metal enclosures and containers are usually grounded by connecting them with a wire going to ground. This wire is called an equipment grounding conductor. Most portable electric tools and appliances are grounded by this means. There is one disadvantage to grounding: a break in the grounding system may occur without the user's knowledge.

Insulation may be damaged by hard usage on the job or simply by aging. If this damage causes the conductors to become exposed, the hazards of shocks, burns, and fire will exist. Double insulation may be used as additional protection on the live parts of a tool, but double insulation does not provide protection against defective cords and plugs or against heavy moisture conditions.

The use of a ground-fault circuit interrupter (GFCI) is one method used to overcome grounding and insulation deficiencies.

101APX25.TIF

WHAT IS A GFCI?

The ground-fault circuit interrupter (GFCI) is a fast-acting circuit breaker which senses small imbalances in the circuit caused by current leakage to ground and, in a fraction of a second, shuts off the electricity. The GFCI continually matches the amount of current going to an electrical device against the amount of current returning from the device along the electrical path. Whenever the amount "going" differs from the amount "returning" by approximately 5 milliamps, the GFCI interrupts the electric power within as little as 1/40 of a second. (See diagram.)

101APX26.TIF

Ground-Fault Circuit Interrupter

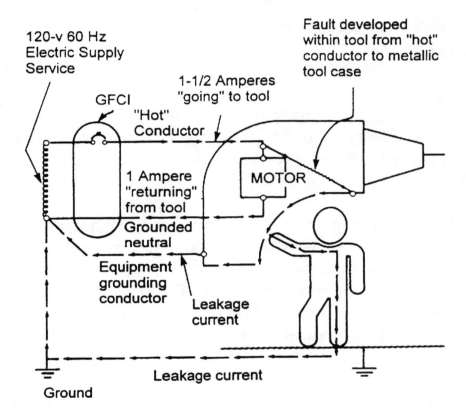

GFCI monitors the difference in current flowing into the "hot" and out to the grounded neutral conductors. The difference (1/2 ampere in this case) will flow back through any available path, such as the equipment grounding conductor, and through a person holding the tool, if the person is in contact with a grounded object.

101APX27.TIF

However, the GFCI will not protect the employee from line-to-line contact hazards (such as a person holding two "hot" wires or a hot and a neutral wire in each hand). It does provide protection against the most common form of electrical shock hazard--the ground fault. It also provides protection against fires, overheating, and destruction of insulation on wiring.

WHAT ARE THE HAZARDS?

With the wide use of portable tools on construction sites, the use of flexible cords often becomes necessary. Hazards are created when cords, cord connectors, receptacles, and cord- and plug-connected equipment are improperly used and maintained.

Generally, flexible cords are more vulnerable to damage than is fixed wiring. Flexible cords must be connected to devices and to fittings so as to prevent tension at joints and terminal screws. Because a cord is exposed, flexible, and unsecured, joints and terminals become more vulnerable. Flexible cord conductors are finely stranded for flexibility, but the strands of one conductor may loosen from under terminal screws and touch another conductor, especially if the cord is subjected to stress or strain.

A flexible cord may be damaged by activities on the job, by door or window edges, by staples or fastenings, by abrasion from adjacent materials, or simply by aging. If the electrical conductors become exposed, there is a danger of shocks, burns, or fire. A frequent hazard on a construction site is a cord assembly with improperly connected terminals.

When a cord connector is wet, hazardous leakage can occur to the equipment grounding conductor and to humans who pick up that connector if they also provide a path to ground. Such leakage is not limited to the face of the connector but also develops at any wetted portion of it.

When the leakage current of tools is below 1 ampere, and the grounding conductor has a low resistance, no shock should be perceived. However, should the resistance

101APX28.TIF

of the equipment grounding conductor increase, the current through the body also will increase. Thus, if the resistance of the equipment grounding conductor is significantly greater than 1 ohm, tools with even small leakages become hazardous.

PREVENTING AND ELIMINATING HAZARDS

GFCIs can be used successfully to reduce electrical hazards on construction sites. Tripping of GFCIs--interruption of current flow--is sometimes caused by wet connectors and tools. It is good practice to limit exposure of connectors and tools to excessive moisture by using watertight or sealable connectors. Providing more GFCIs or shorter circuits can prevent tripping caused by the cumulative leakage from several tools or by leakages from extremely long circuits.

EMPLOYER'S RESPONSIBILITY

OSHA ground-fault protection rules and regulations have been determined necessary and appropriate for employee safety and health. Therefore, it is the employer's responsibility to provide either: (a) ground-fault circuit interrupters on construction sites for receptacle outlets in use and not part of the permanent wiring of the building or structure; or (b) a scheduled and recorded assured equipment grounding conductor program on construction sites, covering all cord sets, receptacles which are not part of the permanent wiring of the building or structure, and equipment connected by cord and plug which are available for use or used by employees.

GROUND-FAULT CIRCUIT INTERRUPTERS

The employer is required to provide approved ground-fault circuit interrupters for all 120-volt, single-phase, 15- and 20-ampere receptacle outlets on construction sites which are not a part of the permanent wiring of the building or structure and which are in use by employees. Receptacles on the ends of extension cords are not part of the permanent wiring and, therefore, must be protected by GFCIs whether or not the extension cord is plugged into permanent wiring. These GFCIs monitor the current-to-the-load for leakage to ground. When this leakage exceeds 5 mA ± 1 mA,

101APX29.TIF

the GFCI interrupts the current. They are rated to trip quickly enough to prevent electrocution. This protection is required in addition to, not as a substitute for, the grounding requirements of OSHA safety and health rules and regulations, 29 CFR 1926. The requirements which employers must meet, if they choose the GFCI option, are stated in 29 CFR 1926.404(b)(1)(ii). (See appendix.)

ASSURED EQUIPMENT GROUNDING CONDUCTOR PROGRAM

The assured equipment grounding conductor program covers all cord sets, receptacles which are not a part of the permanent wiring of the building or structure, and equipment connected by cord and plug which are available for use or used by employees. The requirements which the program must meet are stated in 29 CFR 1926.404(b)(1)(iii), but employers may provide additional tests or procedures. (See appendix.) OSHA requires that a written description of the employer's assured equipment grounding conductor program, including the specific procedures adopted, be kept at the jobsite. This program should outline the employer's specific procedures for the required equipment inspections, tests, and test schedule.

The required tests must be recorded, and the record maintained until replaced by a more current record. The written program description and the recorded tests must be made available, at the jobsite, to OSHA and to any affected employee upon request. The employer is required to designate one or more **competent persons** to implement the program.

Electrical equipment noted in the assured equipment grounding conductor program must be visually inspected for damage or defects before each day's use. Any damaged or defective equipment must not be used by the employee until repaired.

Two tests are required by OSHA. One is a continuity test to ensure that the equipment grounding conductor is electrically continuous. It must be performed on all cord sets, receptacles which are not part of the permanent wiring of the building or structure, and on cord- and plug-connected equipment which is required to be grounded. This test may be performed using a simple continuity tester, such as a

101APX30.TIF

lamp and battery, a bell and battery, an ohmmeter, or a receptacle tester.

The other test must be performed on receptacles and plugs to ensure that the equipment grounding conductor is connected to its proper terminal. This test can be performed with the same equipment used in the first test.

These tests are required before first use, after any repairs, after damage is suspected to have occurred, and at 3-month intervals. Cord sets and receptacles which are essentially fixed and not exposed to damage must be tested at regular intervals not to exceed 6 months. Any equipment which fails to pass the required tests shall not be made available or used by employees.

SUMMARY

This discussion provides information to help guide employers and employees in protecting themselves against 120-volt electrical hazards on the construction site, through the use of ground-fault circuit interrupters or through an assured equipment grounding conductor program.

When planning your program, remember to use the OSHA rules and regulations as a guide to ensure employee safety and health. Following these rules and regulations will help reduce the number of injuries and accidents from electrical hazards. Work disruptions should be minor, and the necessary inspections and maintenance should require little time.

An effective safety and health program requires the cooperation of both the employer and employees.

If you need additional information planning your program, contact the OSHA office nearest you.

101APX31.TIF

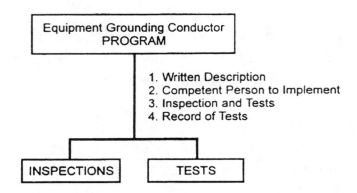

Visual inspection of following:

1. cord sets

2. cap, plug and receptacle of cord sets

3. equipment connected by cord and plug

Exceptions:

• receptacles and cord sets which are fixed and not exposed to damage

Frequency of Inspections:

• before each day's use

Conduct tests for:

1. continuity of equipment grounding conductor

2. proper terminal connection of equipment grounding conductor

Frequency of Tests:

• before first use

• after repair, and before placing back in service

• before use, after suspected damage

• every 3 months, except that cord sets and receptacles that are fixed and not exposed to damage must be tested at regular intervals not to exceed 6 months.

101APX32.TIF

APPENDIX

Construction Safety and Health Regulations Part 1926 Subpart K (Partial)

§1926.404 Wiring design and protection.

(b) Branch circuits--(1) Ground-fault protection--(i) General.
The employer shall use either ground-fault circuit interrupters as specified in paragraph (b)(l)(ii) of this section or an assured equipment grounding conductor program as specified in paragraph (b)(l)(iii) of this section to protect employees on construction sites. These requirements are in addition to any other requirements for equipment grounding conductors.

(ii) Ground-fault circuit interrupters. All 120-volt, single-phase, 15- and 20-ampere receptacle outlets on construction sites, which are not a part of the permanent wiring of the building or structure and which are in use by employees, shall have approved ground-fault circuit interrupters for personnel protection. Receptacles on a two-wire, single-phase portable or vehicle-mounted generator rated not more than 5kW, where the circuit conductors of the generator are insulated from the generator frame and all other grounded surfaces, need not be protected with ground-fault circuit interrupters.

(iii) Assured equipment grounding conductor program. The employer shall establish and implement an assured equipment grounding conductor program on construction sites covering cord sets, receptacles which are not a part of the building or structure, and equipment connected by cord and plug which are available for use or used by employees. This program shall comply with the following minimum requirements:

(A) A written description of the program, including the specific procedures adopted by the employer, shall be available at the jobsite for inspection and copying by the Assistant Secretary and any affected employee.

(B) The employer shall designate one or more competent persons [as defined in §1926.32(f)] to implement the program.

101APX33.TIF

(C) Each cord set, attachment cap, plug and receptacle of cord sets, and any equipment connected by cord and plug, except cord sets and receptacles which are fixed and not exposed to damage, shall be visually inspected before each day's use for external defects, such as deformed or missing pins or insulation damage, and for indications of possible internal damage. Equipment found damaged or defective shall not be used until repaired.

(D) The following tests shall be performed on all cord sets, receptacles which are not a part of the permanent wiring of the building or structure, and cord- and plug-connected equipment required to be grounded:

(1) All equipment grounding conductors shall be tested for continuity and shall be electrically continuous.

(2) Each receptacle and attachment cap or plug shall be tested for correct attachment of the equipment grounding conductor. The equipment grounding conductor shall be connected to its proper terminal.

(E) All required tests shall be performed:

(1) Before first use;

(2) Before equipment is returned to service following any repairs;

(3) Before equipment is used after any incident which can be reasonably suspected to have caused damage (for example, when a cord set is run over); and

(4) At intervals not to exceed 3 months, except that cord sets and receptacles which are fixed and not exposed to damage shall be tested at intervals not exceeding 6 months.

101APX34.TIF

(F) The employer shall not make available or permit the use by employees of any equipment which has not met the requirements of this paragraph (b)(1)(iii) of this section.

(G) Tests performed as required in this paragraph shall be recorded. This test record shall identify each receptacle, cord set, and cord- and plug-connected equipment that passed the test and shall indicate the last date it was tested or the interval for which it was tested. This record shall be kept by means of logs, color coding, or other effective means and shall be maintained until replaced by a more current record. The record shall be made available on the jobsite for inspection by the Assistant Secretary and any affected employee.

101APX35.TIF

The NCCER makes every effort to keep these manuals up-to-date and free of technical errors. We appreciate your help in this process. If you have an idea for improving this manual, or if you find an error, a typographical mistake, or an inaccuracy in the NCCER's Craft Training Manuals, please write us, using this form or a photocopy. Be sure to include the exact module number, page number, a description of the problem, and the correction, if possible. Your input will be brought to the attention of the Technical Review Committee. Thank you for your assistance.

Instructors – If you found that additional materials were necessary in order to teach this module effectively, please let us know so that we may include them in the Equipment/Materials list in the Instructor's Guide.

Write: Curriculum Development and Revision Department
National Center for Construction Education and Research
P.O. Box 141104
Gainesville, FL 32614-1104

Fax: 352-334-0932

Craft _____ Module Name _____

Copyright Date _____ Module Number _____ Page Number(s) _____

Description of Problem

_____

_____

_____

_____

(Optional) Correction of Problem

_____

_____

_____

(Optional) Your Name and Address

_____

_____

_____

Hand Bending

Module 26102

**NATIONAL
CENTER FOR
CONSTRUCTION
EDUCATION AND
RESEARCH**

HAND BENDING

OBJECTIVES

Upon completion of this module, the trainee will be able to:

1. Identify the methods of hand bending conduit.
2. Identify the various methods used to install conduit.
3. Use math formulas to determine conduit bends.
4. Make 90° bends, back-to-back bends, offsets, kicks, and saddle bends using a hand bender.
5. Cut, ream, and thread conduit.

Prerequisites

Successful completion of the following Task Modules is required before beginning study of this Task Module: Core Curricula; Electrical Level 1, Module 26101.

Required Trainee Materials

1. Trainee Task Module
2. Copy of the latest edition of the *National Electrical Code*
3. *OSHA Electrical Safety Guidelines* (pocket guide)
4. Appropriate Personal Protective Equipment

Note: The designations "National Electrical Code," "NE Code," and "NEC," where used in this document, refer to the *National Electrical Code®*, which is a registered trademark of the National Fire Protection Association, Quincy, MA. *All National Electrical Code (NEC) references in this module refer to the 1999 edition of the NEC.*

Course Map

This course map shows all of the task modules in the first level of the Electrical curricula. The suggested training order begins at the bottom and proceeds up. Skill levels increase as a trainee advances on the course map. The training order may be adjusted by the local Training Program Sponsor.

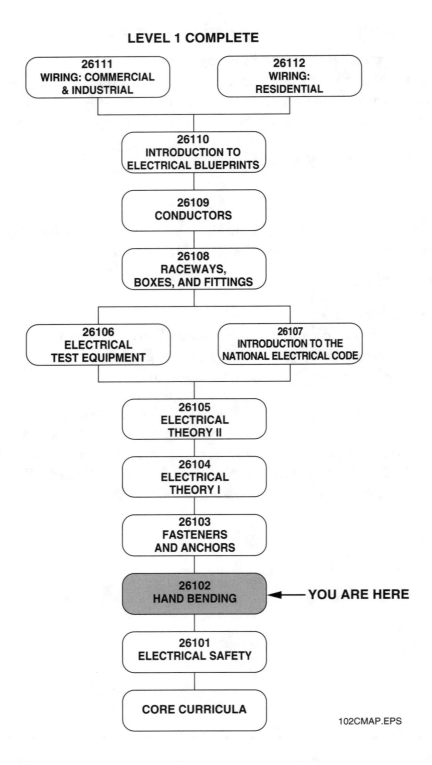

LEVEL 1 COMPLETE

26111
WIRING: COMMERCIAL & INDUSTRIAL

26112
WIRING: RESIDENTIAL

26110
INTRODUCTION TO ELECTRICAL BLUEPRINTS

26109
CONDUCTORS

26108
RACEWAYS, BOXES, AND FITTINGS

26106
ELECTRICAL TEST EQUIPMENT

26107
INTRODUCTION TO THE NATIONAL ELECTRICAL CODE

26105
ELECTRICAL THEORY II

26104
ELECTRICAL THEORY I

26103
FASTENERS AND ANCHORS

26102
HAND BENDING ◄─── YOU ARE HERE

26101
ELECTRICAL SAFETY

CORE CURRICULA

102CMAP.EPS

TABLE OF CONTENTS

Terms Introduced In This Module

90° bend: A bend that changes the direction of the conduit by 90°.

Back-to-back bend: Any bend formed by two 90° bends with a straight section of conduit between the bends.

Concentric bends: Making 90° bends in two or more parallel runs of conduit and increasing the radius of each conduit from inside of the run toward the outside.

Developed length: The actual length of the conduit that will be bent.

Gain: Because a conduit bends in a radius and not at right angles, the length of conduit needed for a bend will not equal the total determined length. Gain is the distance saved by the arc of a 90° bend.

Offsets: An offset (kick) is two bends placed in a piece of conduit to change elevation to go over or under obstructions or for proper entry into boxes, cabinets, etc.

Rise: The length of the bent section of conduit measured from the bottom, centerline, or top of the straight section to the end of the bent section.

Segment bend: A large bend formed by multiple short bends or *shots*.

Stub-up: Another name for the rise in a section of conduit. Also, a term used for conduit penetrating a slab or the ground.

1.0.0   INTRODUCTION

The art of conduit bending is dependent upon the skills of the electrician and requires a working knowledge of basic terms and proven procedures. Practice, knowledge, and training will help you gain the skills necessary for proper conduit bending and installation. You will be able to practice conduit bending in the lab and in the field under the supervision of experienced coworkers. In this module, the techniques for using hand-operated and step conduit benders such as the hand bender and the hickey will be covered. The process of hand bending conduit, and cutting, reaming, and threading conduit will also be explained.

2.0.0   HAND BENDING EQUIPMENT

Figure 1 shows a hand bender. Hand benders are convenient to use on the job because they are portable and no electrical power is required. Hand benders have a shape that supports the walls of the conduit being bent.

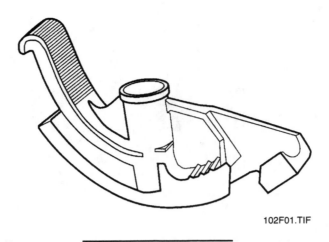

102F01.TIF

Figure 1. Hand Bender

These benders are used to make various bends in smaller-size conduit (½" to 1¼"). Most hand benders are sized to bend rigid conduit and electrical metallic tubing (EMT) of corresponding sizes. For example, a single hand bender can bend either ¾" EMT or ½" rigid conduit. The next larger size of hand bender will bend either 1" EMT or ¾" rigid conduit. This is because the corresponding sizes of conduit have nearly equal outside diameters.

The first step in making a good bend is familiarizing yourself with the bender. The manufacturer of the bender will typically provide documentation indicating starting points, distance between **offsets**, **gains**, and other important values associated with that particular bender. There is no substitute for taking the time to review this information. It will make the job go faster and result in better bends.

CAUTION: When making bends, be sure you have a firm grip on the handle to avoid slippage and possible injury.

When performing a bend, it is important to keep the conduit on a stable, firm, flat surface for the entire duration of the bend. Hand benders are designed to have force applied using one foot and the hands. See *Figure 2*. It is important to use constant foot pressure as well as force on the handle to achieve uniform bends. Allowing the conduit to rise up or performing the bend on soft ground can result in distorting the conduit outside the bender.

Note: Bends should be made in accordance with the guidelines of **NEC Article 345** (intermediate metal conduit or IMC), **Article 346** (rigid metal conduit), **Article 347** (rigid nonmetallic conduit such as polyvinyl chloride or PVC), or **Article 348** (electrical metallic tubing).

A hickey should not be confused with a hand bender. The hickey, which is used for rigid conduit only, functions quite differently. See *Figure 3*.

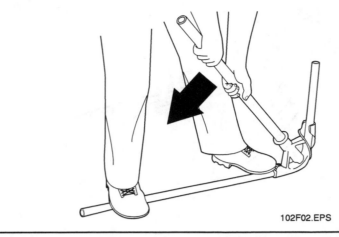

Figure 2. Pushing Down On The Bender To Complete The Bend

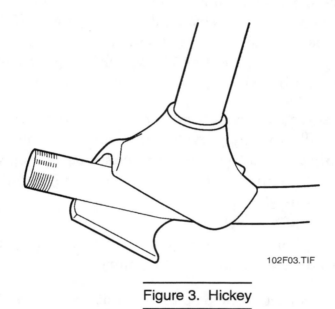

Figure 3. Hickey

When you use a hickey to bend conduit, you are forming the bend as well as the radius. When using a hickey, be careful not to flatten or kink the conduit. Hickeys should only be used with rigid conduit because very little support is given to the walls of the conduit being bent. A hickey is a segment bending device. First, a small bend of about 10° is made. Then, the hickey is moved to a new position and another small bend is made. This process is continued until the bend is completed. A hickey can be used for conduit **stub-ups** in slabs and decks.

PVC conduit is bent using a heating unit (*Figure 4*). The PVC must be rotated regularly while it is in the heater so that it heats evenly. Once heated, the PVC is removed, and the bending is performed by hand. Some units use an electric heating element, while others use liquid propane (LP). After bending, a damp sponge or cloth is often used so that the PVC sets up faster.

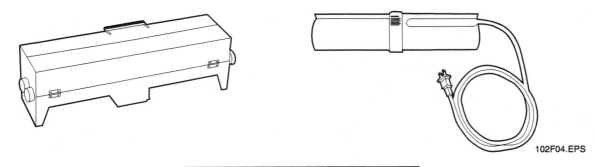

Figure 4. Typical PVC Heating Units

CAUTION: Avoid contact with the case of the heating unit; it can become very hot and cause burns. Also, to avoid a fire hazard, ensure that the unit is cool before storage. If using an LP unit, keep a fire extinguisher nearby.

When bending PVC that is 2" or larger in diameter, there is a risk of wrinkling or flattening the bend. A plug set eliminates this problem (*Figure 5*). A plug is inserted into each end of the piece of PVC being bent. Then, a hand pump is used to pressurize the conduit before bending it. The pressure is about 3 to 5 psi.

Note: The plugs must remain in place until the pipe is cooled and set.

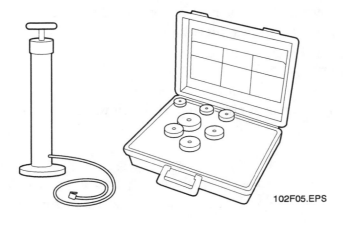

Figure 5. Typical Plug Set

2.1.0   GEOMETRY REQUIRED TO MAKE A BEND

Bending conduit requires that you use some basic geometry. You may already be familiar with most of the concepts needed; however, here is a review of the concepts directly related to this task. A right triangle is defined as any triangle with a 90° angle. The side directly opposite the 90° angle is called the *hypotenuse,* and the side on which the triangle sits is the *base.* The vertical side is called the *height.* On the job, you will apply the relationships in a right triangle when making an offset bend. The offset forms the hypotenuse of a right triangle (*Figure 6*).

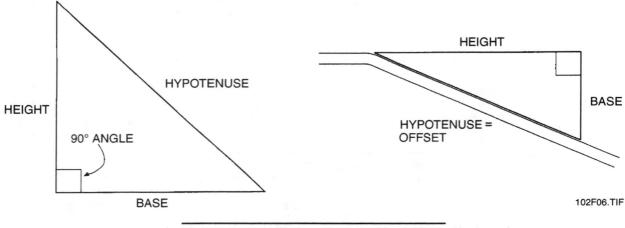

Figure 6. Right Triangle And Offset Bend

Note: There are reference tables for sizing offset bends based on these relationships (see *Appendix A*).

A circle is defined as a closed curved line whose points are all the same distance from its center. The distance from the center point to the edge of the circle is called the *radius*. The length from one edge of the circle to the other edge is the *diameter*. The distance around the circle is called the *circumference*. A circle can be divided into four equal quadrants. Each quadrant accounts for 90°, making a total of 360°. When you make a **90° bend**, you will use ¼ of a circle, or one quadrant. Concentric circles are circles that have a common center but different radii. The concept of concentric circles can be applied to **concentric bends** in conduit. The angle of each bend is 90°. Such bends have the same center point, but the radius of each is different. See *Figure 7*.

To calculate the circumference of a circle, use the following formula:

$$C = \pi \times D \text{ or } C = \pi D$$

In this formula, C = circumference, D = diameter, and π = 3.14. Another way of stating the formula for circumference is $C = 2\pi R$, where R equals the radius or ½ the diameter.

To figure the arc of a quadrant use:

$$\text{Length of arc} = (.25)\, 2\pi R = 1.57R$$

For this formula, the arc of a quadrant equals ¼ the circumference of the circle or 1.57 times the radius.

A bending radius table is included in *Appendix B*.

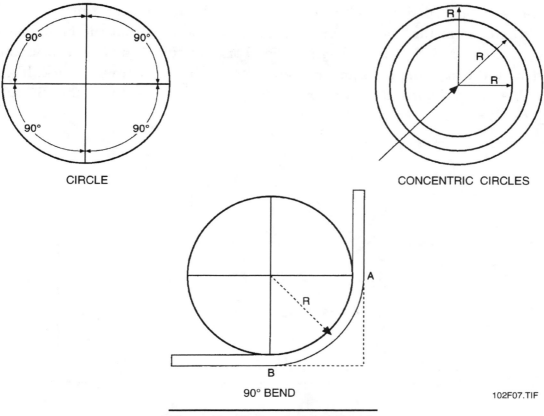

CIRCLE

CONCENTRIC CIRCLES

90° BEND

102F07.TIF

Figure 7. Circles And 90° Bends

2.2.0 MAKING A 90° BEND

The 90° stub bend is probably the most basic bend of all. The stub bend is used much of the time, regardless of the type of conduit being installed. Before beginning to make the bend, you need to know two measurements:

- Desired **rise** or stub-up
- Take-up distance of the bender

The desired rise is the height of the stub-up. The *take-up* is the amount of conduit the bender will use to form the bend. Take-up distances are usually listed in the manufacturer's instruction manual. Typical bender take-up distances are shown in *Table 1*.

EMT	Rigid/IMC	Take-Up
$\frac{1}{2}$"	—	5"
$\frac{3}{4}$"	$\frac{1}{2}$"	6"
1"	$\frac{3}{4}$"	8"
$1\frac{1}{4}$"	1"	11"

Table 1. Typical Bender Take-Up Distances

Once you have determined the take-up, subtract it from the stub-up height. Mark that distance on the conduit (all the way around) at that distance from the end. The mark will indicate the point at which you will begin to bend the conduit. Line up the starting point on the conduit with the starting point on the bender. Most benders have a mark, like an arrow, to indicate this point. *Figure 8* shows the take-up required to achieve an 18" stub-up on a piece of ½" EMT.

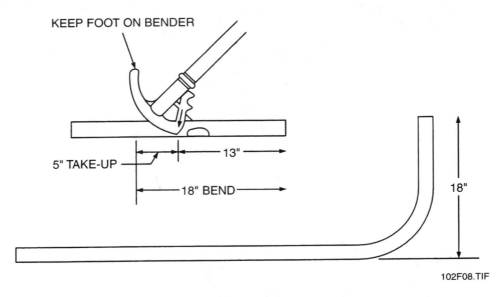

Figure 8. Bending An 18-Inch Stub-Up

Once you have lined up the bender, use one foot to hold the conduit steady. Keep your heel on the floor for balance. Apply pressure on the bender foot pedal with your other foot. Make sure you hold the bender handle level, as far up as possible, to get maximum leverage. Then, bend the conduit in one smooth motion, pulling as steadily as possible. Avoid over-stretching.

Note: When bending conduit using the take-up method, always place the bender on the conduit and make the bend facing the hook of the conduit from which the measurements were taken.

After finishing the bend, check to make sure you have the correct angle and measurement. Use the following steps to check a 90° bend:

Step 1   With the back of the bend on the floor, measure to the end of the conduit stub-up to make sure it is the right length.

Step 2   Check the 90° angle of the bend with a square or at the angle formed by the floor and a wall. A torpedo level may also be used.

Note: If you overbend a conduit slightly past the desired angle, you can use the bender to bend the conduit back to the correct angle.

ELECTRICAL — TRAINEE TASK MODULE 26102

The above procedure will produce a 90° *one-shot bend*. That means that it took a single bend to form the conduit bend. A **segment bend** is any bend that is formed by a series of bends of a few degrees each, rather than a single one-shot bend. A shot is actually one bend in a segment bend. Segment or sweep bends must conform to the provisions of the NEC.

2.3.0   GAIN

The gain is the distance saved by the arc of a 90° bend. Knowing the gain can help you to pre-cut, ream, and pre-thread both ends of the conduit before you bend it. This will make your work go more quickly because it is easier to work with conduit while it is straight. *Figure 9* shows that the overall **developed length** of a piece of conduit with a 90° bend is less than the sum of the horizontal and vertical distances when measured square to the corner. This is shown by the following equation:

Developed length = (A + B) − gain

An example of a manufacturer's gain table is also shown. These tables are used to determine the gain for a certain size conduit.

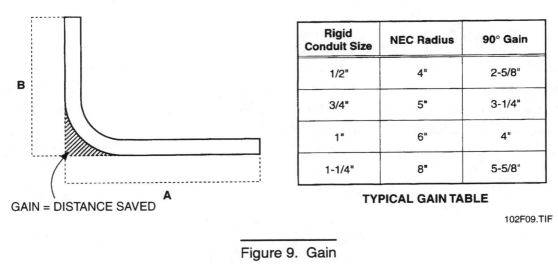

Rigid Conduit Size	NEC Radius	90° Gain
1/2"	4"	2-5/8"
3/4"	5"	3-1/4"
1"	6"	4"
1-1/4"	8"	5-5/8"

TYPICAL GAIN TABLE

102F09.TIF

GAIN = DISTANCE SAVED

Figure 9. Gain

2.4.0   BACK-TO-BACK 90° BENDS

A **back-to-back bend** consists of two 90° bends made on the same piece of conduit and placed back-to-back (*Figure 10*).

To make a back-to-back bend, make the first bend (labeled *X* in *Figure 10*) in the usual manner. To make the second bend, measure the required distance between the bends from the back of the first bend. This distance is labeled *L* in the figure. Reverse the bender on the conduit, as shown in *Figure 10*. Place the bender's back-to-back indicating mark at point Y on the conduit. Note that outside measurements from point X to point Y are used. Holding the bender in the reverse position and properly aligned, apply foot pressure and complete the second bend.

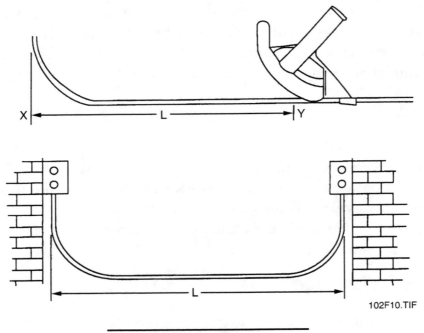

Figure 10. Back-To-Back Bend

2.5.0   MAKING AN OFFSET

Many situations require that the conduit be bent so that it can pass over objects such as beams and other conduits, or enter meter cabinets and junction boxes. Bends used for this purpose are called *offsets (kicks)*. To produce an offset, two equal bends of less than 90° are required, a specified distance apart, as shown in *Figure 11*.

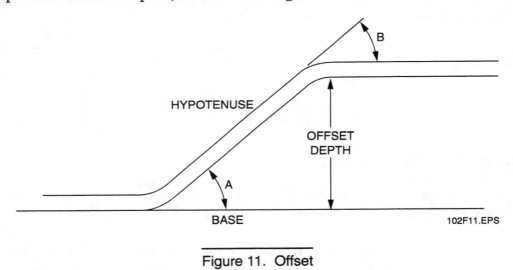

Figure 11. Offset

Offsets are a tradeoff between space and the effort it will take to pull the wire. The larger the degree of bend, the harder it will be to pull the wire. The smaller the degree of bend, the easier it will be to pull the wire. Use the shallowest degree of bend that will still allow the conduit to bypass the obstruction and fit in the given space.

ELECTRICAL — TRAINEE TASK MODULE 26102

When conduit is offset, some of the conduit length is used. If the offset is made into the area, an allowance must be made for this shrinkage. If the offset angle is away from the obstruction, the shrinkage can be ignored. *Table 2* shows the amount of shrinkage per inch of rise for common offset angles.

Offset Angle	Multiplier	Shrinkage (per inch of rise)
10° x 10°	6	$\frac{1}{16}$"
22½° x 22½°	2.6	$\frac{3}{16}$"
30° x 30°	2	$\frac{1}{4}$"
45° x 45°	1.4	$\frac{3}{8}$"
60° x 60°	1.2	$\frac{1}{2}$"

Table 2. Shrinkage Calculation

The formula for figuring the distance between bends is as follows:

Distance between bends = depth of offset x multiplier

The distance between the offset bends can generally be found in the manufacturer's documentation for the bender. *Table 3* shows the distance between bends for the most common offset angles.

Offset Depth	22½°		30°		45°		60°	
	Between Bends	Shrinkage	Between Bends	Shrinkage	Between Bends	Shrinkage	Between Bends	Shrinkage
2	5¼	$\frac{3}{8}$	—	—	—	—	—	—
3	7¾	$\frac{9}{16}$	6	$\frac{3}{4}$	—	—	—	—
4	10½	$\frac{3}{4}$	8	1	—	—	—	—
5	13	$1\frac{5}{16}$	10	1¼	7	1⅞	—	—
6	15½	1⅛	12	1½	8½	2¼	7¼	3
7	18¼	$1\frac{5}{16}$	14	1¾	9¾	2⅝	8⅜	3½
8	20¾	1½	16	2	11¼	3	9⅝	4
9	23½	1¾	18	2¼	12½	3⅜	10⅞	4½
10	26	1⅞	20	2½	14	3¾	12	5

Table 3. Common Offset Factors (In Inches)

Calculations related to offsets are derived from the branch of mathematics known as *trigonometry*, which deals with triangles. The multipliers shown in *Table 2* represent the *cosecant* (COS) of the related offset angle. The multiplier is determined by dividing the depth of the offset by the hypotenuse of the triangle created by the offset (*Figure 11*).

Basic trigonometry (trig) functions are briefly covered in *Appendix A*. As you will see in the next section, the *tangent* (TAN) of the offset angle is also used in calculating parallel offsets. Understanding trig functions will help you understand how offsets are determined. If you have a scientific calculator and understand these functions, you can calculate offset angles when you know the dimensions of the triangle created by the offset and the obstacle.

2.6.0 PARALLEL OFFSETS

Often, multiple pieces of conduit must be bent around a common obstruction. In this case, parallel offsets are made. Since the bends are laid out along a common radius, an adjustment must be made to ensure that the ends do not come out uneven, as shown in *Figure 12*.

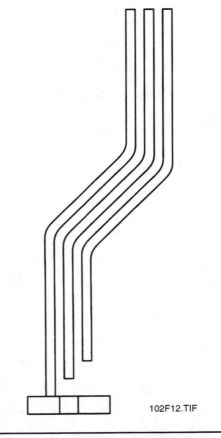

102F12.TIF

Figure 12. Incorrect Parallel Offsets

The center of the first bend of the innermost conduit is found first, as shown in *Figure 13*. Each successive conduit must have its centerline moved farther away from the end of the pipe, as shown in *Figure 14*. The amount to add is calculated as follows:

Amount added = center-to-center spacing x tangent (TAN) of ½ offset angle

Tangents can be found using the trig tables provided in *Appendix A*.

ELECTRICAL — TRAINEE TASK MODULE 26102

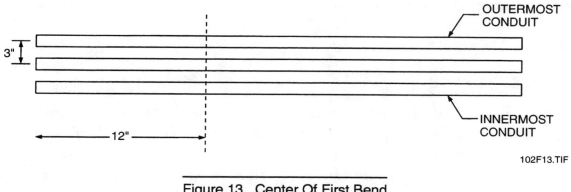

Figure 13. Center Of First Bend

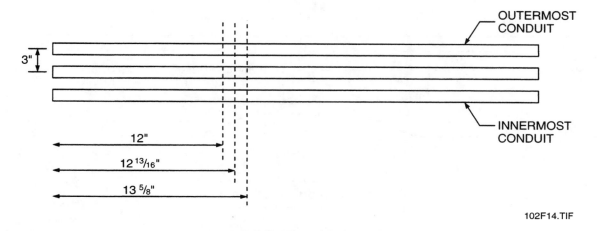

Figure 14. Successive Centerlines

For example, *Figure 15* shows three pipes laid out as parallel and offset. The angle of the offset is 30°. The center-to-center spacing is 3". The start of the innermost pipe's first bend is 12".

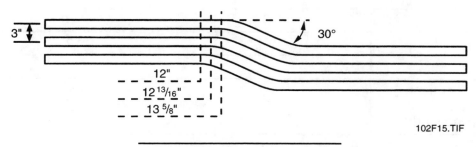

Figure 15. Parallel Offset Pipes

The starting point of the second pipe will be:

12" + [center-to-center spacing x TAN (½ offset angle)]

12" + (3" x TAN 15°) = 12" + (3" x .2679) = 12" + .8037"

This is approximately $12\frac{13}{16}$".

The starting point for the outermost pipe is:

$$12\tfrac{13}{16}" + \tfrac{13}{16}" = 13\tfrac{5}{8}"$$

2.7.0 SADDLE BENDS

A saddle bend is used to go around obstructions. *Figure 16* illustrates an example of a saddle bend that is required to clear a pipe obstruction. Making a saddle bend will cause the center of the saddle to shorten $\tfrac{3}{16}"$ for every inch of saddle depth (see *Table 4*). For example, if the pipe diameter is 2 inches, this would cause a $\tfrac{3}{8}"$ shortening of the conduit on each side of the bend.

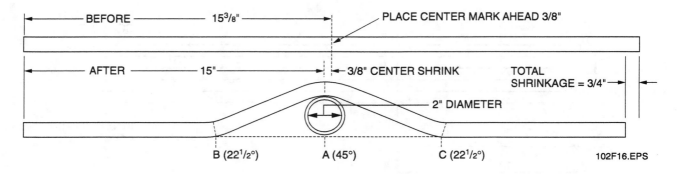

Figure 16. Saddle Measurement

Obstruction Depth	Shrinkage Amount (Move Center Mark Forward)	Make Outside Marks from *New* Center Mark
1	$\tfrac{3}{16}"$	$2\tfrac{1}{2}"$
2	$\tfrac{3}{8}"$	$5"$
3	$\tfrac{9}{16}"$	$7\tfrac{1}{2}"$
4	$\tfrac{3}{4}"$	$10"$
5	$\tfrac{15}{16}"$	$12\tfrac{1}{2}"$
6	$1\tfrac{1}{8}"$	$15"$
For each additional inch, add	$\tfrac{3}{16}"$	$2\tfrac{1}{2}"$

Table 4. Shrinkage Chart For Saddle Bends With A 45° Center Bend And Two 22½° Bends

When making saddle bends, the following steps should apply:

Step 1   Locate the center mark A on the conduit by using the size of the obstruction (i.e., pipe diameter) and calculate the shrink rate of the obstruction (for example, if the pipe diameter is 2 inches, $\tfrac{3}{8}"$ of conduit will be lost on each side of the bend for a total shrinkage of $\tfrac{3}{4}"$). This figure will be added to the measurement from the end of the conduit to the centerline of the obstruction (for example, if the distance measured from the conduit end and the obstruction centerline was 15", the distance to A would be $15\frac{3}{8}$").

Step 2 Locate marks B and C on the conduit by measuring $2\frac{1}{2}$" for every 1" of saddle depth *from* the A mark (i.e., for the saddle depth of 2 inches, the B mark would be 5" before the A mark and the C mark would be 5" after the A mark). See *Figure 17*.

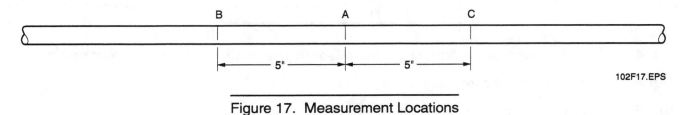

Figure 17. Measurement Locations

Step 3 Refer to *Figure 18* and make a 45° bend at point A, make a $22\frac{1}{2}$° bend at point B, and make a $22\frac{1}{2}$° bend at point C. (Be sure to check the manufacturer's specifications.)

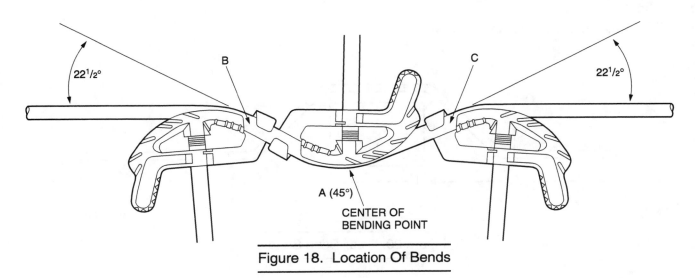

Figure 18. Location Of Bends

2.8.0 FOUR-BEND SADDLES

Four-bend saddles can be difficult. The reason is that four bends must be aligned exactly on the same plane. Extra time spent laying it out and performing the bends will pay off in not having to scrap the whole piece and start over.

Figure 19 illustrates that the four-bend saddle is really two offsets formed back-to-back. Working left to right, the procedure for forming this saddle is as follows:

Step 1 Determine the height of the offset.

Step 2 Determine the correct spacing for the first offset and mark the conduit.

Step 3 Bend the first offset.

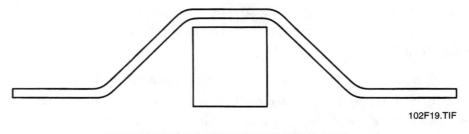

102F19.TIF

Figure 19. Typical Four-Bend Saddle

Step 4   Mark the start point for the second offset at the trailing edge of the obstruction.

Step 5   Mark the spacing for the second offset.

Step 6   Bend the second offset.

Using *Figure 20* as an example, a four-bend saddle using ½" EMT is laid out as follows:

- Height of the box = 6"
- Width of the box = 8"
- Distance to the obstruction = 36"

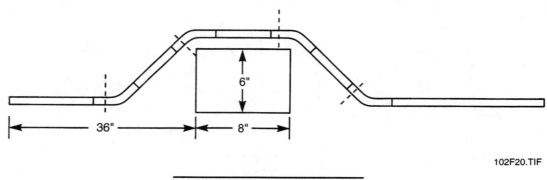

102F20.TIF

Figure 20. Four-Bend Saddle

Two 30° offsets will be used to form the saddle. It is created as follows:

Step 1   See *Figure 21*. Working from left to right, calculate the start point for the first bend. The distance to the obstruction is 36", the offset is 6", and the 30° multiplier from *Table 2* is 2.00:

Distance to the obstruction − (offset x constant for the angle) + shrinkage = distance to the first bend

36" − (6" x 2.00) + 1½" = 25½"

Step 2   Determine where the second bend will end to ensure the conduit clears the obstruction. See *Figure 22*.

Distance to the first bend + distance to second bend + shrinkage = total length of the first offset

25½" + 12" + 1½" = 39"

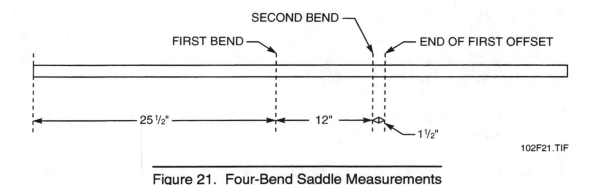

Figure 21. Four-Bend Saddle Measurements

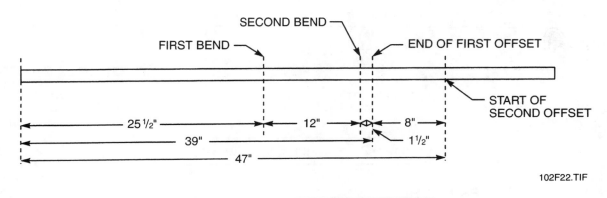

Figure 22. Bend And Offset Measurements

Step 3 Determine the start point of the second offset. The width of the box is 8"; therefore, the start point of the second offset should be 8" beyond the end of the first offset.

8" + 39" = 47"

Step 4 Determine the spacing for the second offset. Since the first and second offsets have the same rise and angle, the distance between bends will be the same, or 12".

3.0.0 CUTTING, REAMING, AND THREADING CONDUIT

Rigid conduit, IMC, and EMT are available in standard 10-foot lengths. When installing conduit, it is cut to fit the job requirements.

3.1.0 HACKSAW METHOD OF CUTTING CONDUIT

Conduit is normally cut using a hacksaw. To cut conduit with a hacksaw:

Step 1 Inspect the blade of the hacksaw and replace it, if needed. A blade with 18, 24, or 32 cutting teeth per inch is recommended for conduit. Use a higher tooth count for EMT and a lower tooth count for rigid conduit and IMC. If the blade needs to be replaced, point the teeth toward the front of the saw when installing the new blade.

Step 2 Secure the conduit in a pipe vise.

Step 3   Rest the middle of the hacksaw blade on the conduit where the cut is to be made. Position the saw so the end of the blade is pointing slightly down and the handle is pointing slightly up. Push forward gently until the cut is started. Make even strokes until the cut is finished.

CAUTION: To avoid bruising your knuckles on the newly-cut pipe, use gentle strokes for the final cut.

Step 4   Check the cut. The end of the conduit should be straight and smooth. *Figure 23* shows correct and incorrect cuts. Ream the conduit.

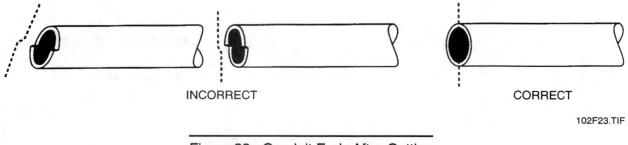

INCORRECT   CORRECT

102F23.TIF

Figure 23. Conduit Ends After Cutting

3.2.0 PIPE CUTTER METHOD

A pipe cutter can also be used to cut rigid IMC conduit. To use a pipe cutter:

Step 1   Secure the conduit in a pipe vise and mark a place for the cut.

Step 2   Open the cutter and place it over the conduit with the cutter wheel on the mark.

Step 3   Tighten the cutter by rotating the screw handle.

CAUTION: Do not overtighten the cutter. Overtightening can break the cutter wheel and distort the wall of the conduit.

Step 4   Rotate the cutter counterclockwise to start the cut. *Figure 24* shows the proper way to rotate the cutter.

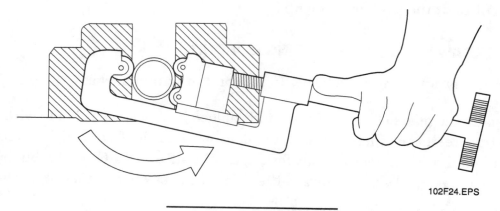

102F24.EPS

Figure 24. Cutter Rotation

Step 5   Tighten the cutter handle ¼ turn for each full turn around the conduit. Again, make sure that you do not overtighten it.

Step 6   Add a few drops of cutting oil to the groove and continue cutting. Avoid skin contact with the oil.

Step 7   When the cut is almost finished, stop cutting and snap the conduit to finish the cut. This reduces the ridge that can be formed on the inside of the conduit.

Step 8   Clean the conduit and cutter with a shop towel rag.

Step 9   Ream the conduit.

3.3.0   REAMING CONDUIT

When the conduit is cut, the inside edge is sharp. This edge will damage the insulation of the wire when it is pulled through. To avoid this damage, the inside edge must be smoothed or *reamed* using a reamer (*Figure 25*).

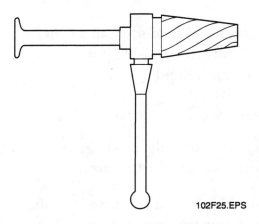

102F25.EPS

Figure 25. Rigid Conduit Reamer

To ream the inside edge of a piece of conduit using a hand reamer, proceed as follows:

Step 1   Place the conduit in a pipe vise.

Step 2   Insert the reamer tip in the conduit.

Step 3   Apply light forward pressure and start rotating the reamer. *Figure 26* shows the proper way to rotate the reamer. It should be rotated using a downward motion. The reamer can be damaged if you rotate it in the wrong direction. The reamer should bite as soon as you apply the proper pressure.

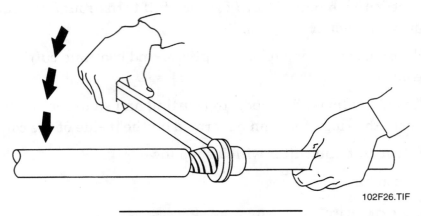

Figure 26. Reamer Rotation

Step 4   Remove the reamer by pulling back on it while continuing to rotate it. Check the progress and then reinsert the reamer. Rotate the reamer until the inside edge is smooth. You should stop when all burrs have been removed.

Note: If a conduit reamer is not available, use a half-round file (tang of file must have a handle attached). EMT may be reamed using the nose of diagonal cutters or small hand reamers.

3.4.0   THREADING CONDUIT

After conduit is cut and reamed, it is usually threaded so it can be properly joined. Only rigid conduit and IMC have walls thick enough for threading.

The tool used to cut threads in conduit is called a *die*. Conduit dies are made to cut a taper of $\frac{3}{4}$ inch per foot. The number of threads per inch varies from 8 to 18, depending upon the diameter of the conduit. A thread gauge is used to measure how many threads per inch are cut.

The threading dies are contained in a die head. The die head can be used with a hand-operated ratchet threader (*Figure 27*) or with a portable power drive.

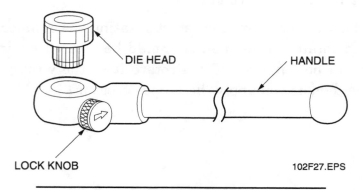

Figure 27. Hand-Operated Ratchet Threader

ELECTRICAL — TRAINEE TASK MODULE 26102

To thread conduit using a hand-operated threader, proceed as follows:

Step 1   Insert the conduit in a pipe vise. Make sure the vise is fastened to a strong surface. Place supports, if necessary, to help secure the conduit.

Step 2   Determine the correct die and head. Inspect the die for damage such as broken teeth. Never use a damaged die.

Step 3   Insert the die securely in the head. Make sure the proper die is in the appropriately numbered slot on the head.

Step 4   Determine the correct thread length to cut for the conduit size used (match the manufacturer's thread length).

Step 5   Lubricate the die with cutting oil at the beginning and throughout the threading operation. Avoid skin contact with the oil.

Step 6   Cut threads to the proper length. Make sure that the conduit enters the tapered side of the die. Apply pressure and start turning the head. You should back off the head each $\frac{1}{4}$ turn to clear away chips.

Step 7   Remove the die when the proper cut is made. Threads should be cut only to the length of the die. Overcutting will leave the threads exposed to corrosion.

Step 8   Inspect the threads to make sure they are clean, sharp, and properly made. Use a thread gauge to measure the threads. The finished end should allow for a wrench-tight fit with one or two threads exposed.

Note: The conduit should be reamed again after threading to remove any burrs and edges. Cutting oil must be swabbed from the inside and outside of the conduit. Use a sandbox or drip pan under the threader to collect drips and shavings.

Die heads can also be used with portable power drives. You will follow the same steps when using a portable power drive. Threading machines are often used on larger conduit and where frequent threading is required. Threading machines hold and rotate the conduit while the die is fed onto the conduit for cutting. When using a threading machine, make sure you secure the legs properly and follow the manufacturer's instructions.

3.5.0   CUTTING AND JOINING PVC CONDUIT

PVC conduit may be easily cut with a fine-tooth handsaw. To ensure square cuts, a miter box or similar device is recommended for cutting 2" and larger PVC. You can deburr the cut ends using a pocket knife. Smaller diameter PVC conduit, up to $1\frac{1}{2}$", may be cut using a PVC cutter.

Use the following steps to join PVC conduit sections or attachments to plastic boxes:

Step 1   Wipe all the contacting surfaces clean and dry.

Step 2   Apply a coat of cement (a brush or aerosol can is recommended) to the male end to be attached.

Step 3   Press the conduit and fitting together and rotate about a half-turn to evenly distribute the cement.

Note: Cementing the PVC must be done quickly. The aerosol spray cans of cement or the cement/brush combination are usually provided by the PVC manufacturer. Make sure you use the recommended cement.

Forming PVC in the field requires a special tool called a *hot box* or other specialized methods. PVC may not be threaded when it is used for electrical applications.

CAUTION: Solvents and cements used with PVC are hazardous. Wear gloves and follow the product instructions.

SUMMARY

You must choose a conduit bender to suit the kind of conduit being installed and the type of bend to be made. Some knowledge of the geometry of right triangles and circles needs to be mastered to make the necessary calculations. You must be able to calculate, lay out, and perform bending operations on a single run of conduit and also on two or more parallel runs of conduit. At times, data tables for the figures may be consulted for the calculations. All work must conform to the requirements of the NEC.

References

For advanced study of topics covered in this task module, the following books are suggested:

Benfield Conduit Bending Manual, Latest Edition, McGraw-Hill Publishing Company, New York, NY.

National Electrical Code Handbook, Latest Edition, National Fire Protection Association, Quincy, MA.

Tom Henry's Conduit Bending Package (includes video, book, and bending chart), Code Electrical Classes, Inc., Winter Park, FL.

REVIEW QUESTIONS

1. The field bending of PVC requires a _____.
    a. hickey
    b. heating unit
    c. segmented bender
    d. one-shot bender

2. After bending PVC, the bend can be set by using _____.
    a. a damp sponge or cloth
    b. dry ice
    c. ice cold water
    d. a blow dryer

3. A hickey is used for bending _____.
    a. rigid conduit
    b. EMT and IMC
    c. PVC conduit
    d. rigid conduit, EMT, IMC, and PVC conduit

4. A plug set is typically used to prevent _____ PVC when bending it.
    a. overpressurizing
    b. flattening
    c. corroding
    d. cutting

5. What is the key to accurate bending with a hand bender?
    a. Correct size and length of handle
    b. Constant foot pressure on the back piece
    c. Using only the correct brand of bender
    d. Correct inverting of the conduit bender

6. In a right triangle, the side directly opposite the 90° angle is called the _____.
    a. right side
    b. hypotenuse
    c. altitude
    d. base

7. The formula for calculating the circumference of a circle is _____.
    a. $\pi \times R^2$
    b. $2\pi \times R^2$
    c. $\pi \times D$
    d. $2\pi \times D$

8. Prior to making a 90° bend, what two measurements must be known?
   a. Length of conduit and size of conduit
   b. Desired rise and length of conduit
   c. Size of bender and size of conduit
   d. Stub-up distance and take-up distance

9. A back-to-back bend is _____.
   a. a two-shot 90° bend
   b. two 90° bends made back-to-back
   c. an offset with four bends
   d. a segmented bend

10. To ensure the conduit enters straight into the junction box, a(n) _____ may be required.
   a. back-to-back bend
   b. saddle bend
   c. offset
   d. take-up

11. To prevent the ends of the conduit from being staggered, what additional information must be used when making parallel offset bends?
   a. Center-to-center spacing and tangent of ½ the offset angle
   b. Length of conduit and size of conduit
   c. Stub-up distance and take-up distance
   d. Offset angle and length of conduit

12. When making a saddle bend, the center of the saddle will cause the conduit to shrink _____ for every inch of saddle depth.
   a. $\frac{3}{8}$"
   b. $\frac{3}{16}$"
   c. $\frac{3}{4}$"
   d. $\frac{3}{32}$"

13. When using a pipe cutter, always rotate the cutter _____ to start the cut.
   a. in a clockwise direction
   b. with the grain
   c. in a counterclockwise direction
   d. against the grain

14. You would use _____ to smooth the sharp inside edge of metal conduit after it has been cut.
   a. a flat file
   b. rough sandpaper
   c. a reamer
   d. a pocket knife

15. What tool is used to cut threads in rigid conduit or IMC?
    a.   A thread gauge
    b.   A cutter
    c.   A tap
    d.   A die

ANSWERS TO REVIEW QUESTIONS

Answer		**Section**
1.	b	2.0.0
2.	a	2.0.0
3.	a	2.0.0
4.	b	2.0.0
5.	b	2.0.0
6.	b	2.1.0
7.	c	2.1.0
8.	d	2.2.0
9.	b	2.4.0
10.	c	2.5.0
11.	a	2.6.0
12.	b	2.7.0
13.	c	3.2.0
14.	c	3.3.0
15.	d	3.4.0

USING TRIGONOMETRY TO DETERMINE OFFSET ANGLES AND MULTIPLIERS

You do not have to be a mathematician to use trigonometry. Understanding the basic trig functions and how to use them can help you calculate unknown distances or angles. Assume that the right triangle below represents a conduit offset. If you know the length of one side and the angle, you can calculate the length of the other sides, or if you know the length of any two of the sides of the triangle, you can then find the offset angle using one or more of these trig functions. You can use a trig table such as that shown on the following pages or a scientific calculator to determine the offset angle. For example, if the cosecant of angle A is 2.6, the trig table tells you that the offset angle is $22\frac{1}{2}°$.

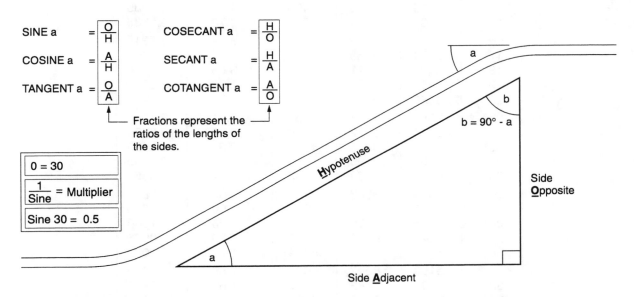

To determine the multiplier for the distance between bends in an offset:

1. Determine the angle of the offset: 30°

2. Find the sine of the angle: 0.5

3. Find the inverse (reciprocal) of the sine: $\frac{1}{0.5}$ = 2. This is also listed in trig tables as the cosecant of the angle.

4. This number multiplied by the height of the offset gives the hypotenuse of the triangle, which is equal to the distance between bends.

102APX01.EPS

TRIGONOMETRY TABLE

Angle	Sine	Cosine	Tangent	Cotangent	Cosecant
1°	.0175	.9998	.0175	57.3	57.3065
2°	.0349	.9994	.0349	28.6	28.6532
3°	.0523	.9986	.0524	19.1	19.1058
4°	.0698	.9976	.0699	14.3	14.3348
5°	.0872	.9962	.0875	11.4	11.4731
6°	.1045	.9945	.1051	9.51	9.5666
7°	.1219	.9925	.1228	8.14	8.2054
8°	.1392	.9903	.1405	7.12	7.1854
9°	.1564	.9877	.1584	6.31	6.3926
10°	.1736	.9848	.1763	5.67	5.7587
11°	.1908	.9816	.1944	5.14	5.2408
12°	.2079	.9781	.2126	4.70	4.8097
13°	.2250	.9744	.2309	4.33	4.4454
14°	.2419	.9703	.2493	4.01	4.1335
15°	.2588	.9659	.2679	3.73	3.8636
16°	.2756	.9613	.2867	3.49	3.5915
17°	.2924	.9563	.3057	3.27	3.4203
18°	.3090	.9511	.3249	3.08	3.2360
19°	.3256	.9455	.3443	2.90	3.0715
20°	.3420	.9397	.3640	2.75	2.9238
21°	.3584	.9336	.3839	2.61	2.7904
22°	.3744	.9272	.4040	2.48	2.6694
23°	.3907	.9205	.4245	2.36	2.5593
24°	.4067	.9135	.4452	2.25	2.4585
25°	.4226	.9063	.4663	2.14	2.3661
26°	.4384	.8988	.4877	2.05	2.2811
27°	.4540	.8910	.5095	1.96	2.2026
28°	.4695	.8829	.5317	1.88	2.1300
29°	.4848	.8746	.5543	1.80	2.0626
30°	.5000	.8660	.5774	1.73	2.0000
31°	.5150	.8572	.6009	1.66	1.9415
32°	.5299	.8480	.6249	1.60	1.8870
33°	.5446	.8387	.6494	1.54	1.8360
34°	.5592	.8290	.6745	1.48	1.7883
35°	.5736	.8192	.7002	1.43	1.7434
36°	.5878	.8090	.7265	1.38	1.7012
37°	.6018	.7986	.7536	1.33	1.6616
38°	.6157	.7880	.7813	1.28	1.6242
39°	.6293	.7771	.8098	1.23	1.5890
40°	.6428	.7660	.8391	1.19	1.5557
41°	.6561	.7547	.8693	1.15	1.5242
42°	.6691	.7431	.9004	1.11	1.4944
43°	.6820	.7314	.9325	1.07	1.4662
44°	.6947	.7193	.9657	1.04	1.4395
45°	.7071	.7071	1.0000	1.00	1.4142

102APX02.TIF

TRIGONOMETRY TABLE (Continued)

Angle	Sine	Cosine	Tangent	Cotangent	Cosecant
46°	.7193	.6947	1.035	.966	1.4395
47°	.7314	.6820	1.0724	.933	1.3673
48°	.7431	.6691	1.1106	.900	1.3456
49°	.7547	.6561	1.1504	.869	1.3250
50°	.7660	.6428	1.1918	.839	1.3054
51°	.7771	.6293	1.2349	.810	1.2867
52°	.7880	.6157	1.2799	.781	1.2690
53°	.7986	.6018	1.3270	.754	1.2521
54°	.8090	.5878	1.3764	.727	1.2360
55°	.8192	.5736	1.4281	.700	1.2207
56°	.8290	.5592	1.4826	.675	1.2062
57°	.8387	.5446	1.5399	.649	1.1923
58°	.8480	.5299	1.6003	.625	1.1791
59°	.8572	.5150	1.6643	.601	1.1666
60°	.8660	.5000	1.7321	.577	1.1547
61°	.8746	.4848	1.8040	.554	1.1433
62°	.8829	.4695	1.8807	.532	1.1325
63°	.8910	.4540	1.9626	.510	1.1223
64°	.8988	.4384	2.0503	.488	1.1126
65°	.9063	.4226	2.1445	.466	1.1033
66°	.9135	.4067	2.2460	.445	1.0946
67°	.9205	.3907	2.3559	.424	1.0863
68°	.9272	.3746	2.4751	.404	1.0785
69°	.9336	.3584	2.6051	.384	1.0711
70°	.9397	.3420	2.7475	.364	1.0641
71°	.9455	.3256	2.9042	.344	1.0576
72°	.9511	.3090	3.0777	.325	1.0514
73°	.9563	.2924	3.2709	.306	1.0456
74°	.9613	.2756	3.4874	.287	1.0402
75°	.9659	.2588	3.7321	.268	1.0352
76°	.9703	.2419	4.0108	.249	1.0306
77°	.9744	.2250	4.3315	.231	1.0263
78°	.9781	.2079	4.7046	.213	1.0223
79°	.9816	.1908	5.1446	.194	1.0187
80°	.9848	.1736	5.6713	.176	1.0154
81°	.9877	.1564	6.3138	.158	1.0124
82°	.9903	.1392	7.1154	.141	1.0098
83°	.9925	.1219	8.1443	.123	1.0075
84°	.9945	.1045	9.5144	.105	1.0055
85°	.9962	.0872	11.4300	.088	1.0038
86°	.9976	.0698	14.3010	.070	1.0024
87°	.9986	.0523	19.0810	.052	1.0013
88°	.9994	.0349	28.6360	.035	1.0006
89°	.9998	.0175	57.2900	.018	1.0001
90°	1.0000	.0000	00	.000	1.0000

102APX03.TIF

BENDING RADIUS TABLE

RADIUS (INCHES)	RADIUS INCREMENTS (INCHES)									
	0	1	2	3	4	5	6	7	8	9
0	0.00	1.57	3.14	4.71	6.28	7.85	.942	10.99	12.56	14.13
10	15.70	17.27	18.84	20.41	21.98	23.85	25.12	26.69	28.26	29.83
20	31.40	32.97	34.54	36.11	37.68	39.25	40.82	42.39	43.96	45.83
30	47.10	48.67	50.24	51.81	53.38	54.95	56.52	58.09	59.66	61.23
40	62.80	64.37	65.94	67.50	69.03	70.65	72.22	73.79	75.36	76.93
50	87.50	80.07	81.64	83.21	84.78	86.35	87.92	89.49	91.06	92.63
60	94.20	95.77	97.34	98.91	100.48	102.05	103.62	105.19	106.76	108.33
70	109.90	111.47	113.04	114.61	116.18	117.75	119.32	120.89	122.46	124.03
80	125.60	127.17	128.74	130.31	131.88	133.45	135.02	136.59	138.16	139.73
90	141.30	142.87	144.44	146.01	147.58	149.15	150.72			

Developed length for following angles use fraction of 90° chart.

For	15°	22-1/2°	30°	45°	60°	67-1/2°	75°	90°
Take	1/6	1/4	1/3	1/2	2/3	3/4	5/6	See Chart

For any other degrees: Developed length = .01744 x radius x degrees.

102APX04.TIF

The NCCER makes every effort to keep these manuals up-to-date and free of technical errors. We appreciate your help in this process. If you have an idea for improving this manual, or if you find an error, a typographical mistake, or an inaccuracy in the NCCER's Craft Training Manuals, please write us, using this form or a photocopy. Be sure to include the exact module number, page number, a description of the problem, and the correction, if possible. Your input will be brought to the attention of the Technical Review Committee. Thank you for your assistance.

Instructors – If you found that additional materials were necessary in order to teach this module effectively, please let us know so that we may include them in the Equipment/Materials list in the Instructor's Guide.

Write: Curriculum Development and Revision Department
National Center for Construction Education and Research
P.O. Box 141104
Gainesville, FL 32614-1104

Fax: 352-334-0932

Craft _____ Module Name _____

Copyright Date _____ Module Number _____ Page Number(s) _____

Description of Problem _____

_____

_____

_____

(Optional) Correction of Problem _____

_____

_____

(Optional) Your Name and Address _____

_____

_____

Fasteners and Anchors

Module 26103

**NATIONAL
CENTER FOR
CONSTRUCTION
EDUCATION AND
RESEARCH**

FASTENERS AND ANCHORS

OBJECTIVES

Upon completion of this module, the trainee will be able to:

1. Identify and explain the use of threaded fasteners.
2. Identify and explain the use of non-threaded fasteners.
3. Identify and explain the use of anchors.
4. Demonstrate the correct applications for fasteners and anchors.
5. Install fasteners and anchors.

Prerequisites

Successful completion of the following Task Modules is required before beginning study of this Task Module: Core Curricula; Electrical Level 1, Modules 26101 and 26102.

Required Trainee Materials

1. Trainee Task Module
2. Copy of the latest edition of the *National Electrical Code*
3. Appropriate Personal Protective Equipment

Note: The designations "National Electrical Code," "NE Code," and "NEC," where used in this document, refer to the *National Electrical Code*®, which is a registered trademark of the National Fire Protection Association, Quincy, MA. *All National Electrical Code (NEC) references in this module refer to the 1999 edition of the NEC.*

Course Map

This course map shows all of the task modules in the first level of the Electrical curricula. The suggested training order begins at the bottom and proceeds up. Skill levels increase as a trainee advances on the course map. The training order may be adjusted by the local Training Program Sponsor.

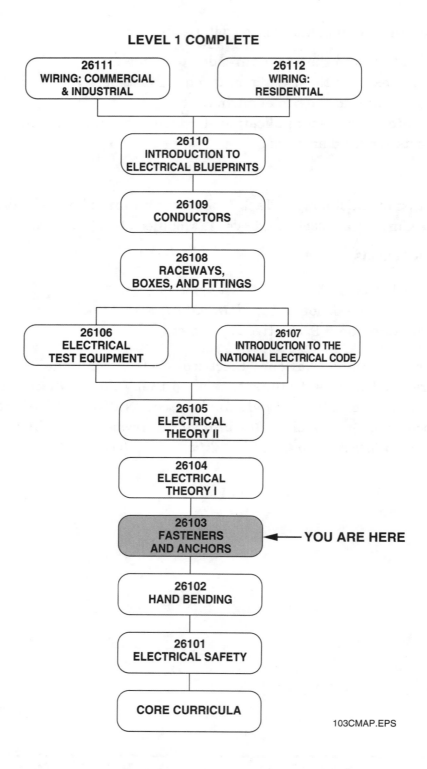

LEVEL 1 COMPLETE

26111
WIRING: COMMERCIAL
& INDUSTRIAL

26112
WIRING:
RESIDENTIAL

26110
INTRODUCTION TO
ELECTRICAL BLUEPRINTS

26109
CONDUCTORS

26108
RACEWAYS,
BOXES, AND FITTINGS

26106
ELECTRICAL
TEST EQUIPMENT

26107
INTRODUCTION TO THE
NATIONAL ELECTRICAL CODE

26105
ELECTRICAL
THEORY II

26104
ELECTRICAL
THEORY I

26103
FASTENERS
AND ANCHORS   ◄— YOU ARE HERE

26102
HAND BENDING

26101
ELECTRICAL SAFETY

CORE CURRICULA

103CMAP.EPS

TABLE OF CONTENTS

TABLE OF CONTENTS (Continued)

Trade Terms Introduced In This Module

American Society for Testing of Materials (ASTM): An organization that publishes specifications and standards relating to fasteners.

Clearance: The amount of space between the threads of bolts and their nuts.

Foot pounds (ft. lbs.): The normal method used for measuring the amount of torque being applied to bolts or nuts.

Inch pounds (in. lbs.): A method of measuring the amount of torque applied to small bolts or nuts that require measurement in smaller increments than foot pounds.

Key: A machined metal part that fits into a keyway and prevents parts such as gears or pulleys from rotating on a shaft.

Keyway: A machined slot in a shaft and on parts such as gears and pulleys that accepts a key.

Nominal size: A means of expressing the size of a bolt or screw. It is the approximate diameter of a bolt or screw.

Society of Automotive Engineers (SAE): An organization that publishes specifications and standards relating to fasteners.

Thread classes: Threads are distinguished by three classifications according to the amount of tolerance the threads provide between the bolt and nut.

Thread identification: Standard symbols used to identify threads.

Thread standards: An established set of standards for machining threads.

Tolerance: The amount of difference allowed from a standard.

Torque: The turning force applied to a fastener.

Unified National Coarse (UNC) thread: A standard type of coarse thread.

Unified National Extra Fine (UNEF) thread: A standard type of extra-fine thread.

Unified National Fine (UNF) thread: A standard type of fine thread.

1.0.0 INTRODUCTION

Fasteners are used to assemble and install many different types of equipment, parts, and materials. Fasteners include screws, bolts, nuts, pins, clamps, retainers, tie wraps, rivets, and **keys**. Fasteners are used extensively in the electrical craft. You need to be familiar with the many different types of fasteners in order to identify, select, and properly install the correct fastener for a specific application.

The two primary categories of fasteners are:

- Threaded fasteners
- Non-threaded fasteners

Within each of these two categories, there are numerous different types and sizes of fasteners. Each type of fastener is designed for a specific application. The kind of fastener used for a job may be listed in the project specifications, or you may have to select an appropriate fastener.

Failure of fasteners can result in a number of different problems. To perform quality electrical work, it is important to use the correct type and size of fastener for the particular job. It is equally important that the fastener be installed properly.

In this module, you will be introduced to fasteners commonly used in electrical work.

2.0.0 THREADED FASTENERS

Threaded fasteners are the most commonly used type of fastener. Many threaded fasteners are assembled with nuts and washers. The following sections describe standard threads used on threaded fasteners, as well as different types of bolts, screws, nuts, and washers. *Figure 1* shows several types of threaded fasteners.

2.1.0 THREAD STANDARDS

There are many different types of threads used for manufacturing fasteners. The different types of threads are designed to be used for different jobs. Threads used on fasteners are manufactured to industry established standards for uniformity. The most common **thread standard** is the Unified standard, sometimes referred to as the *American standard*. Unified standards are used to establish thread series and classes.

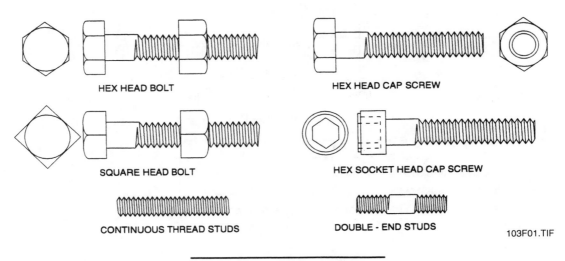

HEX HEAD BOLT

HEX HEAD CAP SCREW

SQUARE HEAD BOLT

HEX SOCKET HEAD CAP SCREW

CONTINUOUS THREAD STUDS

DOUBLE - END STUDS

103F01.TIF

Figure 1. Threaded Fasteners

2.1.1 Thread Series

Unified standards are established for three series of threads, depending on the number of threads per inch for a certain diameter of fastener. These three series are:

- *Unified National Coarse (UNC) thread* – Used for bolts, screws, nuts, and other general purposes. Fasteners with UNC threads are commonly used for rapid assembly or disassembly of parts and where corrosion or slight damage may occur.
- *Unified National Fine (UNF) thread* – Used for bolts, screws, nuts, and other applications where a finer thread for a tighter fit is desired.
- *Unified National Extra Fine (UNEF) thread* – Used on thin-walled tubes, nuts, ferrules, and couplings.

2.1.2 Thread Classes

The Unified standards also establish **thread classes**. Classes 1A, 2A, and 3A apply to external threads only. Classes 1B, 2B, and 3B apply to internal threads only. Thread classes are distinguished from each other by the amounts of **tolerance** provided. Classes 3A and 3B provide a minimum **clearance** and classes 1A and 1B provide a maximum clearance.

Classes 2A and 2B are the most commonly used. Classes 3A and 3B are used when close tolerances are needed. Classes 1A and 1B are used where quick and easy assembly is needed and a large tolerance is acceptable.

2.1.3 Thread Identification

Thread identification is done using a standard method. *Figure 2* shows how screw threads are designated for a common fastener.

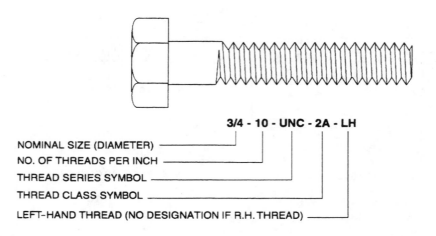

NOMINAL SIZE (DIAMETER)
NO. OF THREADS PER INCH
THREAD SERIES SYMBOL
THREAD CLASS SYMBOL
LEFT-HAND THREAD (NO DESIGNATION IF R.H. THREAD)

3/4 - 10 - UNC - 2A - LH

103F02.TIF

Figure 2. Screw Thread Designations

- *Nominal size* – The **nominal size** is the approximate diameter of the fastener.
- *Number of threads per inch (TPI)* – The TPI is standard for all diameters.
- *Thread series symbol* – The Unified standard thread type (UNC, UNF, or UNEF).
- *Thread class symbol* – The closeness of fit between the bolt threads and nut threads.
- *Left-hand thread symbol* – Specified by the symbol LH. Unless threads are specified with the LH symbol, the threads are right-hand threads.

2.1.4   Grade Markings

Special markings on the head of a bolt or screw can be used to determine the quality of the fastener. The **Society of Automotive Engineers (SAE)** and the **American Society for Testing of Materials (ASTM)** have developed the standards for these markings. These grade or line markings for steel bolts and screws are shown in *Figure 3*.

Generally, the higher-quality steel fasteners have a greater number of marks on the head. If the head is unmarked, the fastener is usually considered to be made of mild steel (having low carbon content).

2.2.0   BOLT AND SCREW TYPES

Bolts and screws are made in many different sizes and shapes and from a variety of materials. They are usually identified by the head type or other special characteristics. The following sections describe several different types of bolts and screws.

2.2.1   Machine Screws

Machine screws are used for general assembly work. They come in a variety of types with slotted or recessed heads. Machine screws are generally available in diameters ranging from 0 (0.060") to ½ (0.500"). The length of machine screws typically varies from ⅛" to 3".

ASTM AND SAE GRADE MARKINGS FOR STEEL BOLTS & SCREWS

GRADE MARKING	SPECIFICATION	MATERIAL
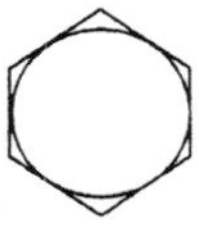 (hex bolt, plain)	SAE-GRADE 0	STEEL
	SAE-GRADE 1 / ASTM-A 307	LOW CARBON STEEL
	SAE-GRADE 2	LOW CARBON STEEL
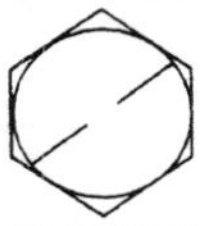 (hex bolt, one line)	SAE-GRADE 3	MEDIUM CARBON STEEL, COLD WORKED
 (hex bolt, A 449)	SAE-GRADE 5	MEDIUM CARBON STEEL, QUENCHED AND TEMPERED
	ASTM-A 449	
 (hex bolt, A 325)	ASTM-A 325	MEDIUM CARBON STEEL, QUENCHED AND TEMPERED
 (hex bolt, BB)	ASTM-A 354 GRADE BB	LOW ALLOY STEEL, QUENCHED AND TEMPERED
 (hex bolt, BC)	ASTM-A 354 GRADE BC	LOW ALLOY STEEL, QUENCHED AND TEMPERED
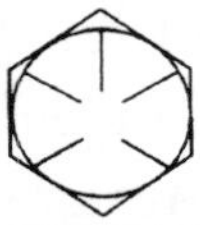 (hex bolt, six lines)	SAE-GRADE 7	MEDIUM CARBON ALLOY STEEL, QUENCHED AND TEMPERED ROLL THREADED AFTER HEAT TREATMENT
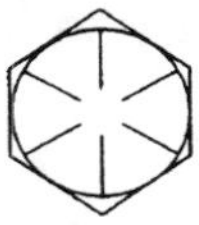 (hex bolt, six lines)	SAE-GRADE 8	MEDIUM CARBON ALLOY STEEL, QUENCHED AND TEMPERED
	ASTM-A 354 GRADE BD	ALLOY STEEL, QUENCHED AND TEMPERED
 (hex bolt, A 490)	ASTM-A 490	ALLOY STEEL, QUENCHED AND TEMPERED

ASTM SPECIFICATIONS

A 307 - LOW CARBON STEEL EXTERNALLY AND INTERNALLY THREADED STANDARD FASTENERS.
A 325 - HIGH STRENGTH STEEL BOLTS FOR STRUCTURAL STEEL JOINTS, INCLUDING SUITABLE NUTS AND PLAIN HARDENED WASHERS.
A 449 - QUENCHED AND TEMPERED STEEL BOLTS AND STUDS.
A 354 - QUENCHED AND TEMPERED ALLOY STEEL BOLTS AND STUDS WITH SUITABLE NUTS.
A 490 - HIGH STRENGTH ALLOY STEEL BOLTS FOR STRUCTURAL STEEL JOINTS, INCLUDING SUITABLE NUTS AND PLAIN HARDENED WASHERS.

SAE SPECIFICATION

J 429 - MECHANICAL AND QUALITY REQUIREMENTS FOR THREADED FASTENERS.

103F03.TIF

Figure 3. Grade Markings For Steel Bolts And Screws

Machine screws are also manufactured in metric sizes. *Figure 4* shows different types of machine screws.

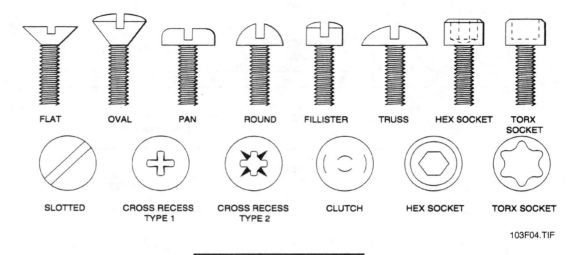

Figure 4. Machine Screws

As shown, the heads of machine screws are made in different shapes and with slots made to fit various kinds of manual and power tool screwdrivers. Flat head screws are used in a countersunk hole and tightened so that the head is flush with the surface. Oval head screws are also used in a countersunk hole in applications where a more decorative finish is desired. Pan and round head screws are general use fastening screws. Fillister, hex socket, and torx socket screws are typically used in confined space applications on machined assemblies that need a finished appearance. They are often installed in a recessed hole. Truss screws are a low-profile screw generally used without a washer. To prevent damage when tightening and removing machine screws (regardless of head type), make sure to use a screwdriver or power tool bit with the proper tip to drive them.

2.2.2 Machine Bolts

Machine bolts are generally used to assemble parts where close tolerances are not required. Machine bolts have square or hexagonal heads and are generally available in diameters ranging from $\frac{1}{4}$" to 3". The length of machine bolts typically varies from $\frac{1}{2}$" to 30". Nuts used with machine bolts are similar in shape to the bolt heads. The nuts are usually purchased at the same time as the bolts. *Figure 5* shows two different types of machine bolts.

2.2.3 Cap Screws

Cap screws are often used on high-quality assemblies requiring a finished appearance. The cap screw passes through a clearance hole in one of the assembly parts and screws into a threaded hole in the other part. The clamping action occurs by tightening the cap screw.

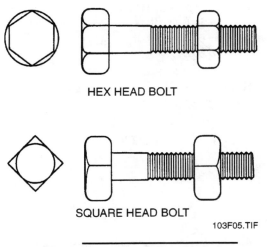

HEX HEAD BOLT

SQUARE HEAD BOLT

103F05.TIF

Figure 5. Machine Bolts

Cap screws are made to close tolerances and are provided with a machined or semi-finished bearing surface under the head. They are normally made in coarse and fine thread series and in diameters from $\frac{1}{4}$" to 2". Lengths may range from $\frac{3}{8}$" to 10". Metric sizes are also available. *Figure 6* shows typical cap screws.

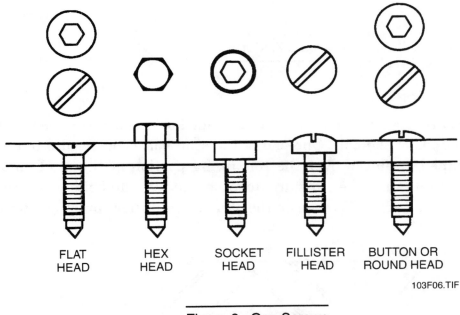

| FLAT HEAD | HEX HEAD | SOCKET HEAD | FILLISTER HEAD | BUTTON OR ROUND HEAD |

103F06.TIF

Figure 6. Cap Screws

2.2.4 Set Screws

Heat-treated steel is normally used to make set screws. Common uses of set screws include preventing pulleys from slipping on shafts, holding collars in place on shafts, and holding shafts in place. The head style and point style are typically used to classify set screws. *Figure 7* shows several set screw heads and point styles.

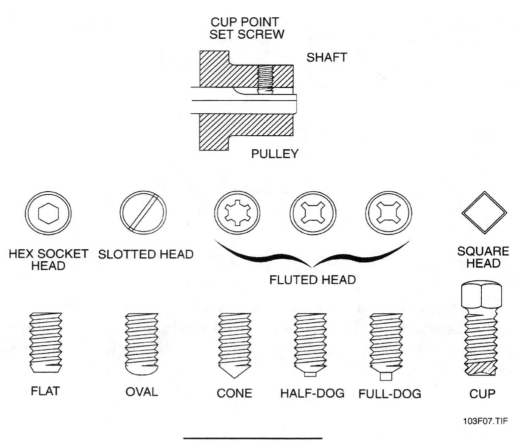

Figure 7. Set Screws

2.2.5 Stud Bolts

Stud bolts (*Figure 8*) are headless bolts that are threaded over the entire length of the bolt or for a length on both ends of the bolt. One end of the stud bolt is screwed into a tapped hole. The part to be clamped is placed over the remaining portion of the stud, and a nut and washer are screwed on to clamp the two parts together. Other stud bolts have machine-screw threads on one end and lag-screw threads on the other so that they can be screwed into wood.

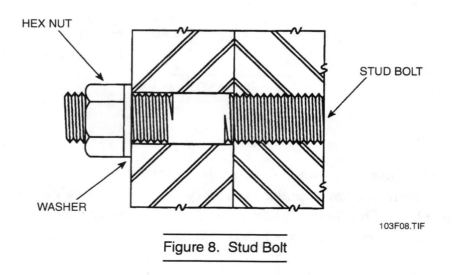

Figure 8. Stud Bolt

Stud bolts are used for several purposes, including holding together inspection covers on equipment and bearing caps.

2.3.0 NUTS

Most nuts used with threaded fasteners are hexagonal or square. They are usually used with bolts having the same shaped head. *Figure 9* shows several different types of nuts that are used with threaded fasteners.

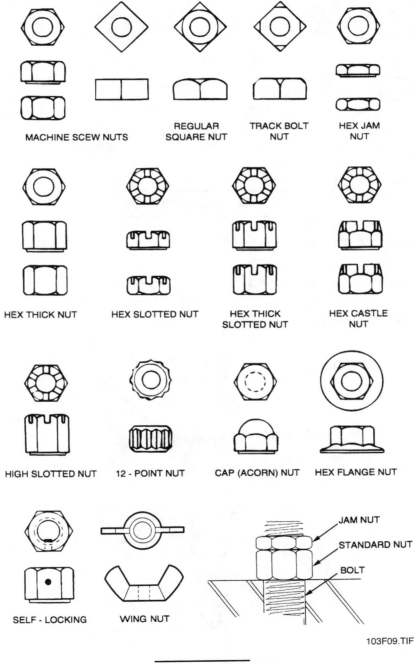

103F09.TIF

Figure 9. Nuts

Nuts are typically classified as regular, semi-finished, or finished. The only machining done on regular nuts is to the threads. In addition to the threads, semi-finished nuts are also machined on the bearing face. Machining the bearing face makes a truer surface for fitting the washer. The only difference between semi-finished and finished nuts is that finished nuts are made to closer tolerances.

The standard machine screw nut has a regular finish. Regular and semi-finished nuts are shown in *Figure 10*.

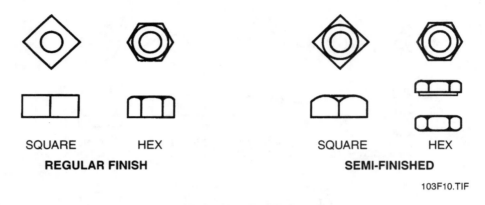

103F10.TIF

Figure 10. Nut Finishes

2.3.1 Jam Nuts

A jam nut is used to lock a standard nut in place. A jam nut is a thin nut installed on top of the standard nut. *Figure 11* shows an example of a jam nut installation. Note that a regular nut can also be used as a jam nut.

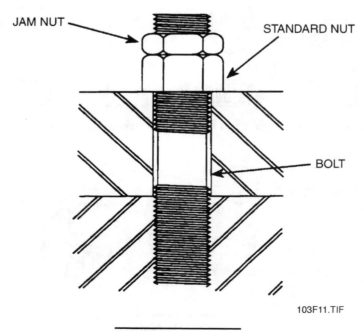

103F11.TIF

Figure 11. Jam Nut

2.3.2 Castellated, Slotted, And Self-Locking Nuts

Castellated (castle) and slotted nuts are slotted across the flat part of the nut. They are used with specially-manufactured bolts in applications where little or no loosening of the fastener can be tolerated. After the nut has been tightened, a cotter pin is fitted in through a hole in the bolt and one set of slots in the nut. The cotter pin keeps the nut from loosening under working conditions.

Self-locking nuts are also used in many applications where loosening of the fastener cannot be tolerated. Self-locking nuts are designed with nylon inserts, or they are deliberately deformed in such a manner so they cannot work loose. An advantage of self-locking nuts is that no hole in the bolt is needed. *Figure 12* shows typical castellated, slotted, and self-locking nuts.

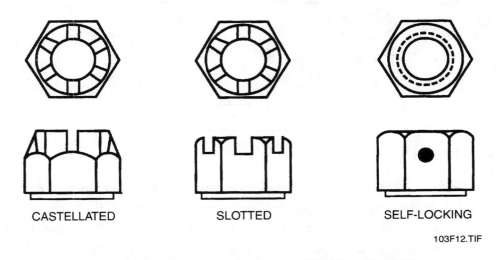

CASTELLATED   SLOTTED   SELF-LOCKING

103F12.TIF

Figure 12. Castellated, Slotted, And Self-Locking Nuts

2.3.3 Acorn Nuts

When appearance is important or exposed, sharp thread edges on the fastener must be avoided, acorn (cap) nuts are used. The acorn nut tightens on the bolt and covers the ends of the threads. The tightening capability of an acorn nut is limited by the depth of the nut. *Figure 13* shows a typical acorn nut.

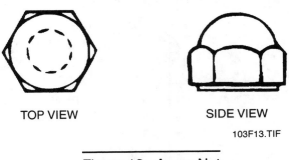

TOP VIEW   SIDE VIEW

103F13.TIF

Figure 13. Acorn Nut

2.3.4 Wing Nuts

Wing nuts are designed to allow rapid loosening and tightening of the fastener without the need for a wrench. They are used in applications where limited **torque** is required and where frequent adjustments and service are necessary. *Figure 14* shows a typical wing nut.

Note: Wing nuts should be used for applications where hand tightening is sufficient.

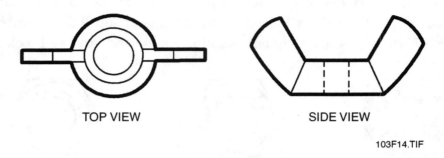

TOP VIEW   SIDE VIEW

103F14.TIF

Figure 14. Wing Nut

2.4.0 WASHERS

There are several different types and sizes of washers. They fit over a bolt or screw to provide an enlarged surface for bolt heads and nuts. Washers also serve to distribute the fastener load over a larger area and to prevent marring of the surfaces. Standard washers are made in light, medium, heavy-duty, and extra heavy-duty series. *Figure 15* shows different types of washers.

Note: The threads of the bolt or screw should have minimal clearance from the hole in the washer.

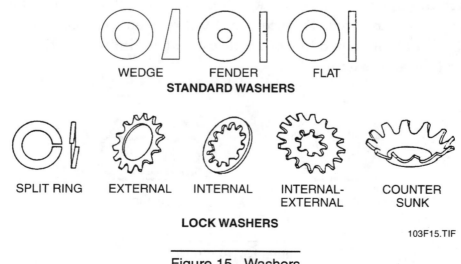

WEDGE   FENDER   FLAT
STANDARD WASHERS

SPLIT RING   EXTERNAL   INTERNAL   INTERNAL-   COUNTER
                                         EXTERNAL   SUNK

LOCK WASHERS

103F15.TIF

Figure 15. Washers

2.4.1 Lock Washers

Lock washers are designed to keep bolts or nuts from working loose. There are various types of lock washers for different applications.

- *Split-ring* – Commonly used with bolts and cap screws.
- *External* – Used for the greatest resistance.
- *Internal* – Used with small screws.
- *Internal-external* – Used for oversized mounting holes.
- *Countersunk* – Used with flat or oval-head screws.

2.4.2 Flat And Fender Washers

Flat washers are used under bolts or nuts to spread the load over a larger area and protect the surface. Common flat washers are made to fit bolt or screw sizes ranging from No. 6 up to 1" with outside diameters ranging from $\frac{3}{8}$" to 2", respectively.

Fender washers are wide-surfaced washers made to bridge oversized holes or other wide clearances to keep bolts or nuts from pulling through the material being fastened. They are flat washers that have a larger diameter and surface area than regular washers. They may also be thinner than a regular washer. Fender washers are typically made to fit bolt or screw sizes ranging from $\frac{3}{16}$" to $\frac{1}{2}$" with outside diameters ranging from $\frac{3}{4}$" to 2", respectively.

2.5.0 INSTALLING FASTENERS

Different types of fasteners require different installation techniques. However, all installations require knowing the proper installation methods, tightening sequence, and torque specifications for the type of fastener being used. Some bolts and nuts require that special safety wires or pins be installed to keep them from working loose.

Most fastener manufacturers provide charts that specify the size hole that should be drilled into the base material for use with each of their products (*Figure 16*). The charts typically show the proper size drill bit to use if it is necessary to first drill and tap holes for use with machine bolts, screws, or other threaded fasteners. They also show the proper size drill to use for drilling pilot holes used with metal and wood screws. (Various kinds of screws are described in detail later in this module.)

DRILL THIS SIZE HOLE		To Tap For This Size Bolt or Screw	For This Size Wood Screw Pilot in Hard Wood
Drill Size	Dec. Equiv.		
60	.0400		
59	.0410		
58	.0420		
57	.0430		
56	.0465	0 x 80	
3/64	.0469		
55	.0520		
54	.0550	1 x 56	No. 3
53	.0595	1 x 64-72	
1/16	.0625		
52	.0635		No. 4
51	.0670		
50	.0700	2 x 56-64	
49	.0730		No. 5
48	.0760		
5/64	.0781		
47	.0785	3 x 48	No. 6
46	.0810		
45	.0820	3 x 56	
44	.0860	4 x 36	No. 7
43	.0890	4 x 40	
42	.0935	4 x 48	
3/32	.0937		
41	.0960		
40	.0980	5 x 36	No. 8
39	.0995		
38	.1015	5 x 40	
37	.1040	5 x 44	No. 9
36	.1069		
7/64	.1094		
35	.1100	6 x 32	
34	.1110	6 x 36	
33	.1130	6 x 40	No. 10
32	.1160		
31	.1200		No. 11
1/8	.1250	7 x 36	
30	.1285	8 x 30	No. 12
29	.1360	8 x 32-36	
28	.1405	8 x 40	
27	.1440	9 x 30	
26	.1470	3/16 x 24	
25	.1495	10 x 24	No. 14
24	.1520		
23	.1540	10 x 28	
5/32	.1562		
22	.1570	10 x 30	
21	.1590	10 x 32	
20	.1610	3/16 x 32	
19	.1660		
18	.1695		No. 16
11/64	.1719		
17	.1730		
16	.1770	12 x 24	
15	.1800		
14	.1820	12 x 28	
13	.1850	12 x 32	
3/16	.1875		No. 18
12	.1890		

DRILL THIS SIZE HOLE		To Tap For This Size Bolt or Screw	For This Size Wood Screw Pilot in Hard Wood
Drill Size	Dec. Equiv.		
11	.1910		
10	.1935	15 x 20	
9	.1960		
8	.1990		
7	.2010	1/4 x 20	
13/64	.2031		
6	.2040		
5	.2055		
4	.2090	1/4 x 24	No. 20
3	.2130	1/4 x 28	
7/32	.2187	1/4 x 32	
2	.2210		
1	.2280		No. 24
A	.2340		
15/64	.2344		
B	.2380		
C	.2420		
D	.2460		
1/4	.2500		

DRILL THIS SIZE HOLE		To Tap For This Size Bolt or Screw
Drill Size	Dec. Equiv.	
E	.2500	
F	.2570	5/16 x 18
G	.2610	
17/64	.2656	5/16 x 18
H	.2660	
I	.2720	
J	.2770	5/16 x 24-32*
K	.2810	
9/32	.2812	5/16 x 24-32*
L	.2900	
M	.2950	
19/64	.2969	
N	.3020	
5/16	.3125	3/8* x 16-1/8* P
O	.3160	
P	.3230	
21/64	.3281	3/8 x 20-24
Q	.3332	
R	.3390	
11/32	.3437	
S	.3480	
T	.3580	
23/64	.3594	
U	.3680	
3/8	.3750	7/16 x 14
V	.3770	
W	.3860	
25/64	.3906	7/16 x 14
X	.3970	
Y	.4040	
13/32	.4062	
Z	.4130	
27/64	.4219	1/2 x 12-13
7/16	.4375	1/4* Pipe
29/64	.4531	1/2 x 20-24
15/32	.4687	1/2 x 27
31/64	.4844	9/16 x 12
1/2	.5000	

* All tap drill sizes are for 75% full thread except asterisked sizes which are 60% full thread.

103F16.EPS

Figure 16. Fastener Hole Guide Chart

ELECTRICAL — TRAINEE TASK MODULE 26103

2.5.1 Torque Tightening

To properly tighten a threaded fastener, two primary factors must be considered:

- The strength of the fastener material
- The degree to which the fastener is tightened

A torque wrench is used to control the degree of tightness. The torque wrench measures how much a fastener is being tightened. *Torque* is the turning force applied to the fastener. Torque is normally expressed in **inch pounds (in. lbs.)** or **foot pounds (ft. lbs.)**. A one-pound force applied to a wrench that is one-foot long exerts one foot pound, or twelve inch pounds, of torque. The torque reading is shown on the indicator on the torque wrench as the fastener is being tightened. *Figure 17* shows two types of torque wrenches.

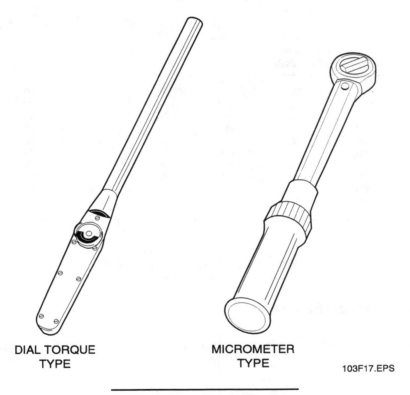

DIAL TORQUE
TYPE

MICROMETER
TYPE

103F17.EPS

Figure 17. Torque Wrenches

Different types of bolts, nuts, and screws are torqued to different values depending on the application. Always check the project specifications and the manufacturer's manual to determine the proper torque for a particular type of fastener. *Figure 18* shows selected torque values for various graded steel bolts.

TORQUE IN FOOT POUNDS

FASTENER DIAMETER	THREADS PER INCH	MILD STEEL	STAINLESS STEEL 18-8	ALLOY STEEL
1/4	20	4	6	8
5/16	18	8	11	16
3/8	16	12	18	24
7/16	14	20	32	40
1/2	13	30	43	60
5/8	11	60	92	120
3/4	10	100	128	200
7/8	9	160	180	320
1	8	245	285	490

SUGGESTED TORQUE VALUES FOR GRADED STEEL BOLTS

GRADE		SAE 1 OR 2	SAE 5	SAE 6	SAE 8
TENSILE STRENGTH		64000 PSI	105000 PSI	130000 PSI	150000 PSI
GRADE MARK		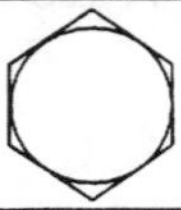	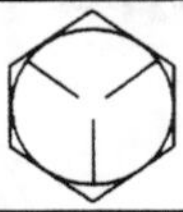	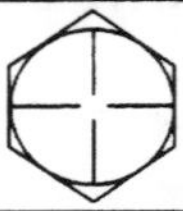	
BOLT DIAMETER	THREADS PER INCH	FOOT POUNDS TORQUE			
1/4	20	5	7	10	10
5/16	18	9	14	19	22
3/8	16	15	25	34	37
7/16	14	24	40	55	60
1/2	13	37	60	85	92
9/16	12	53	88	120	132
5/8	11	74	120	169	180
3/4	10	120	200	280	296
7/8	9	190	302	440	473
1	8	282	466	660	714

103F18.TIF

Figure 18. Torque Value Chart

2.5.2 Installing Threaded Fasteners

The following general procedure can be used to install threaded fasteners in a variety of applications.

Note: When installing threaded fasteners for a specific job, make sure to check all installation requirements.

WARNING! Follow all safety precautions.

Step 1 Select the proper bolts or screws for the job.

Step 2 Check for damaged or dirty internal and external threads.

Step 3 Clean the bolt or screw threads. Do not lubricate the threads if a torque wrench is to be used to tighten the nuts.

Step 4   Insert the bolts through the pre-drilled holes and tighten the nuts by hand. Or, insert the screws through the holes and start the threads by hand.

Note: Turn the nuts or screws several turns by hand and check for cross threading.

Step 5   Following the proper tightening sequence, tighten the bolts or screws snugly.

Step 6   Check the torque specification. Following the proper tightening sequence, tighten each bolt, nut, or screw several times approaching the specified torque. Tighten to the final torque specification.

Step 7   If required to keep the bolts or nuts from working loose, install jam nuts, cotter pins, or safety wire. *Figure 19* shows fasteners with a safety wire installed.

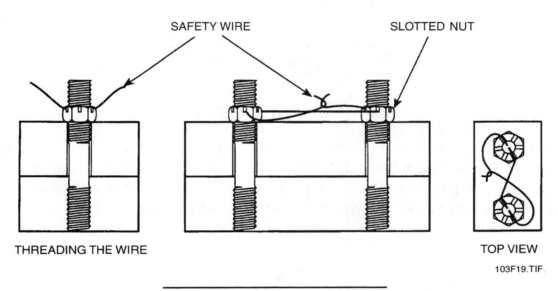

Figure 19. Safety-Wired Fasteners

3.0.0   NON-THREADED FASTENERS

Non-threaded fasteners have many uses in the electrical field. Different types of non-threaded fasteners include retainers, keys, pins, clamps, washers, rivets, and tie wraps.

3.1.0   RETAINER FASTENERS

Retainer fasteners, also called *retaining rings*, are used for both internal and external applications. Some retaining rings are seated in grooves in the fastener. Other types of retainer fasteners are self-locking and do not require a groove. To easily remove internal and external retainer rings without damaging the ring or the fastener, special pliers are used. *Figure 20* shows several types of retainer fasteners.

Note: External retainer fasteners are sometimes called *clips*.

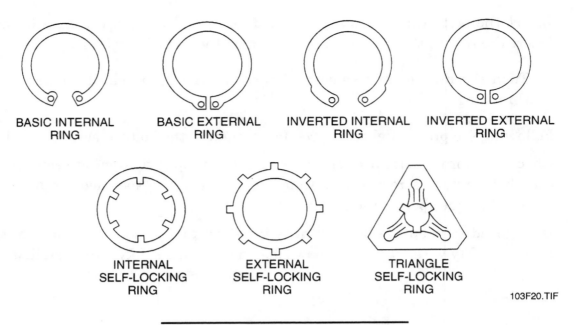

BASIC INTERNAL RING

BASIC EXTERNAL RING

INVERTED INTERNAL RING

INVERTED EXTERNAL RING

INTERNAL SELF-LOCKING RING

EXTERNAL SELF-LOCKING RING

TRIANGLE SELF-LOCKING RING

103F20.TIF

Figure 20. Retainer Fasteners (Rings)

3.2.0   KEYS

To prevent a gear or pulley from rotating on a shaft, keys are inserted. Half of the key fits into a keyseat on the shaft. The other half fits into a **keyway** in the hub of the gear or pulley. The key fastens the two parts together, stopping the gear or pulley from turning on the shaft. *Figure 21* shows several types of keys and keyways and their uses.

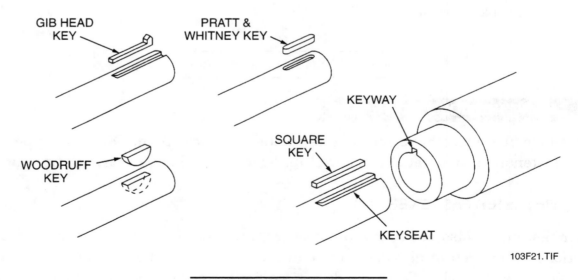

GIB HEAD KEY

PRATT & WHITNEY KEY

WOODRUFF KEY

SQUARE KEY

KEYWAY

KEYSEAT

103F21.TIF

Figure 21. Keys And Keyways

Some different types of keys include:

- *Square key* – Usually one-quarter of the shaft diameter. It may be slightly tapered on the top for easier fitting.
- *Pratt and Whitney key* – Similar to the square key, but rounded at both ends. It fits into a keyseat of the same shape.
- *Gib head key* – Interchangeable with the square key. The head design allows easy removal from the assembly.
- *Woodruff key* – Semicircular shape that fits into a keyseat of the same shape. The top of the key fits into the keyway of the mating part.

3.3.0 PIN FASTENERS

Pin fasteners come in several types and sizes. They have a variety of applications. Common uses of pin fasteners include holding moving parts together, aligning mating parts, fastening hinges, holding gears and pulleys on shafts, and securing slotted nuts. *Figure 22* shows several pin fasteners.

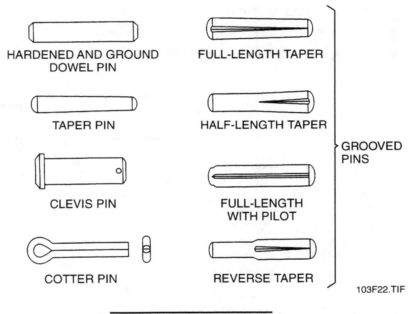

HARDENED AND GROUND DOWEL PIN

FULL-LENGTH TAPER

TAPER PIN

HALF-LENGTH TAPER

CLEVIS PIN

FULL-LENGTH WITH PILOT

GROOVED PINS

COTTER PIN

REVERSE TAPER

103F22.TIF

Figure 22. Pin Fasteners

3.3.1 Dowel Pins

Dowel pins are fit into holes to position mating parts. They may also support a portion of the load placed on the parts. *Figure 23* shows an application of dowel pins used to position mating parts.

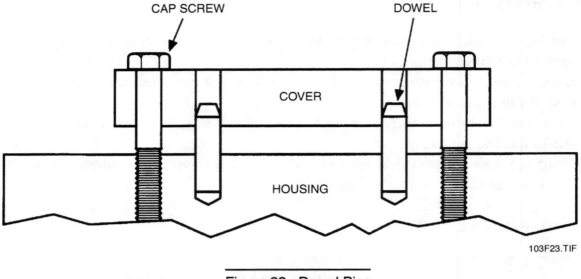

Figure 23. Dowel Pins

3.3.2 Taper And Spring Pins

Taper and spring pins are used to fasten gears, pulleys, and collars to a shaft. *Figure 24* shows how taper and spring pins are used to attach a component to a shaft. The groove in a spring pin allows it to compress against the walls in a spring-like fashion.

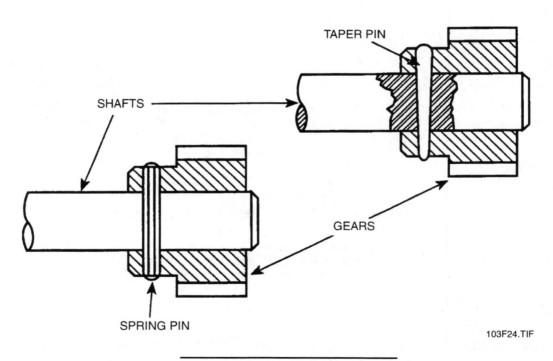

Figure 24. Taper And Spring Pins

ELECTRICAL — TRAINEE TASK MODULE 26103

3.3.3 Cotter Pins

There are several different types of cotter pins used as a locking device for a variety of applications. Cotter pins are often inserted through a hole drilled crosswise through a shaft to prevent parts from slipping on or off the shaft. They are also used to keep slotted nuts from working loose. Standard cotter pins are general use pins. When installed, the extended prong is normally bent back over the nut to provide the locking action. If it is ever removed, throw it away and replace it with a new one. The humped, cinch, and hitch-type cotter pins are self-locking pins. The humped and cinch type should also be thrown away and replaced with a new one if removed. The hitch pin, also called a *hair pin*, is a reusable pin made to be installed and removed quickly. *Figure 25* shows several common types of cotter pins.

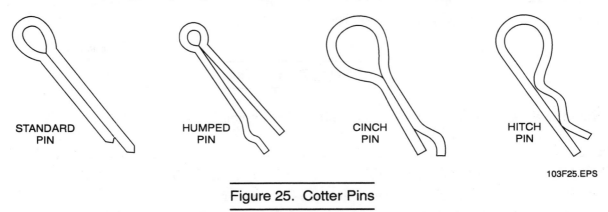

STANDARD PIN   HUMPED PIN   CINCH PIN   HITCH PIN

103F25.EPS

Figure 25. Cotter Pins

3.4.0 BLIND/POP RIVETS

When only one side of a joint can be reached, blind rivets can be used to fasten the parts together. Some applications of blind rivets include fastening light to heavy gauge sheet metal, fiberglass, plastics, and belting. Blind rivets are made of a variety of materials and come in several sizes and lengths. They are installed using special riveting tools. *Figure 26* shows a typical blind rivet installation.

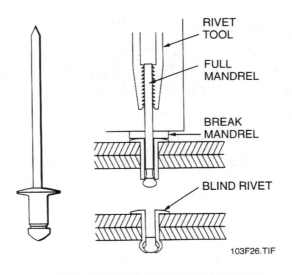

RIVET TOOL

FULL MANDREL

BREAK MANDREL

BLIND RIVET

103F26.TIF

Figure 26. Blind Rivet Installation

Blind rivets are installed through drilled or punched holes using a special blind (pop) rivet gun. *Figure 27* shows a typical rivet gun.

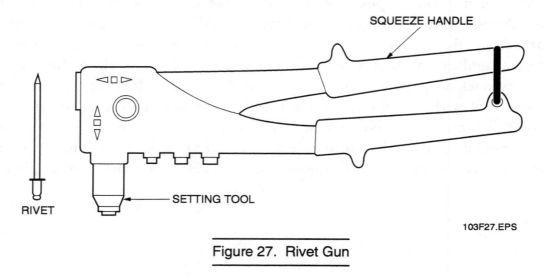

Figure 27. Rivet Gun

Use the following general procedure to install blind rivets.

WARNING!	Follow all safety precautions. Make sure to wear proper eye and face protection when riveting.

Step 1   Select the correct length and diameter of blind rivet to be used.

Step 2   Select the appropriate drill bit for the size of rivet being used.

Step 3   Drill a hole through both parts being connected.

Step 4   Inspect the rivet gun for any defects that might make it unsafe for use.

Step 5   Place the rivet mandrel into the proper size setting tool.

Step 6   Insert the rivet end into the pre-drilled hole.

Step 7   Install the rivet by squeezing the handle of the rivet gun, causing the jaws in the setting tool to grip the mandrel. The mandrel is pulled up, expanding the rivet until it breaks at the shear point. *Figure 28* shows the rivet and tool positioned for joining parts together.

Step 8   Inspect the rivet to make sure the pieces are firmly riveted together and that the rivet is properly installed. *Figure 29* shows a properly-installed blind rivet.

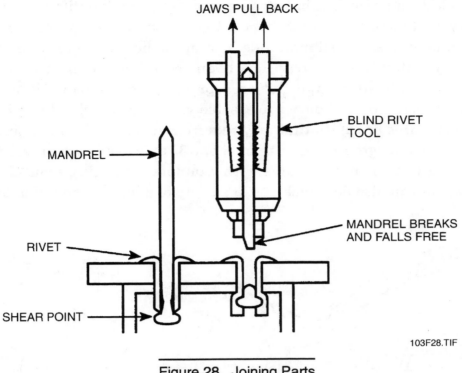

Figure 28. Joining Parts

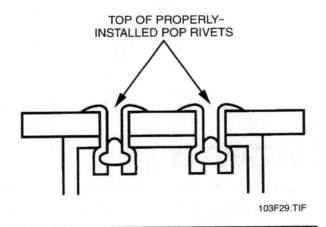

Figure 29. Properly-Installed Blind (Pop) Rivets

3.5.0 TIE WRAPS

A tie wrap is a one-piece, self-locking cable tie, usually made of nylon, that is used to fasten a bundle of wires and cables together. Tie wraps can be quickly installed either manually or using a special installation tool. Black tie wraps resist ultraviolet light and are recommended for outdoor use.

Tie wraps are made in standard, cable strap and clamp, and identification configurations (*Figure 30*). All types function to clamp bundled wires or cables together. In addition, the cable strap and clamp has a molded mounting hole in the head used to secure the tie with a rivet, screw, or bolt after the tie wrap has been installed around the wires or cable. Identification tie wraps have a large flat area provided for imprinting or writing cable identification information. There is also a releasable version available. It is a non-permanent tie used for bundling wires or cables that may require frequent additions or deletions. Cable ties are made in various lengths ranging from about 3" to 30", allowing them to be used for fastening wires and cables into bundles with diameters ranging from about ½" to 9", respectively. Tie wraps can also be attached to a variety of adhesive mounting bases made for that purpose.

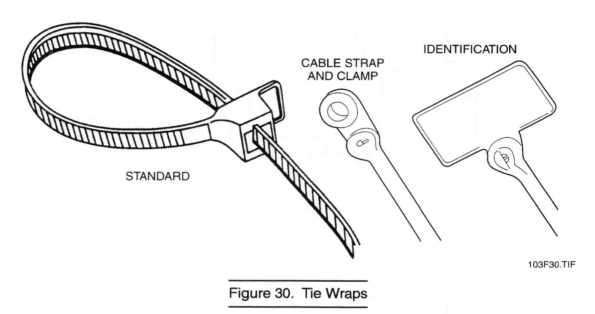

Figure 30. Tie Wraps

4.0.0   SPECIAL THREADED FASTENERS

Special threaded fasteners consist of hardware manufactured in several shapes and sizes and designed to perform specific jobs. Certain types of nuts may be considered special threaded fasteners if they are designed especially for a particular application. In the electrical craft, special threaded fasteners are used on a number of different jobs.

The three types of special threaded fasteners described below are eye bolts, toggle bolts, and J-bolts.

4.1.0   EYE BOLTS

Eye bolts get their name from the eye or loop at one end. The other end of an eye bolt is threaded. There are many types of eye bolts. The eye on some eye bolts is formed and welded while the eye on other types is forged. Shoulder forged eye bolts are commonly used as lifting devices and guides for wires, cables, and cords. *Figure 31* shows some typical eye bolts.

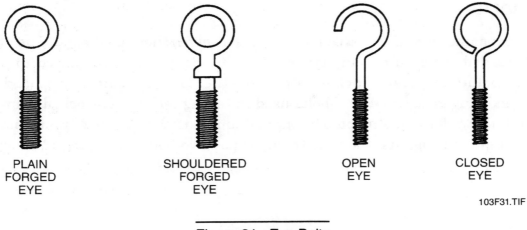

PLAIN
FORGED
EYE

SHOULDERED
FORGED
EYE

OPEN
EYE

CLOSED
EYE

103F31.TIF

Figure 31. Eye Bolts

4.2.0 ANCHOR BOLTS

An anchor bolt is used to fasten parts, machines, and equipment to concrete or masonry foundations, floors, and walls. There are several types of anchor bolts designed for different applications. *Figure 32* shows a type of anchor bolt for use in wet concrete. If the concrete has already hardened, expansion anchor bolts are used. These are covered later in this module.

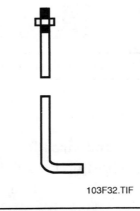

103F32.TIF

Figure 32. Anchor Bolt

One common method used to install anchor bolts in wet concrete involves making a wooden template to locate the anchor bolts. The template positions the anchor bolts so that they correspond to those in the equipment to be fastened.

4.3.0 J-BOLTS

J-bolts get their name from the curve on one end that gives them a J shape. The other end of a J-bolt is threaded. There are many types of J-bolts. Some J-bolts are used to hold tubing bundles and include a plastic jacket to protect the tubing. Others are used to attach equipment to existing grating. Most J-bolts used in tubing racks are attached using two nuts. The upper nut allows for adjustment. The tubing bundle is clamped firmly, but not flattened. Both nuts are tightened against the tube track for positive holding. *Figure 33* shows a typical J-bolt.

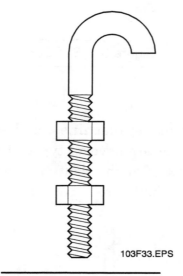

103F33.EPS

Figure 33. Typical J-Bolt

5.0.0 SCREWS

Screws are made in a variety of shapes and sizes for different fastening jobs. The finish or coating used on a screw determines whether it is for interior or exterior use, corrosion resistant, etc. Screws of all types have heads with different shapes and slots similar to those previously described for machine screws. Some have machine threads and are self-drilling. The size or diameter of a screw body or shank is given in gauge numbers ranging from No. 0 to No. 24, and in fractions of an inch for screws with diameters larger than ¼". The higher the gauge number, the larger the diameter of the shank. Screw lengths range from ¼" to 6", measured from the tip to the part of the head that is flush to the surface when driven in. When choosing a screw for an application, you must consider the type and thickness of the materials to be fastened, the size of the screw, the material it is made of, the shape of its head, and the type of driver. Because of the wide diversity in the types of screws and their application, always follow the manufacturer's recommendation to select the right screw for the job. To prevent damage to the screw head or the material being fastened, always use a screwdriver or power driver bit with the proper size and shape tip to fit the screw.

Some of the more common types of screws are:

- Wood screws
- Lag screws
- Masonry/concrete screws
- Thread-forming and thread-cutting screws
- Deck screws
- Drywall screws
- Drive screws

5.1.0   WOOD SCREWS

Wood screws (*Figure 34*) are typically used to fasten boxes, panel enclosures, etc. to wood framing or structures where greater holding power is needed than can be provided by nails. They are also used to fasten equipment to wood in applications where it may occasionally need to be unfastened and removed. Wood screws are commonly made in lengths from ¼" to 4", with shank gauge sizes ranging from 0 to 24. The shank size used is normally determined by the size hole provided in the box, panel, etc. to be fastened. When determining the length of a wood screw to use, a good rule-of-thumb is to select screws long enough to allow about ⅔ of the screw length to enter the piece of wood that is being gripped.

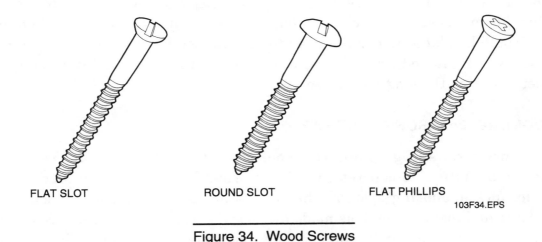

FLAT SLOT   ROUND SLOT   FLAT PHILLIPS

103F34.EPS

Figure 34. Wood Screws

5.2.0   LAG SCREWS AND SHIELDS

Lag screws (*Figure 35*) or lag bolts are heavy-duty wood screws with square- or hex-shaped heads that provide greater holding power. Lag screws with diameters ranging between ¼ and ½" and lengths ranging from 1" to 6" are common. They are typically used to fasten heavy equipment to wood, but can also be used to fasten equipment to concrete when a lag shield is used.

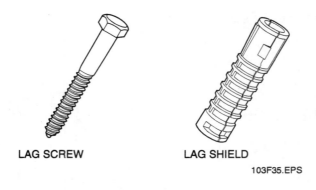

LAG SCREW   LAG SHIELD

103F35.EPS

Figure 35. Lag Screw And Shield

A lag shield is a lead tube that is split lengthwise but remains joined at one end. It is placed in a pre-drilled hole in the concrete. When a lag screw is screwed into the lag shield, the shield expands in the hole, firmly securing the lag screw. In hard masonry, short lag shields (typically 1" to 2" long) may be used to minimize drilling time. In soft or weak masonry, long lag shields (typically 1½" to 3" long) should be used to achieve maximum holding strength.

Make sure to use the proper length lag screw to achieve proper expansion. The length of the lag screw used should be equal to the thickness of the component being fastened plus the length of the lag shield. Also, drill the hole in the masonry to a depth approximately ½" longer than the shield being used. If the head of a lag screw rests directly on wood when installed, a flat washer should be placed under the head to prevent the head from digging into the wood as the lag screw is tightened down. Be sure to take the thickness of any washers used into account when selecting the length of the screw.

5.3.0   CONCRETE/MASONRY SCREWS

Concrete/masonry screws (*Figure 36*), commonly called *self-threading anchors*, are used to fasten a device or fixture to concrete, block, or brick. No anchor is needed. To provide a matched tolerance anchoring system, the screws are installed using specially-designed carbide drill bits and installation tools made for use with the screws. These tools are typically used with a standard rotary drill hammer. The installation tool, along with an appropriate drive socket or bit, is used to drive the screws directly into pre-drilled holes that have a diameter and depth specified by the screw manufacturer. When being driven into the concrete, the widely-spaced threads on the screws cut into the walls of the hole to provide a tight friction fit. Most types of concrete/masonry screws can be removed and reinstalled to allow for shimming and leveling of the fastened device.

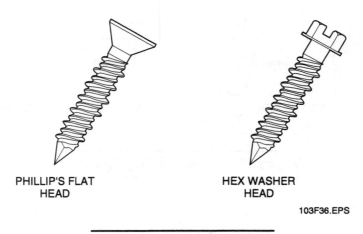

PHILLIP'S FLAT
HEAD

HEX WASHER
HEAD

103F36.EPS

Figure 36. Concrete Screws

5.4.0 THREAD-FORMING AND THREAD-CUTTING SCREWS

Thread-forming screws (*Figure 37*), commonly called *sheet metal screws*, are made of hard metal. They form a thread as they are driven into the work. This thread-forming action eliminates the need to tap a hole before installing the screw. To achieve proper holding, it is important to make sure to use the proper size bit when drilling pilot holes for thread-forming screws. The correct drill bit size used for a specific size screw is usually marked on the box containing the screws. Some types of thread-forming screws also drill their own holes, eliminating drilling, punching, and aligning parts. Thread-forming screws are primarily used to fasten light gauge metal parts together. They are made in the same diameters and lengths as wood screws.

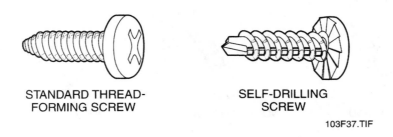

STANDARD THREAD-
FORMING SCREW

SELF-DRILLING
SCREW

103F37.TIF

Figure 37. Thread-Forming Screws

Hardened steel thread-cutting metal screws with blunt points and fine threads (*Figure 38*) are used to join heavy-gauge metals, metals of different gauges, and nonferrous metals. They are also used to fasten sheet metal to building structural members. These screws are made of hardened steel that is harder than the metal being tapped. They cut threads by removing and cutting a portion of the metal as they are driven into a pilot hole and through the material.

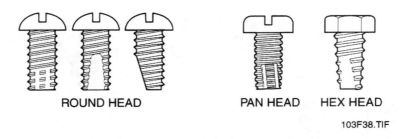

ROUND HEAD   PAN HEAD   HEX HEAD

103F38.TIF

Figure 38. Thread-Cutting Screws

5.5.0 DECK SCREWS

Deck screws (*Figure 39*) are made in a wide variety of shapes and sizes for different indoor and outdoor applications. Some are made to fasten pressure-treated and other types of wood decking to wood framing. Self-drilling types are made to fasten wood decking to different gauges of metal support structures. Similarly, other self-drilling kinds are made to fasten metal decking and sheeting to different gauges and types of metal structural support members. Because of their wide diversity, it is important to follow the manufacturer's recommendations for selection of the proper screw for a particular application. Many manufacturers make a stand-up installation tool used for driving their deck screws. Use of this tool eliminates angle driving, underdriven or overdriven screws, screw wobble, etc. It also reduces operator fatigue.

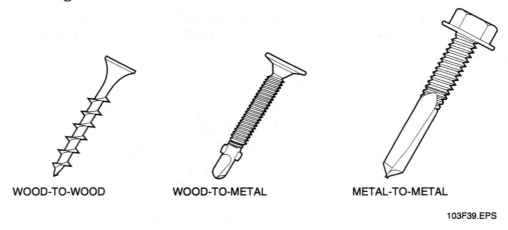

WOOD-TO-WOOD   WOOD-TO-METAL   METAL-TO-METAL

103F39.EPS

Figure 39. Typical Deck Screws

5.6.0 DRYWALL SCREWS

Drywall screws (*Figure 40*) are thin, self-drilling screws with bugle-shaped heads. Depending on the type of screw, it cuts through the wallboard and anchors itself into wood and/or metal studs, holding the wallboard tight to the stud. Coarse thread screws are normally used to fasten wallboard to wood studs. Fine thread and high and low thread types are generally used for fastening to metal studs. Some screws are made for use in either wood or metal. A Phillips or Robertson drive head allows the drywall screw to be countersunk without tearing the surface of the wallboard.

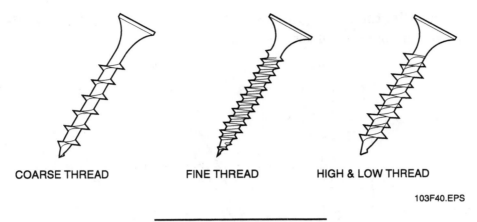

COARSE THREAD   FINE THREAD   HIGH & LOW THREAD

103F40.EPS

Figure 40. Drywall Screws

5.7.0 DRIVE SCREWS

Drive screws do not require that the hole be tapped. They are installed by hammering the screw into a drilled or punched hole of the proper size. Drive screws are mostly used to fasten parts that will not be exposed to much pressure. A typical use of drive screws is to attach permanent name plates on electric motors and other types of equipment. *Figure 41* shows typical drive screws.

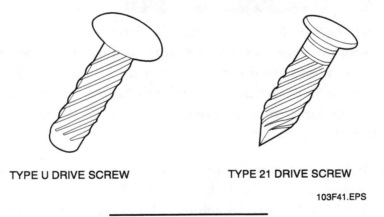

TYPE U DRIVE SCREW   TYPE 21 DRIVE SCREW

103F41.EPS

Figure 41. Drive Screws

6.0.0 HAMMER-DRIVEN PINS AND STUDS

Hammer-driven pins or threaded studs (*Figure 42*) can be used to fasten wood or steel to concrete or block without the need to pre-drill holes. The pin or threaded stud is inserted into a hammer-driven tool designed for its use. The pin or stud is inserted in the tool point end out with the washer seated in the recess. The pin or stud is then positioned against the base material where it is to be fastened and the drive rod of the tool tapped lightly until the striker pin contacts the pin or stud. Following this, the tool's drive rod is struck using heavy blows with about a two-pound engineer's hammer. The force of the hammer blows is transmitted through the tool directly to the head of the fastener, causing it to be driven into the concrete or block. For best results, the drive pin or stud should be embedded a minimum of ½ " in hard concrete to 1¼" in softer concrete block.

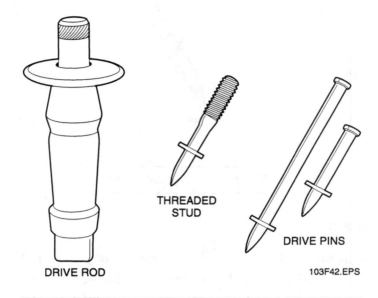

THREADED STUD

DRIVE PINS

DRIVE ROD

103F42.EPS

Figure 42. Hammer-Driven Pins And Installation Tool

7.0.0 POWDER-ACTUATED TOOLS AND FASTENERS

Powder-actuated tools (*Figure 43*) can be used to drive a wide variety of specially-designed pin and threaded stud-type fasteners into masonry and steel. These tools look and fire like a gun and use the force of a detonated gunpowder load (typically 22, 25, or 27 caliber) to drive the fastener into the material. The depth to which the pin or stud is driven is controlled by the density of the base material in which the pin or stud is being installed and by the power level or strength of the cased powder load.

Powder loads and their cases are designed for use with specific types and/or models of powder-actuated tools and are not interchangeable. Typically, powder loads are made in 12 increasing power or load levels used to achieve the proper penetration. The different power levels are identified by a color-code system and load case types. Note that different manufacturers may use different color codes to identify load strength. Power level 1 is the lowest power level while 12 is the highest. Higher number power levels are used when driving into hard materials or when a deeper penetration is needed. Powder loads are available as single-shot units for use with single-shot tools. They are also made in multi-shot strips or disks for semi-automatic tools.

WARNING!   Powder-actuated fastening tools are to be used only by trained and licensed operators and in accordance with the tool operator's manual. You must carry your license with you whenever you are using a powder-actuated tool.

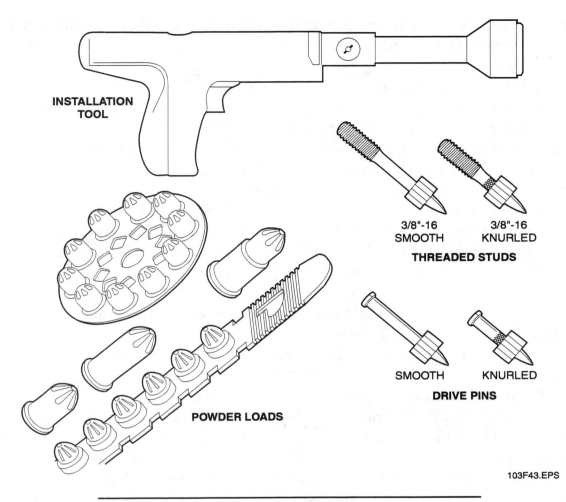

INSTALLATION TOOL

3/8"-16 SMOOTH   3/8"-16 KNURLED
THREADED STUDS

SMOOTH   KNURLED
DRIVE PINS

POWDER LOADS

103F43.EPS

Figure 43. Powder-Actuated Installation Tool And Fasteners

OSHA Standard 29 CFR 1926.302(e) governs the use of powder-actuated tools and requires only operators who have been trained in the operation of the particular tool in use be allowed to operate a powder-actuated tool. Authorized instructors available from the various powder-actuated tool manufacturers generally provide such training and licensing. Trained operators must take precautions to protect both themselves and others in the area when using a powder-actuated driver tool:

- Always use the tool in accordance with the published tool operation instructions. The instructions should be kept with the tool. Never attempt to override the safety features of the tool.
- Never place your hand or other body parts over the front muzzle end of the tool.
- Use only fasteners, powder loads, and tool parts specifically made for use with the tool. Use of other materials can cause improper and unsafe functioning of the tool.
- Operators and bystanders must wear eye and hearing protection along with hard hats. Other personal safety gear, as required, must also be used.
- Always post warning signs which state *Powder-Actuated Tool in Use* within 50 feet of the area where tools are used.

- Prior to using a tool, make sure it is unloaded and perform a proper function test. Check the functioning of the unloaded tool as described in the published tool operation instructions.
- Do not guess before fastening into any base material; always perform a center punch test.
- Always make a test firing into a suitable base material with the lowest power load recommended for the tool being used. If this does not set the fastener, try the next higher power level. Continue this procedure until the proper fastener penetration is obtained.
- Always point the tool away from operators or bystanders.
- Never use the tool in an explosive or flammable area.
- Never leave a loaded tool unattended. Do not load the tool until you are prepared to complete the fastening. Should you decide not to make a fastening after the tool has been loaded, always remove the powder load first, then the fastener. Always unload the tool before cleaning, servicing, or when changing parts, prior to work breaks, and when storing the tool.
- Always hold the tool perpendicular to the work surface and use the spall (chip or fragment) guard or stop spall whenever possible.
- Always follow the required spacing, edge distance, and base material thickness requirements.
- Never fire through an existing hole or into a weld area.
- In the event of a misfire, always hold the tool depressed against the work surface for at least 30 seconds. If the tool still does not fire, follow the published tool instructions. Never carelessly discard or throw unfired powder loads into a trash receptacle.
- Always store the powder loads and unloaded tool under lock and key.

8.0.0 MECHANICAL ANCHORS

Mechanical anchors are devices used to give fasteners a firm grip in a variety of materials, where the fasteners by themselves would otherwise have a tendency to pull out. Anchors can be classified in many ways by different manufacturers. In this module, anchors have been divided into five broad categories:

- One-step anchors
- Bolt anchors
- Screw anchors
- Self-drilling anchors
- Hollow-wall anchors

8.1.0 ONE-STEP ANCHORS

One-step anchors are designed so that they can be installed through the mounting holes in the component to be fastened. This is because the anchor and the drilled hole into which it is installed have the same size diameter. They come in various diameters ranging from ¼" to 1¼" with lengths ranging from 1¾" to 12". Wedge, stud, sleeve, one-piece, screw, and nail anchors (*Figure 44*) are common types of one-step anchors.

ELECTRICAL — TRAINEE TASK MODULE 26103

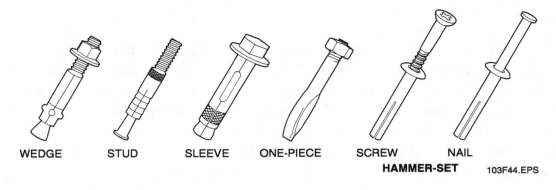

Figure 44. One-Step Anchors

8.1.1 Wedge Anchors

Wedge anchors are heavy-duty anchors supplied with nuts and washers. The drill bit size used to drill the hole is the same diameter as the anchor. The depth of the hole is not critical as long as the minimum length recommended by the manufacturer is drilled. After blowing the hole clean of dust and other material, the anchor is inserted into the hole and driven with a hammer far enough so that at least six threads are below the top surface of the component. Then, the component is fastened by tightening the anchor nut to expand the anchor and tighten it in the hole.

8.1.2 Stud Bolt Anchors

Stud bolt anchors are heavy-duty threaded anchors. Because this type of anchor is made to bottom in its mounting hole, it is a good choice to use when jacking or leveling of the fastened component is needed. The depth of the hole drilled in the masonry must be as specified by the manufacturer in order to achieve proper expansion. After blowing the hole clean of dust and other material, the anchor is inserted in the hole with the expander plug end down. Following this, the anchor is driven into the hole with a hammer (or setting tool) to expand the anchor and tighten it in the hole. The anchor is fully set when it can no longer be driven into the hole. The component is fastened using the correct size and thread bolt for use with the anchor stud.

8.1.3 Sleeve Anchors

Sleeve anchors are multi-purpose anchors. The depth of the anchor hole is not critical as long as the minimum length recommended by the manufacturer is drilled. After blowing the hole clean of dust and other material, the anchor is inserted into the hole and tapped until flush with the component. Then, the anchor nut or screw is tightened to expand the anchor and tighten it in the hole.

8.1.4 One-Piece Anchors

One-piece anchors are multi-purpose anchors. They work on the principle that as the anchor is driven into the hole, the spring force of the expansion mechanism is compressed and flexes to fit the size of the hole. Once set, it tries to regain its original shape. The depth of the hole drilled in the masonry must be at least ½" deeper than the required embedment. The proper depth is crucial. Overdrilling is as bad as underdrilling. After blowing the hole clean of dust and other material, the anchor is inserted through the component and driven with a hammer into the hole until the head is firmly seated against the component. It is important to make sure that the anchor is driven to the proper embedment depth. Note that manufacturers also make specially-designed drivers and manual tools that are used instead of a hammer to drive one-piece anchors. These tools allow the anchors to be installed in confined spaces and help prevent damage to the component from stray hammer blows.

8.1.5 Hammer-Set Anchors

Hammer-set anchors are made for use in concrete and masonry. There are two types: nail and screw. An advantage of the screw-type anchors is that they are removable. Both types have a diameter the same size as the anchoring hole. For both types, the anchor hole must be drilled to the diameter of the anchor and to a depth of at least ¼" deeper than that required for embedment. After blowing the hole clean of dust and other material, the anchor is inserted into the hole through the mounting holes in the component to be fastened, then the screw or nail is driven into the anchor body to expand it. It is important to make sure that the head is seated firmly against the component and is at the proper embedment.

8.2.0 BOLT ANCHORS

Bolt anchors are designed to be installed flush with the surface of the base material. They are used in conjunction with threaded machine bolts or screws. In some types, they can be used with threaded rod. Drop-in, single and double expansion, and caulk-in anchors (*Figure 45*) are commonly used types of bolt anchors.

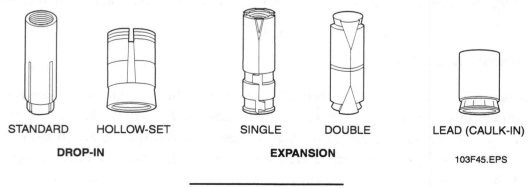

STANDARD   HOLLOW-SET   SINGLE   DOUBLE   LEAD (CAULK-IN)

DROP-IN   **EXPANSION**

103F45.EPS

Figure 45. Bolt Anchors

8.2.1   Drop-In Anchors

Drop-in anchors are typically used as heavy-duty anchors. There are two types of drop-in anchors. The first type, made for use in solid concrete and masonry, has an internally-threaded expansion anchor with a pre-assembled internal expander plug. The anchor hole must be drilled to the specific diameter and depth specified by the manufacturer. After blowing the hole clean of dust and other material, the anchor is inserted into the hole and tapped until it is flush with the surface. Following this, a setting tool supplied with the anchor is driven into the anchor to expand it. The component to be fastened is positioned in place and fastened by threading and tightening the correct size machine bolt or screw into the anchor.

The second type, called a *hollow set drop-in anchor*, is made for use in hollow concrete and masonry base materials. Hollow set drop-in anchors have a slotted, tapered expansion sleeve and a serrated expansion cone. They come in various lengths compatible with the outer wall thickness of most hollow base materials. They can also be used in solid concrete and masonry. The anchor hole must be drilled to the specific diameter specified by the manufacturer. When installed in hollow base materials, the hole is drilled into the cell or void. After blowing the hole clean of dust and other material, the anchor is inserted into the hole and tapped until it is flush with the surface. Following this, the component to be fastened is positioned in place, then the proper size machine bolt or screw is threaded into the anchor and tightened to expand the anchor in the hole.

8.2.2   Single- And Double-Expansion Anchors

Single- and double-expansion anchors are both made for use in concrete and other masonry. The double-expansion anchor is used mainly when fastening into concrete or masonry of questionable strength. For both types, the anchor hole must be drilled to the specific diameter and depth specified by the manufacturer. After blowing the hole clean of dust and other material, the anchor is inserted into the hole, threaded cone end first. It is then tapped until it is flush with the surface. Following this, the component to be fastened is positioned in place, then the proper size machine bolt or screw is threaded into the anchor and tightened to expand the anchor in the hole.

8.2.3   Lead (Caulk-In) Anchors

Lead (caulk-in) anchors are a cast-type anchor used in concrete and masonry. They consist of an internally-threaded expander cone with a series of vertical internal ribs and a lead sleeve. The vertical internal ribs prevent the cone from turning in the sleeve as the anchor is tightened. The anchor hole must be drilled to the specific diameter and depth specified by the manufacturer. However, in weak or soft masonry, a slightly deeper hole can be drilled to countersink the anchor below the surface. After blowing the hole clean of dust and other material, the anchor is inserted into the hole, threaded cone end first. Following this, a setting tool supplied with the anchor is driven into the anchor to expand it. The component to be fastened is positioned in place and fastened by threading and tightening the correct size machine bolt or screw into the anchor.

8.3.0 SCREW ANCHORS

Screw anchors are lighter-duty anchors made to be installed flush with the surface of the base material. They are used in conjunction with sheet metal, wood, or lag screws depending on the anchor type. Fiber, lead, and plastic anchors are common types of screw anchors (*Figure 46*). The lag shield anchor used with lag screws was described earlier in this module.

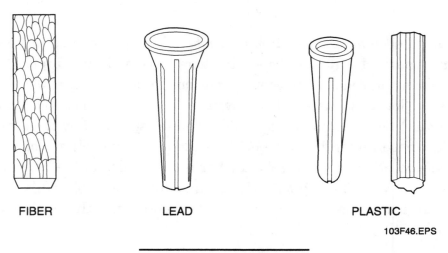

FIBER   LEAD   PLASTIC

103F46.EPS

Figure 46. Screw Anchors

Fiber, lead, and plastic anchors are typically used in concrete and masonry. Plastic anchors are also commonly used in wallboard and similar base materials. The installation of all types is simple. The anchor hole must be drilled to the diameter specified by the manufacturer. The minimum depth of the hole must equal the anchor length. After blowing the hole clean of dust and other material, the anchor is inserted into the hole and tapped until it is flush with the surface. Following this, the component to be fastened is positioned in place, then the proper type and size screw is driven through the component mounting hole and into the anchor to expand the anchor in the hole.

8.4.0 SELF-DRILLING ANCHORS

Some anchors made for use in masonry are self-drilling anchors. *Figure 47* is typical of those in common use. This fastener has a cutting sleeve that is first used as a drill bit and later becomes the expandable fastener itself. A rotary hammer is used to drill the hole in the concrete using the anchor sleeve as the drill bit. After the hole is drilled, the anchor is pulled out and the hole cleaned. This is followed by inserting the anchor's expander plug into the cutting end of the sleeve. The anchor sleeve and expander plug are driven back into the hole with the rotary hammer until they are flush with the surface of the concrete. As the fastener is hammered down, it hits the bottom where the tapered expander causes the fastener to expand and lock into the hole. The anchor is then snapped off at the shear point with a quick lateral movement of the hammer. The component to be fastened can then be attached to the anchor using the proper size bolt.

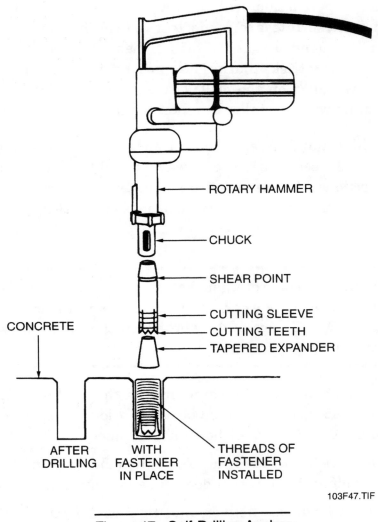

103F47.TIF

Figure 47. Self-Drilling Anchor

8.5.0 GUIDELINES FOR DRILLING ANCHOR HOLES IN HARDENED CONCRETE OR MASONRY

When selecting masonry anchors, regardless of the type, always take into consideration and follow the manufacturer's recommendations pertaining to hole diameter and depth, minimum embedment in concrete, maximum thickness of material to be fastened, and the pullout and shear load capacities.

When installing anchors and/or anchor bolts in hardened concrete, make sure the area where the equipment or component is to be fastened is smooth so that it will have solid footing. Uneven footing might cause the equipment to twist, warp, not tighten properly, or vibrate when in operation. Before starting, carefully inspect the rotary hammer or hammer drill and the drill bit(s) to ensure they are in good operating condition. Be sure to use the type of carbide-tipped masonry or percussion drill bits recommended by the drill/hammer or anchor manufacturer because these bits are made to take the higher impact of the masonry materials. Also, it is recommended that the drill or hammer tool depth gauge be set to the depth of the hole needed. The trick to using masonry drill bits is not to force them into the material by pushing down hard on the drill. Use a little pressure and let the drill do the work. For large holes, start with a smaller bit, then change to a larger bit.

The methods for installing the different types of anchors in hardened concrete or masonry were briefly described in the sections above. Always install the selected anchors according to the manufacturer's directions. Here is an example of a typical procedure used to install many types of expansion anchors in hardened concrete or masonry. Refer to *Figure 48* as you study the procedure.

WARNING!   Drilling in concrete generates noise, dust, and flying particles. Always wear safety goggles, ear protectors, and gloves. Make sure other workers in the area also wear protective equipment.

Step 1 Drill the anchor bolt hole the same size as the anchor bolt. The hole must be deep enough for six threads of the bolt to be below the surface of the concrete (see *Figure 48, Step 1*). Clean out the hole using a squeeze bulb.

Step 2 Drive the anchor bolt into the hole using a hammer (*Figure 48, Step 2*). Protect the threads of the bolt with a nut that does not allow any threads to be exposed.

Step 3 Put a washer and nut on the bolt, and tighten the nut with a wrench until the anchor is secure in the concrete (*Figure 48, Step 3*).

8.6.0 HOLLOW-WALL ANCHORS

Hollow-wall anchors are used in hollow materials such as concrete plank, block, structural steel, wallboard, and plaster. Some types can also be used in solid materials. Toggle bolts, sleeve-type wall anchors, wallboard anchors, and metal drive-in anchors are common anchors used when fastening to hollow materials.

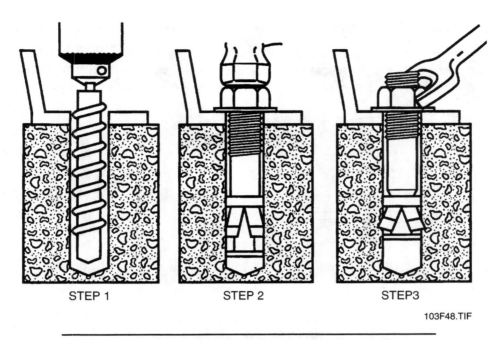

Figure 48. Installing An Anchor Bolt In Hardened Concrete

103F48.TIF

When installing anchors in hollow walls or ceilings, regardless of the type, always follow the manufacturer's recommendations pertaining to use, hole diameter, wall thickness, grip range (thickness of the anchoring material), and the pullout and shear load capacities.

8.6.1 Toggle Bolts

Toggle bolts (*Figure 49*) are used to fasten equipment, hangers, supports, and similar items into hollow surfaces such as walls and ceilings. They consist of a slotted bolt or screw and spring-loaded wings. When inserted through the item to be fastened, then through a pre-drilled hole in the wall or ceiling, the wings spring apart and provide a firm hold on the inside of the hollow wall or ceiling as the bolt is tightened. Note that the hole drilled in the wall or ceiling should be just large enough for the compressed wing-head to pass through. Once the toggle bolt is installed, be careful not to completely unscrew the bolt because the wings will fall off, making the fastener useless. Screw-actuated plastic toggle bolts are also made. These are similar to metal toggle bolts, but they come with a pointed screw and do not require as large a hole. Unlike the metal version, the plastic wings remain in place if the screw is removed.

Toggle bolts are used to fasten a part to hollow block, wallboard, plaster, panel, or tile. The following general procedure can be used to install toggle bolts.

WARNING!   Follow all safety precautions.

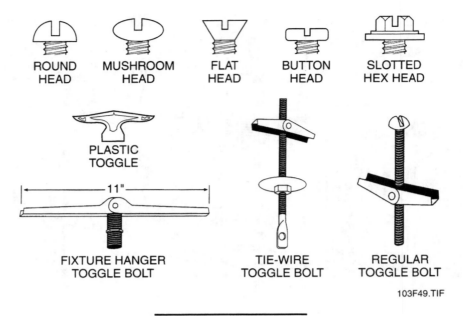

ROUND HEAD
MUSHROOM HEAD
FLAT HEAD
BUTTON HEAD
SLOTTED HEX HEAD

PLASTIC TOGGLE

11"

FIXTURE HANGER TOGGLE BOLT
TIE-WIRE TOGGLE BOLT
REGULAR TOGGLE BOLT

103F49.TIF

Figure 49. Toggle Bolts

Step 1   Select the proper size drill bit or punch and toggle bolt for the job.

Step 2   Check the toggle bolt for damaged or dirty threads or a malfunctioning wing mechanism.

Step 3   Drill a hole completely through the surface to which the part is to be fastened.

Step 4   Insert the toggle bolt through the opening in the item to be fastened.

Step 5   Screw the toggle wing onto the end of the toggle bolt, ensuring that the flat side of the toggle wing is facing the bolt head.

Step 6   Fold the wings completely back and push them through the drilled hole until the wings spring open.

Step 7   Pull back on the item to be fastened in order to hold the wings firmly against the inside surface to which the item is being attached.

Step 8   Tighten the toggle bolt with a screwdriver until it is snug.

8.6.2   Sleeve-Type Wall Anchors

Sleeve-type wall anchors (*Figure 50*) are suitable for use in concrete, block, plywood, wallboard, hollow tile, and similar materials. The two types made are standard and drive. The standard type is commonly used in walls and ceilings and is installed by drilling a mounting hole to the required diameter. The anchor is inserted into the hole and tapped until the gripper prongs embed in the base material. Following this, the anchor's screw is tightened to draw the anchor tight against the inside of the wall or ceiling. Note that the drive-type anchor is hammered into the material without the need for drilling a mounting hole. After the anchor is installed, the anchor screw is removed, the component being fastened is positioned in place, then the screw is reinstalled through the mounting hole in the component and into the anchor. The screw is tightened into the anchor to secure the component.

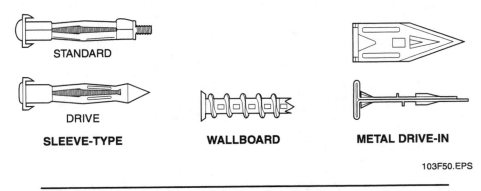

STANDARD

DRIVE

SLEEVE-TYPE   WALLBOARD   METAL DRIVE-IN

103F50.EPS

Figure 50. Sleeve-Type, Wallboard, And Metal Drive-In Anchors

8.6.3   Wallboard Anchors

Wallboard anchors (*Figure 50*) are self-drilling medium- and light-duty anchors used for fastening in wallboard. The anchor is driven into the wall with a Phillips head manual or cordless screwdriver until the head of the anchor is flush with the wall or ceiling surface. Following this, the component being fastened is positioned over the anchor, then secured with the proper size sheet metal screw driven into the anchor.

8.6.4   Metal Drive-In Anchors

Metal drive-in anchors (*Figure 50*) are used to fasten light to medium loads to wallboard. They have two pointed legs that stay together when the anchor is hammered into a wall and spread out against the inside of the wall when a No. 6 or 8 sheet metal screw is driven in.

9.0.0   EPOXY ANCHORING SYSTEMS

Epoxy resin compounds can be used to anchor threaded rods, dowels, and similar fasteners in solid concrete, hollow wall, and brick. For one manufacturer's product, a two-part epoxy is packaged in a two-chamber cartridge that keeps the resin and hardener ingredients separated until use. This cartridge is placed into a special tool similar to a caulking gun. When the gun handle is pumped, the epoxy resin and hardener components are mixed within the gun, then the epoxy is ejected from the gun nozzle.

To use the epoxy to install an anchor in solid concrete (*Figure 51*), a hole of the proper size is drilled in the concrete and cleaned using a nylon (not metal) brush. Following this, a small amount of epoxy is dispensed from the gun to make sure that the resin and hardener have mixed properly. This is indicated by the epoxy being of a uniform color. The gun nozzle is then placed into the hole, and the epoxy is injected into the hole until half the depth of the hole is filled. Following this, the selected fastener is pushed into the hole with a slow twisting motion to make sure that the epoxy fills all voids and crevices, then is set to the required plumb (or level) position. After the recommended cure time for the epoxy has elapsed, the fastener nut can be tightened to secure the component or fixture in place.

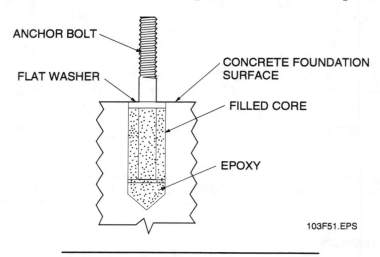

ANCHOR BOLT

FLAT WASHER

CONCRETE FOUNDATION SURFACE

FILLED CORE

EPOXY

103F51.EPS

Figure 51. Fastener Anchored In Epoxy

The procedure for installing a fastener in a hollow wall or brick using epoxy is basically the same as described above. The difference is that the epoxy is first injected into an anchor screen to fill the screen, then the anchor screen is installed into the drilled hole. Use of the anchor screen is necessary to hold the epoxy intact in the hole until the anchor is inserted into the epoxy.

SUMMARY

Fasteners and anchors are used for a variety of tasks in the electrical craft. In this module, you learned about various types of fasteners and anchors and their uses. Basic installation procedures were also included. Selecting the correct fastener or anchor for a particular job is required to perform high-quality work. In the electrical craft, it is important to be familiar with the correct terms used to describe fasteners and anchors. Using the proper technical terms helps avoid confusion and improper selection. Installation techniques for fasteners and anchors may vary depending on the job. Make sure to check the project specifications and manufacturer's information when installing any fastener or anchor.

New fasteners and anchors are being developed every day. Your local distributor/manufacturer is an excellent source of information.

References

For advanced study of topics covered in this task module, the following books are suggested:

American Electrician's Handbook, Latest Edition, McGraw-Hill, New York, NY.
Carpentry, Latest Edition, Delmar Publishers, New York, NY.
The Sheet Metal Toolbox Manual, Latest Edition, Prentice Hall, New York, NY.

REVIEW QUESTIONS

1. The quality of some fasteners can be determined by _____ on the head of a bolt or screw.
   a. the number of grooves cut
   b. the number of sides
   c. the length of the lines
   d. the grade markings

2. The purpose of a jam nut is to _____.
   a. hold a piece of material stationary
   b. lock a standard nut in place
   c. stop rotation of a machine quickly for safety reasons
   d. compress a lock washer in place

3. Washers are used to _____.
   a. distribute the load over a larger area
   b. attach an item to a hollow surface
   c. anchor materials that expand due to temperature changes
   d. allow the bolts to expand with temperature changes

4. Torque is normally expressed in _____.
   a. pounds per square inch (psi)
   b. gallons per minute (gpm)
   c. foot pounds (ft. lbs.)
   d. cubic feet per minute (cpm)

5. When tightening bolts, nuts, or screws, always use the proper _____.
   a. tightening sequence
   b. bolt pattern
   c. ratchet wrench
   d. lubrication

6. What type of fasteners are commonly used as lifting devices?
   a. Toggle bolts
   b. Anchor bolts
   c. J-bolts
   d. Eye bolts

7. When installing a panel in concrete using a lag screw, you would also use a _____.
   a. drop-in anchor
   b. lag shield
   c. caulk-in anchor
   d. double-expansion anchor

8. When fastening a device or fixture to concrete using concrete/masonry screws, the screws _____.
   a. need not be installed in pre-drilled holes because they are self-tapping
   b. are installed using specially-designed carbide drill bits and installation tools made for use with the screws
   c. are installed by hammering them into a drilled or punched hole of the proper size
   d. are installed in a hole drilled with a standard masonry drill bit

9. When using powder-actuated tools and fasteners, _____.
   a. higher-numbered power level loads are used to drive into soft materials
   b. use only powder loads that are specifically designed for use with the installation tool
   c. use only fasteners that are specifically designed for use with the installation tool
   d. both b and c

10. Wedge and sleeve anchors are classified as _____ anchors.
   a. bolt
   b. hollow-wall
   c. one-step
   d. screw

ANSWERS TO REVIEW QUESTIONS

<u>Answer</u>	<u>Section</u>
1. d	2.1.4
2. b	2.3.1
3. a	2.4.0
4. c	2.5.1
5. a	2.5.2
6. d	4.1.0
7. b	5.2.0
8. b	5.3.0
9. d	7.0.0
10. c	8.1.0

MECHANICAL ANCHORS AND THEIR USES (PAGE 1 OF 2)

Anchor Type	Typically Used In	Use With Fastener	Typical Working Load Range*
One-Step Anchors			
Wedge	Concrete **Stone	None	Light, medium, and heavy duty
Stud	Concrete **Stone, solid brick and block	None	Light, medium, and and heavy duty
Sleeve	Concrete, solid brick and block **Stone, hollow brick and block	None	Light and medium duty
One-piece	Concrete, solid block **Stone, solid and hollow brick, hollow block	None	Light and medium duty
Hammer-set	Concrete, solid block **Stone, solid and hollow brick, hollow block	None	Light duty
Bolt Anchors			
Drop-in	Concrete **Stone, solid brick	Machine screw or bolt	Light, medium, and heavy duty
Hollow-set drop-in	Concrete, solid brick and block **Stone, hollow brick and block	Machine screw or bolt	Light and medium duty
Single-expansion	Concrete, solid brick and block **Stone, hollow brick and block	Machine screw or bolt	Light and medium duty
Double-expansion	Concrete, solid brick and block **Stone, hollow brick and block	Machine screw or bolt	Light and medium duty
Lead (caulk-in)	Concrete, solid brick and block **Stone, hollow brick and block	Machine screw or bolt	Light and medium duty

Anchor Type	Typically Used In	Use With Fastener	Typical Working Load Range*
Screw Anchors			
Lag shield	Concrete **Stone, solid and hollow brick and block	Lag screw	Light and medium duty
Fiber	Concrete, stone, solid brick and block **Hollow brick and block, wallboard	Wood, sheet metal, or lag screw	Light and medium duty
Lead	Concrete, solid brick and block **Hollow brick and block, wallboard	Wood or sheet metal screw	Light duty
Plastic	Concrete, stone, solid brick and block **Hollow brick and block, wallboard	Wood or sheet metal screw	Light duty
Hollow-Wall Anchors			
Toggle bolts	Concrete, plank, hollow block, wallboard, plywood/ paneling	None	Light and medium duty
Plastic toggle bolts	Wallboard, plywood/ paneling **Hollow block, structural tile	Wood or sheet metal screw	Light duty
Sleeve-type wall	Wallboard, plywood/ paneling **Hollow block, structural tile	None	Light duty
Wallboard	Wallboard	Sheet metal screw	Light duty
Metal drive-in	Wallboard	Sheet metal screw	Light duty

* Anchor working loads given in the table are defined below. These are approximate loads only. Actual allowable loads depend on such factors as the anchor style and size, base material strength, spacing and edge distance, and the type of service load applied. Always consult the anchor manufacturer's product literature to determine the correct type of anchor and size to use for a specific application.

- Light duty – Less than 400 lbs.
- Medium duty – 400 to 4,000 lbs.
- Heavy duty – Above 4,000 lbs.

** Indicates use may be suitable depending on the application.

The NCCER makes every effort to keep these manuals up-to-date and free of technical errors. We appreciate your help in this process. If you have an idea for improving this manual, or if you find an error, a typographical mistake, or an inaccuracy in the NCCER's Craft Training Manuals, please write us, using this form or a photocopy. Be sure to include the exact module number, page number, a description of the problem, and the correction, if possible. Your input will be brought to the attention of the Technical Review Committee. Thank you for your assistance.

Instructors – If you found that additional materials were necessary in order to teach this module effectively, please let us know so that we may include them in the Equipment/Materials list in the Instructor's Guide.

Write: Curriculum Development and Revision Department
National Center for Construction Education and Research
P.O. Box 141104
Gainesville, FL 32614-1104

Fax: 352-334-0932

Craft _____ Module Name _____

Copyright Date _____ Module Number _____ Page Number(s) _____

Description of Problem _____

_____

_____

_____

_____

(Optional) Correction of Problem _____

_____

_____

(Optional) Your Name and Address _____

_____

_____

notes

Electrical Theory I

Module 26104

Electrical Trainee Task Module 26104

ELECTRICAL THEORY I

OBJECTIVES

Upon completion of this module, the trainee will be able to:

1. Recognize what atoms are and how they are constructed.
2. Define voltage and identify the ways in which it can be produced.
3. Explain the difference between conductors and insulators.
4. Define the units of measurement that are used to measure the properties of electricity.
5. Explain how voltage, current, and resistance are related to each other.
6. Using the formula for Ohm's Law, calculate an unknown value.
7. Explain the different types of meters used to measure voltage, current, and resistance.
8. Using the power formula, calculate the amount of power used by a circuit.

Prerequisites

Successful completion of the following Task Modules is required before beginning study of this Task Module: Core Curricula; Electrical Level 1, Modules 26101 through 26103.

Required Trainee Materials

1. Trainee Task Module
2. Copy of the latest edition of the *National Electrical Code*
3. Appropriate Personal Protective Equipment

Note: The designations "National Electrical Code," "NE Code," and "NEC," where used in this document, refer to the *National Electrical Code®*, which is a registered trademark of the National Fire Protection Association, Quincy, MA. *All National Electrical Code (NEC) references in this module refer to the 1999 edition of the NEC.*

Course Map

This course map shows all of the task modules in the first level of the Electrical curricula. The suggested training order begins at the bottom and proceeds up. Skill levels increase as a trainee advances on the course map. The training order may be adjusted by the local Training Program Sponsor.

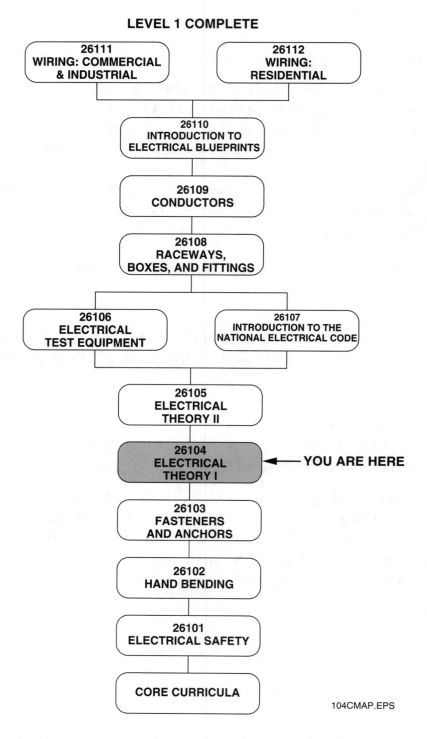

LEVEL 1 COMPLETE

26111
WIRING: COMMERCIAL & INDUSTRIAL

26112
WIRING: RESIDENTIAL

26110
INTRODUCTION TO ELECTRICAL BLUEPRINTS

26109
CONDUCTORS

26108
RACEWAYS, BOXES, AND FITTINGS

26106
ELECTRICAL TEST EQUIPMENT

26107
INTRODUCTION TO THE NATIONAL ELECTRICAL CODE

26105
ELECTRICAL THEORY II

26104
ELECTRICAL THEORY I ◄—— **YOU ARE HERE**

26103
FASTENERS AND ANCHORS

26102
HAND BENDING

26101
ELECTRICAL SAFETY

CORE CURRICULA

104CMAP.EPS

TABLE OF CONTENTS

Trade Terms Introduced In This Module

Ammeter: An instrument for measuring electrical current.

Ampere (A): A unit of electrical current. For example, one volt across one ohm of resistance causes a current flow of one ampere.

Atom: The smallest particle to which an element may be divided and still retain the properties of the element.

Battery: A DC voltage source consisting of two or more cells that convert chemical energy into electrical energy.

Circuit: A complete path for current flow.

Conductor: A material that offers very little resistance to current flow.

Coulomb: 6.25×10^{18} electrons or 6,250,000,000,000,000,000 electrons. A coulomb is the common unit of quantity used for specifying the size of a given charge.

Current: The movement, or flow, of electrons in a circuit. Current (I) is measured in amperes.

Electron: A negatively-charged particle that orbits the nucleus of an atom.

Insulator: A material that offers resistance to current flow.

Joule (J): A unit of measurement that represents one newton-meter (Nm), which is a unit of measure for doing work.

Kilo: A prefix used to indicate one thousand; for example, one kilowatt is equal to one thousand watts.

Matter: Any substance that has mass and occupies space.

Mega: A prefix used to indicate one million; for example, one megawatt is equal to one million watts.

Micro: A prefix used to indicate one-millionth; for example, one microwatt is equal to one-millionth of a watt.

Neutrons: Electrically-neutral particles (neither positive nor negative) that have the same mass as a proton and are found in the nucleus of an atom.

Nucleus: The center of an atom. It contains the protons and neutrons of the atom.

Ohm (Ω): The basic unit of measurement for resistance.

Ohmmeter: An instrument used for measuring resistance.

Ohm's Law: A statement of the relationships among current, voltage, and resistance in an electrical circuit: current (I) equals voltage (E) divided by resistance (R). Generally expressed as a mathematical formula: $I = E/R$.

Power: The rate of doing work or the rate at which energy is used or dissipated. Electrical power is the rate of doing electrical work. Electrical power is measured in watts.

Protons: The smallest positively-charged particles of an atom. Protons are contained in the nucleus of an atom.

Resistance: An electrical property that opposes the flow of current through a circuit. Resistance (R) is measured in ohms.

Resistor: Any device in a circuit that resists the flow of electrons.

Schematic: A type of drawing in which symbols are used to represent the components in a system.

Series circuit: A circuit with only one path for current flow.

Valence shell: The outermost ring of electrons that orbit about the nucleus of an atom.

Volt (V): The unit of measurement for voltage (electromotive force). One volt is equivalent to the force required to produce a current of one ampere through a resistance of one ohm.

Voltage: The driving force that makes current flow in a circuit. Voltage (E) is also referred to as *electromotive force* or *potential*.

Voltage drop: The change in voltage across a component that is caused by the current flowing through it and the amount of resistance opposing it.

Voltmeter: An instrument for measuring voltage. The resistance of the voltmeter is fixed. When the voltmeter is connected to a circuit, the current passing through the meter will be directly proportional to the voltage at the connection points.

Watt (W): The basic unit of measurement for electrical power.

1.0.0 INTRODUCTION TO ELECTRICAL THEORY

As an electrician, you must work with a force that cannot be seen. However, electricity is there on the job, every day of the year. It is necessary that you understand the forces of electricity so that you will be safe on the job. The first step is a basic understanding of the principles of electricity.

The relationships among **current**, **voltage**, **resistance**, and **power** in a basic direct current (DC) **series circuit** are common to all types of electrical **circuits**. This module provides a general introduction to the electrical concepts used in **Ohm's Law**. It also presents the opportunity to practice applying these basic concepts to DC series circuits. In this way, you can prepare for further study in electrical and electronics theory and maintenance techniques. By practicing these techniques for all combinations of DC circuits, you will be prepared to work on any DC circuits you might encounter.

2.0.0 CONDUCTORS AND INSULATORS

2.1.0 THE ATOM

The **atom** is the smallest part of an element that enters into a chemical change, but it does so in the form of a charged particle. These charged particles are called *ions*, and are of two types—positive and negative. A positive ion may be defined as an atom that has become positively charged. A negative ion may be defined as an atom that has become negatively charged. One of the properties of charged ions is that ions of the same charge tend to repel one another, whereas ions of unlike charge will attract one another. The term *charge* can be taken to mean a quantity of electricity that is either positive or negative.

The structure of an atom is best explained by a detailed analysis of the simplest of all atoms, that of the element hydrogen. The hydrogen atom in *Figure 1* is composed of a **nucleus** containing one **proton** and a single orbiting **electron**. As the electron revolves around the nucleus, it is held in this orbit by two counteracting forces. One of these forces is called *centrifugal force*, which is the force that tends to cause the electron to fly outward as it travels around its circular orbit. The second force acting on the electron is *electrostatic force*. This force tends to pull the electron in toward the nucleus and is provided by the mutual attraction between the positive nucleus and the negative electron. At some given radius, the two forces will balance each other, providing a stable path for the electron.

- A proton (+) repels another proton (+).
- An electron (–) repels another electron (–).
- A proton (+) attracts an electron (–).

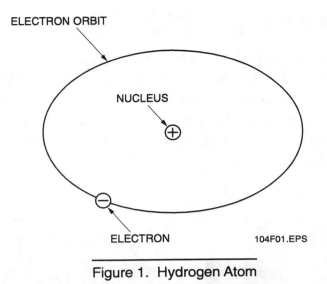

ELECTRON ORBIT

NUCLEUS

ELECTRON   104F01.EPS

Figure 1. Hydrogen Atom

Basically, an atom contains three types of subatomic particles that are of interest in electricity: electrons, protons, and **neutrons**.

The protons and neutrons are located in the center, or nucleus, of the atom, and the electrons travel about the nucleus in orbits.

Because protons are relatively heavy, the repulsive force they exert on one another in the nucleus of an atom has little effect.

The attracting and repelling forces on charged materials occur because of the electrostatic lines of force that exist around the charged materials. In a negatively-charged object, the lines of force of the excess electrons add to produce an electrostatic field that has lines of force coming into the object from all directions. In a positively-charged object, the lines of force of the excess protons add to produce an electrostatic field that has lines of force going out of the object in all directions. The electrostatic fields either aid or oppose each other to attract or repel.

2.1.1   The Nucleus

The nucleus is the central part of the atom. It is made up of heavy particles called protons and **neutrons**. The proton is a charged particle containing the smallest known unit of positive electricity. The neutron has no electrical charge. The number of protons in the nucleus determines how the atom of one element differs from the atom of another element.

Although a neutron is actually a particle by itself, it is generally thought of as an electron and proton combined and is electrically neutral. Since neutrons are electrically neutral, they are not considered important to the electrical nature of atoms.

2.1.2 Electrical Charges

The negative charge of an electron is equal but opposite to the positive charge of a proton. The charges of an electron and a proton are called *electrostatic charges*. The lines of force associated with each particle produce electrostatic fields. Because of the way these fields act together, charged particles can attract or repel one another. The Law of Electrical Charges states that particles with like charges repel each other and those with unlike charges attract each other. This is shown in *Figure 2*.

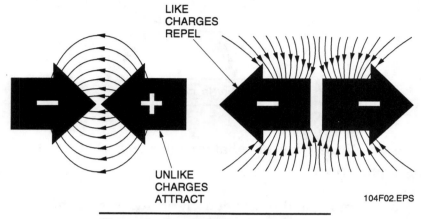

LIKE CHARGES REPEL

UNLIKE CHARGES ATTRACT

104F02.EPS

Figure 2. Law Of Electrical Charges

2.2.0 CONDUCTORS AND INSULATORS

The difference between atoms, with respect to chemical activity and stability, depends on the number and position of the electrons included within the atom. In general, the electrons reside in groups of orbits called *shells*. The shells are arranged in steps that correspond to fixed energy levels.

The number of electrons in the outermost shell determines the valence of an atom. For this reason, the outer shell of an atom is called the **valence shell**, and the electrons contained in this shell are called *valence electrons* (*Figure 3*). The valence of an atom determines its ability to gain or lose an electron, which in turn determines the chemical and electrical properties of the atom. An atom that is lacking only one or two electrons from its outer shell will easily gain electrons to complete its shell, but a large amount of energy is required to free any of its electrons. An atom having a relatively small number of electrons in its outer shell in comparison to the number of electrons required to fill the shell will easily lose these valence electrons.

It is the valence electrons that we are most concerned with in electricity. These are the electrons that are easiest to break loose from their parent atom. Normally, a **conductor** has three or less valence electrons, an **insulator** has five or more valence electrons, and semiconductors usually have four valence electrons.

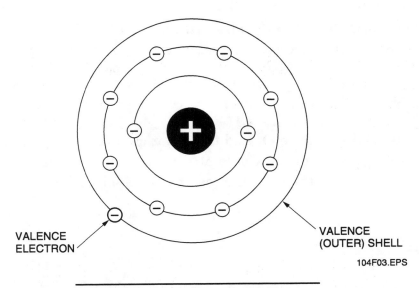

VALENCE
ELECTRON

VALENCE
(OUTER) SHELL

104F03.EPS

Figure 3. Valence Shell And Electrons

All the elements of which **matter** is made may be placed into one of three categories: conductors, insulators, and semiconductors.

Conductors, for example, are elements such as copper and silver that will conduct a flow of electricity very readily. Because of their good conducting abilities, they are formed into wire and used whenever it is desired to transfer electrical energy from one point to another.

Insulators, on the other hand, do not conduct electricity to any great degree and are used when it is desirable to prevent the flow of electricity. Compounds such as porcelain and plastic are good insulators.

Materials such as germanium and silicon are not good conductors but cannot be used as insulators either, since their electrical characteristics fall between those of conductors and those of insulators. These in-between materials are classified as *semiconductors*. As you will learn later in your training, semiconductors play a crucial role in electronic circuits.

3.0.0   ELECTRIC CHARGE AND CURRENT

An electric charge has the ability to do the work of moving another charge by attraction or repulsion. The ability of a charge to do work is called its *potential*. When one charge is different from another, there must be a difference in potential between them. The sum of the difference of potential of all the charges in the electrostatic field is referred to as *electromotive force (emf)* or *voltage*. Voltage is frequently represented by the letter *E*.

Electric charge is measured in **coulombs**. An electron has 1.6×10^{-19} coulombs of charge. Therefore, it takes 6.25×10^{18} electrons to make up one coulomb of charge, as shown below.

$$\frac{1}{1.6 \times 10^{-19}} = 6.25 \times 10^{18} \text{ electrons}$$

If two particles, one having charge Q_1 and the other charge Q_2, are a distance (d) apart, then the force between them is given by Coulomb's Law, which states that the force is directly proportional to the product of the two charges and inversely proportional to the square of the distance between them:

$$\text{Force} = \frac{Q_1 \times Q_2}{d^2}$$

If Q_1 and Q_2 are both positive or both negative, then the force is positive; it is repulsive. If Q_1 and Q_2 are of opposite charges, then the force is negative; it is attractive.

3.1.0 CURRENT FLOW

The movement of the flow of electrons is called *current*. To produce current, the electrons are moved by a potential difference. Current is represented by the letter *I*. The basic unit in which current is measured is the **ampere**, also called the *amp*. The symbol for the ampere is *A*. One ampere of current is defined as the movement of one coulomb past any point of a conductor during one second of time. One coulomb is equal to 6.25×10^{18} electrons; therefore, one ampere is equal to 6.25×10^{18} electrons moving past any point of a conductor during one second of time.

The definition of current can be expressed as an equation:

$$I = \frac{Q}{T}$$

Where:

I = current (amperes)
Q = charge (coulombs)
T = time (seconds)

Charge differs from current in that **Q** is an accumulation of charge, while I measures the intensity of moving charges.

In a conductor, such as copper wire, the free electrons are charges that can be forced to move with relative ease by a potential difference. If a potential difference is connected across two ends of a copper wire, as shown in *Figure 4*, the applied voltage forces the free electrons to move. This current is a flow of electrons from the point of negative charge (–) at one end of the wire, moving through the wire to the positive charge (+), at the other end. The direction of the electron flow is from the negative side of the battery, through the wire, and back to the positive side of the battery. The direction of current flow is therefore from a point of negative potential to a point of positive potential.

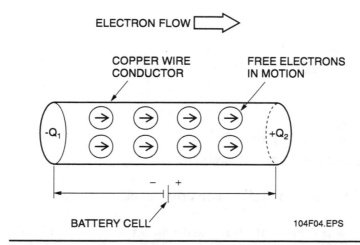

Figure 4. Potential Difference Causing Electric Current

3.2.0   VOLTAGE

The force that causes electrons to move is called *voltage, potential difference,* or *electromotive force (emf).* One **volt (V)** is the potential difference between two points for which one coulomb of electricity will do one **joule (J)** of work. A **battery** is one of several means of creating voltage. It chemically creates a large reserve of free electrons at the negative (–) terminal. The positive (+) terminal has electrons chemically removed and will therefore accept them if an external path is provided from the negative (–) terminal. When a battery is no longer able to chemically deposit electrons at the negative (–) terminal, it is said to be dead, or in need of recharging. Batteries are normally rated in volts. Large batteries are also rated in ampere-hours, where one ampere-hour is a current of one amp supplied for one hour.

4.0.0   RESISTANCE

4.1.0   CHARACTERISTICS OF RESISTANCE

Resistance is directly related to the ability of a material to conduct electricity. Conductors have very low resistance; insulators have very high resistance.

Resistance can be defined as the opposition to current flow. To add resistance to a circuit, electrical components called **resistors** are used. A resistor is a device whose resistance to current flow is a known, specified value. Resistance is measured in ohms and is represented by the symbol R in equations. One **ohm** is defined as that amount of resistance that will limit the current in a conductor to one ampere when the voltage applied to the conductor is one volt. The symbol for an ohm is Ω.

The resistance of a wire is proportional to the length of the wire, inversely proportional to the cross-sectional area of the wire, and dependent upon the kind of material of which the wire is made. The relationship for finding the resistance of a wire is:

$$R = \rho \frac{L}{A}$$

Where:

R = resistance (ohms)
L = length of wire (feet)
A = area of wire (circular mils, CM, or cm²)
ρ = specific resistance (ohm-CM/ft. or microhm-cm)

A *mil* equals 0.001 inch; a circular mil is the cross-sectional area of a wire one mil in diameter.

The specific resistance is a constant that depends on the material of which the wire is made. *Table 1* shows the properties of various wire conductors.

Metal	Specific Resistance (Resistance of 1 CM/ft. in ohms)	
	32°F or 0°C or	75°F or 23.8°C
Silver, pure annealed	8.831	9.674
Copper, pure annealed	9.39	10.351
Copper, annealed	9.59	10.505
Copper, hard-drawn	9.81	10.745
Gold	13.216	14.404
Aluminum	15.219	16.758
Zinc	34.595	37.957
Iron	54.529	62.643

Table 1. Conductor Properties

Table 1 shows that at 75°F, a one-mil diameter, pure annealed copper wire that is one-foot long has a resistance of 10.351 ohms; while a one-mil diameter, one-foot long aluminum wire has a resistance of 16.758 ohms. Temperature is important in determining the resistance of a wire. The hotter a wire, the greater its resistance.

4.2.0 OHM'S LAW

Ohm's Law defines the relationship between current, voltage, and resistance. There are three ways to express Ohm's Law mathematically.

- The current in a circuit is equal to the voltage applied to the circuit divided by the resistance of the circuit:

$$I = \frac{E}{R}$$

- The resistance of a circuit is equal to the voltage applied to the circuit divided by the current in the circuit:

$$R = \frac{E}{I}$$

- The applied voltage to a circuit is equal to the product of the current and the resistance of the circuit:

$$E = I \times R = IR$$

Where:

  I = current (amperes)
  R = resistance (ohms)
  E = voltage or emf (volts)

If any two of the quantities E, I, or R are known, the third can be calculated.

The Ohm's Law equations can be memorized and practiced effectively by using an Ohm's Law circle, as shown in *Figure 5*. To find the equation for E, I, or R when two quantities are known, cover the unknown third quantity. The other two quantities in the circle will indicate how the covered quantity may be found.

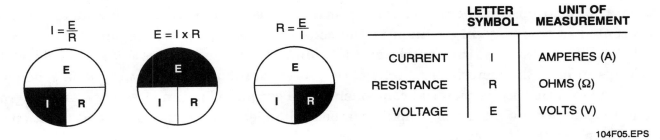

	LETTER SYMBOL	UNIT OF MEASUREMENT
CURRENT	I	AMPERES (A)
RESISTANCE	R	OHMS (Ω)
VOLTAGE	E	VOLTS (V)

104F05.EPS

Figure 5. Ohm's Law Circle

Example 1:

Find I when E = 120V and R = 30Ω.

$$I = \frac{E}{R}$$
$$I = \frac{120V}{30\Omega}$$
$$I = 4A$$

This formula shows that in a DC circuit, current (I) is directly proportional to voltage (E) and inversely proportional to resistance (R).

Example 2:

Find R when E = 240V and I = 20A.

$$R = \frac{E}{I}$$
$$R = \frac{240V}{20A}$$
$$R = 12\Omega$$

Example 3:

Find E when I = 15A and R = 8Ω.

$$E = I \times R$$
$$E = 15A \times 8\Omega$$
$$E = 120V$$

5.0.0   SCHEMATIC REPRESENTATION OF CIRCUIT ELEMENTS

A simple electric circuit is shown in both pictorial and schematic forms in *Figure 6*. The **schematic** diagram is a shorthand way to draw an electric circuit, and circuits are usually represented in this way. In addition to the connecting wire, three components are shown symbolically: the battery, the switch, and the lamp. Note the positive (+) and negative (–) markings in both the pictorial and schematic representations of the battery. The schematic components represent the pictorial components in a simplified manner. A schematic diagram is one that shows, by means of graphic symbols, the electrical connections and functions of the different parts of a circuit.

The standard graphic symbols for commonly used electrical and electronic components are shown in *Figure 7*.

ELECTRICAL — TRAINEE TASK MODULE 26104

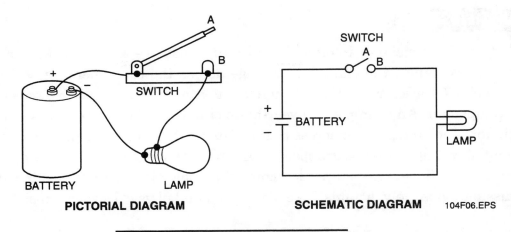

PICTORIAL DIAGRAM   SCHEMATIC DIAGRAM   104F06.EPS

Figure 6. Simple Electrical Symbols

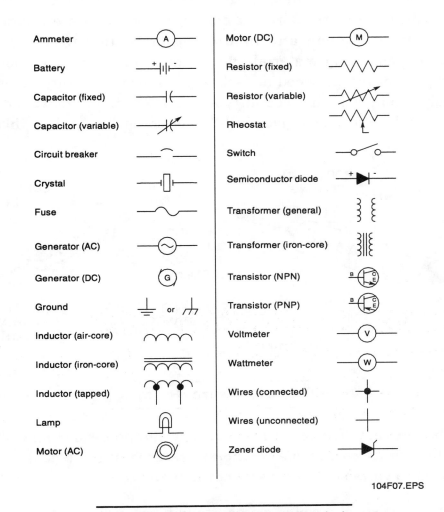

104F07.EPS

Figure 7. Standard Schematic Symbols

6.0.0 RESISTORS

The function of a resistor is to offer a particular resistance to current flow. For a given current and known resistance, the change in voltage across the component, or **voltage drop**, can be predicted using Ohm's Law. Voltage drop refers to a specific amount of voltage used, or developed, by that component. An example is a very basic circuit of a 10V battery and a single resistor in a series circuit. The voltage drop across that resistor is 10V because it is the only component in the circuit and all voltage must be dropped across that resistor. Similarly, for a given applied voltage, the current that flows may be predetermined by selection of the resistor value. The required power dissipation largely dictates the construction and physical size of a resistor.

The two most common types of electronic resistors are *wire-wound* and *carbon composition construction*. A typical wire-wound resistor consists of a length of nickel wire wound on a ceramic tube and covered with porcelain. Low-resistance connecting wires are provided, and the resistance value is usually printed on the side of the component. *Figure 8* illustrates the construction of typical resistors. Carbon composition resistors are constructed by molding mixtures of powdered carbon and insulating materials into a cylindrical shape. An outer sheath of insulating material affords mechanical and electrical protection, and copper connecting wires are provided at each end. Carbon composition resistors are smaller and less expensive than the wire-wound type. However, the wire-wound type is the more rugged of the two and is able to survive much larger power dissipations than the carbon composition type.

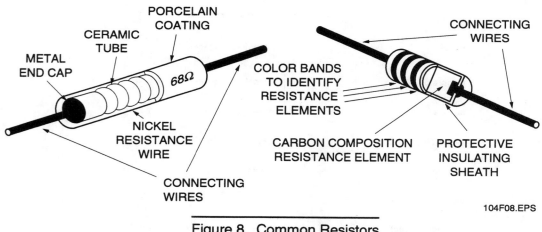

104F08.EPS

Figure 8. Common Resistors

Most resistors have standard fixed values, so they can be termed *fixed resistors*. Variable resistors, also known as *adjustable resistors*, are used a great deal in electronics. Two common symbols for a variable resistor are shown in *Figure 9*.

A variable resistor consists of a coil of closely-wound insulated resistance wire formed into a partial circle. The coil has a low-resistance terminal at each end, and a third terminal is connected to a movable contact with a shaft adjustment facility. The movable contact can be set to any point on a connecting track that extends over one (uninsulated) edge of the coil.

ELECTRICAL — TRAINEE TASK MODULE 26104

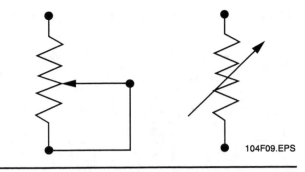

104F09.EPS

Figure 9. Symbols Used For Variable Resistors

Using the adjustable contact, the resistance from either end terminal to the center terminal may be adjusted from zero to the maximum coil resistance.

Another type of variable resistor is known as a *decade resistance box*. This is a laboratory component that contains precise values of switched series-connected resistors.

6.1.0 RESISTOR COLOR CODES

Because carbon composition resistors are physically small (some are less than 1 cm in length), it is not convenient to print the resistance value on the side. Instead, a color code in the form of colored bands is employed to identify the resistance value and tolerance. The color code is illustrated in *Figure 10*. Starting from one end of the resistor, the first two bands identify the first and second digits of the resistance value, and the third band indicates the number of zeros. An exception to this is when the third band is either silver or gold, which indicates a 0.01 or 0.1 multiplier, respectively. The fourth band is always either silver or gold, and in this position, silver indicates a ±10% tolerance and gold indicates a ±5% tolerance. Where no fourth band is present, the resistor tolerance is ±20%.

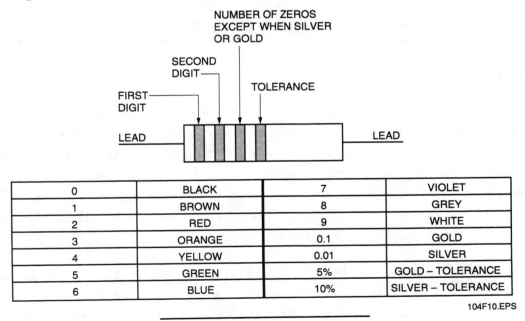

0	BLACK	7	VIOLET
1	BROWN	8	GREY
2	RED	9	WHITE
3	ORANGE	0.1	GOLD
4	YELLOW	0.01	SILVER
5	GREEN	5%	GOLD – TOLERANCE
6	BLUE	10%	SILVER – TOLERANCE

104F10.EPS

Figure 10. Resistor Color Codes

We can put this information to practical use by determining the range of values for the carbon resistor in *Figure 11*.

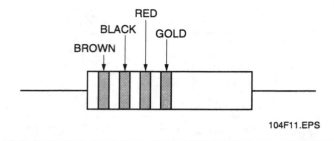

104F11.EPS

Figure 11. Sample Color Codes On A Fixed Resistor

The color code for this resistor is as follows:

- Brown = 1, black = 0, red = 2, gold = a tolerance of ±5%
- First digit = 1, second digit = 0, number of zeros (2) = 1,000Ω

Since this resistor has a value of 1,000Ω ±5%, the resistor can range in value from 950Ω to 1,050Ω.

7.0.0   MEASURING VOLTAGE, CURRENT, AND RESISTANCE

Working with electricity requires making accurate measurements. This section will discuss the basic meters used to measure voltage, current, and resistance: the **voltmeter**, **ammeter**, and **ohmmeter**.

WARNING!   Only qualified individuals may use these meters. Consult your company's safety policy for applicable rules.

7.1.0   BASIC METER OPERATION

When troubleshooting or testing equipment, you will need various meters to check for proper circuit voltages, currents, and resistances and to determine if the wiring is defective. Meters are used in repairing, maintaining, and troubleshooting electrical circuits and equipment. The best and most expensive measuring instrument is of no use to you unless you know what you are measuring and what each reading indicates. Remember that the purpose of a meter is to measure quantities existing within a circuit. For this reason, when the meter is connected to a circuit, it must not change the condition of the circuit.

The three basic electrical quantities discussed in this section are current, voltage, and resistance. Actually, it is really current that causes the meter to respond even when voltage or resistance is being measured. In a basic meter, the measurement of current can be calibrated to indicate almost any electrical quantity based on the principle of Ohm's Law. The amount of current that flows through a meter is determined by the voltage applied to the meter and the resistance of the meter, as stated by $I = E/R$.

For a given meter resistance, different values of applied voltage will cause specific values of current to flow. Although the meter actually measures current, the meter scale can be calibrated in units of voltage. Similarly, for a given applied voltage, different values of resistance will cause specific values of current to flow; therefore, the meter scale can also be calibrated in units of resistance rather than current. The same holds true for power, since power is proportional to current, as stated by $P = EI$. It is on this principle that the meter was developed and its construction allows for the measurement of various parameters by actually measuring current.

You must understand the purpose and function of each individual piece of test equipment and any limitations associated with it. It is also extremely important that you understand how to safely use each piece of equipment. If you understand the capabilities of the test equipment, you can better use the equipment, better understand the indications on the equipment, and know what substitute or backup meters can be used.

7.2.0 VOLTMETER

A simple voltmeter consists of the meter movement in series with the internal resistance of the voltmeter itself. For example, a meter with a 50-microamp (μA) meter movement and a 1,000Ω internal resistance can be used to directly measure voltages up to 0.05V, as shown in *Figure 12*. (The prefix **micro** means one-millionth.) When the meter is placed across the voltage source, a current determined by the internal resistance of the meter flows through the meter movement. A voltmeter's internal resistance is typically high to minimize meter loading effects on the source.

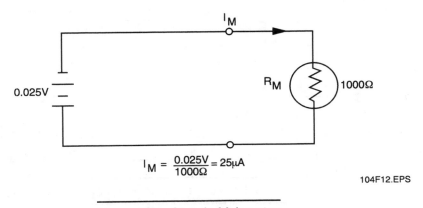

$$I_M = \frac{0.025V}{1000\Omega} = 25\mu A$$

104F12.EPS

Figure 12. Simple Voltmeter

To measure larger voltages, a multiplier resistor is used. This increased series resistance limits the current that can flow through the meter movement, thus extending the range of the meter.

To avoid damage to the meter movement, the following precautions should be observed when using a voltmeter:

- Always set the full-scale voltage of the meter to be larger than the expected voltage to be measured.

- Always ensure that the internal resistance of the voltmeter is much greater than the resistance of the component to be measured. This means that the current it takes to drive the voltmeter (about 50 microamps) should be a negligible fraction of the current flowing through the circuit element being measured.

- If you are unsure of the level of the voltage to be measured, take a reading at the highest range of the voltmeter and progressively (step-by-step) lower the range until the reading is obtained.

In most commercial voltmeters, the internal resistance is expressed by the ohms-per-volt rating of the meter. A typical meter has a rating of 20,000 ohms-per-volt with a 50-microamp movement. This quantity tells what the internal resistance of the meter is on any particular full-scale setting. In general, the meter's internal resistance is the ohms-per-volt rating multiplied by the full-scale voltage. The higher the ohms-per-volt rating, the higher the internal resistance of the meter, and the smaller the effect of the meter on the circuit.

7.3.0 AMMETER

A current meter, usually called an *ammeter*, is used by placing the meter in series with the wire through which the current is flowing. This method of connection is shown in *Figure 13*. Notice how the magnitude of load current will flow through the ammeter. Because of this, an ammeter's internal resistance must be low to minimize the circuit-loading effects as seen by the source. Also, high current magnitudes flowing through an ammeter can damage it. For this reason, ammeter shunts are employed to reduce the ammeter circuit current to a fraction of the current flowing through the load.

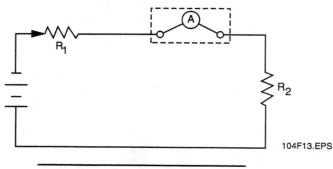

104F13.EPS

Figure 13. Ammeter Connection

To avoid damage to the meter movement, the following precautions should be observed when taking current measurements with an ammeter:

- Always check the polarity of the ammeter. Make certain that the meter is connected to the circuit so that electrons flow into the negative lead and out of the positive lead. It is easy to tell which is the positive lead because it is normally red. The negative lead is usually black.

- Always set the full-scale deflection of the meter to be larger than the expected current. To be safe, set the full-scale current several times larger than the expected current, and then slowly increase the meter sensitivity to the appropriate scale.
- Always connect the ammeter in series with the circuit element through which the current to be measured is flowing. Never connect the ammeter in parallel. When an ammeter is connected across a constant-potential source of appreciable voltage, the low internal resistance of the meter bypasses the circuit resistance. This results in the application of the source voltage directly to the meter terminals. The resulting excess current will burn out the meter coil.

7.4.0 OHMMETER

An ohmmeter is used to measure resistance and check continuity. The deflection of the pointer of an ohmmeter is controlled by the amount of battery current passing through the coil. Current flow depends on the applied voltage and the circuit resistance. By applying a constant source voltage to the circuit under test, the resultant current flow depends only on circuit resistance. This magnitude of current will create meter movement. By knowing the relationship between current and resistance, an ohmmeter's scale can be calibrated to indicate circuit resistance based on the magnitude of current for a constant source voltage. Refer to *Figure 14*, a simple ohmmeter circuit.

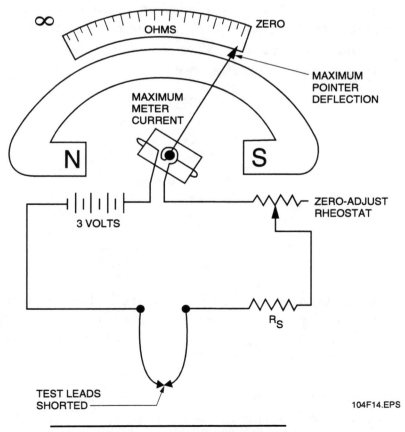

104F14.EPS

Figure 14. Simple Ohmmeter Circuit

8.0.0 ELECTRICAL POWER

Power is defined as the rate of doing work. This is equivalent to the rate at which energy is used or dissipated. Electrons passing through a resistance dissipate energy in the form of heat. In electrical circuits, power is measured in units called **watts (W)**. The power in watts equals the rate of energy conversion. One watt of power equals the work done in one second by one volt of potential difference in moving one coulomb of charge. One coulomb per second is an ampere; therefore, power in watts equals the product of amperes times volts.

The work done in an electrical circuit can be useful work or it can be wasted work. In both cases, the rate at which the work is done is still measured in power. The turning of an electric motor is useful work. On the other hand, the heating of wires or resistors in a circuit is wasted work, since no useful function is performed by the heat.

The unit of electrical work is the joule. This is the amount of work done by one coulomb flowing through a potential difference of one volt. Thus, if five coulombs flow through a potential difference of one volt, five joules of work are done. The time it takes these coulombs to flow through the potential difference has no bearing on the amount of work done.

It is more convenient when working with circuits to think of amperes of current rather than coulombs. As previously discussed, one ampere equals one coulomb passing a point in one second. Using amperes, one joule of work is done in one second when one ampere moves through a potential difference of one volt. This rate of one joule of work in one second is the basic unit of power, and is called a *watt*. Therefore, a watt is the power used when one ampere of current flows through a potential difference of one volt, as shown in *Figure 15*.

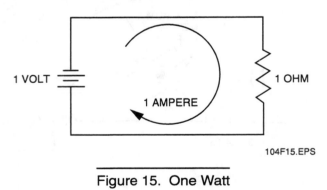

104F15.EPS

Figure 15. One Watt

Mechanical power is usually measured in units of horsepower (hp). To convert from horsepower to watts, multiply the number of horsepower by 746. To convert from watts to horsepower, divide the number of watts by 746. Conversions for common units of power are given in *Table 2*.

The kilowatt-hour (kWh) is commonly used for large amounts of electrical work or energy. (The prefix **kilo** means one thousand.) The amount is calculated simply as the product of the power in kilowatts multiplied by the time in hours during which the power is used. If a light bulb uses 300W or 0.3kW for 4 hours, the amount of energy is 0.3 x 4, which equals 1.2kWh.

1,000 Watts (W) =	1 Kilowatt (kW)
1,000,000 Watts (W) =	1 Megawatt (MW)
1,000 Kilowatts (kW) =	1 Megawatt (MW)
1 Watt (W) =	0.00134 Horsepower (hp)
1 Horsepower (hp) =	746 Watts (W)

Table 2. Conversion Table

Very large amounts of electrical work or energy are measured in megawatts (MW). (The prefix **mega** means one million.)

Electricity usage is figured in kilowatt-hours of energy. The power line voltage is fairly constant at 120V. Suppose the total load current in the main line equals 20A. Then, the power in watts from the 120V line is:

P = 120V x 20A
P = 2,400W or 2.4kW

If this power is used for five hours, then the energy of work supplied equals:

2.4 x 5 = 12kWh

8.1.0 POWER EQUATION

When one ampere flows through a difference of two volts, two watts must be used. In other words, the number of watts used is equal to the number of amperes of current times the potential difference. This is expressed in equation form as:

P = I x E or P = IE

Where:

P = power used in watts
I = current in amperes
E = potential difference in volts

The equation is sometimes called *Ohm's Law for Power*, because it is similar to Ohm's Law. This equation is used to find the power consumed in a circuit or load when the values of current and voltage are known. The second form of the equation is used to find the voltage when the power and current are known:

$$E = \frac{P}{I}$$

The third form of the equation is used to find the current when the power and voltage are known:

$$I = \frac{P}{E}$$

Using these three equations, the power, voltage, or current in a circuit can be calculated whenever any two of the values are already known.

Example 1:

Calculate the power in a circuit where the source of 100V produces 2A in a 50Ω resistance.

$P = IE$
$P = 2 \times 100$
$P = 200W$

This means the source generates 200W of power while the resistance dissipates 200W in the form of heat.

Example 2:

Calculate the source voltage in a circuit that consumes 1,200W at a current of 5A.

$$E = \frac{P}{I}$$

$$E = \frac{1,200}{5}$$

$E = 240V$

Example 3:

Calculate the current in a circuit that consumes 600W with a source voltage of 120V.

$$I = \frac{P}{E}$$

$$I = \frac{600}{120}$$

$I = 5A$

Components that use the power dissipated in their resistance are generally rated in terms of power. The power is rated at normal operating voltage, which is usually 120V. For instance, an appliance that draws 5A at 120V would dissipate 600W. The rating for the appliance would then be 600W/120V.

To calculate I or R for components rated in terms of power at a specified voltage, it may be convenient to use the power formula in different forms. There are actually three basic power formulas, but each can be rearranged into three other forms for a total of nine combinations:

$$P = IE \qquad P = I^2R \qquad P = \frac{E^2}{R}$$

$$I = \frac{P}{E} \qquad R = \frac{P}{I^2} \qquad R = \frac{E^2}{P}$$

$$E = \frac{P}{I} \qquad I = \sqrt{\frac{P}{R}} \qquad E = \sqrt{PR}$$

Note that all of these formulas are based on Ohm's Law (E = IR) and the power formula (P = I x E). *Figure 16* shows all of the applicable power, voltage, resistance, and current equations.

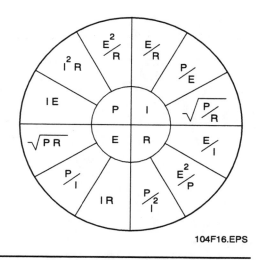

104F16.EPS

Figure 16. Expanded Ohm's Law Circle

8.2.0 POWER RATING OF RESISTORS

If too much current flows through a resistor, the heat caused by the current will damage or destroy the resistor. This heat is caused by I^2R heating, which is power loss expressed in watts. Therefore, every resistor is given a wattage, or power rating, to show how much I^2R heating it can take before it burns out. This means that a resistor with a power rating of one watt will burn out if it is used in a circuit where the current causes it to dissipate heat at a rate greater than one watt.

If the power rating of a resistor is known, the maximum current it can carry is found by using an equation derived from $P = I^2R$:

$P = I^2R$ becomes $I^2 = P/R$, which becomes $I = \sqrt{P/R}$

Using this equation, find the maximum current that can be carried by a 1Ω resistor with a power rating of 4W:

$$I = \sqrt{P/R} = \sqrt{4/1} = 2 \text{ amperes}$$

If such a resistor conducts more than 2 amperes, it will dissipate more than its rated power and burn out.

Power ratings assigned by resistor manufacturers are usually based on the resistors being mounted in an open location where there is free air circulation, and where the temperature is not higher than 104°F (40°C). Therefore, if a resistor is mounted in a small, crowded, enclosed space, or where the temperature is higher than 104°F, there is a good chance it will burn out even before its power rating is exceeded. Also, some resistors are designed to be attached to a chassis or frame that will carry away the heat.

SUMMARY

The relationships among current, voltage, resistance, and power are consistent for all types of DC circuits and can be calculated using Ohm's Law and Ohm's Law for Power. Understanding and being able to apply these concepts is necessary for effective circuit analysis and troubleshooting.

References

For advanced study of topics covered in this task module, the following books are suggested:

Electronics Fundamentals, Latest Edition, Prentice Hall, New York, NY.
Principles of Electric Circuits, Latest Edition, Prentice Hall, New York, NY.

REVIEW QUESTIONS

1. A negative charge that is a type of subatomic particle is a(n) _____.
   a. proton
   b. neutron
   c. electron
   d. atom

2. A proton has a _____ charge.
   a. negative
   b. positive
   c. neutral
   d. both b and c

3. Like charges _____ each other.
   a. attract
   b. repel
   c. have no effect on
   d. complement

4. An electron has _____ coulombs of charge.
   a. 1.6×10^{-9}
   b. 1.6×10^{9}
   c. 1.6×10^{19}
   d. 1.6×10^{-19}

5. The quantity that Ohm's Law does not express a relationship for in an electrical circuit is _____.
   a. charge
   b. resistance
   c. voltage
   d. current

6. The color band that represents tolerance on a resistor is the _____.
   a. 4th band
   b. 3rd band
   c. 2nd band
   d. 1st band

7. A resistor with a color code of red/red/orange indicates a value of _____.
   a. 22,000 ohms
   b. 66 ohms
   c. 223 ohms
   d. 220 ohms

8. An ammeter is placed in _____ with the circuit being tested.
    a. parallel
    b. series
    c. compound
    d. none of the above

9. The basic unit of power is the _____ .
    a. volt
    b. ampere
    c. coulomb
    d. watt

10. The power in a circuit with 120 volts and 5 amps is _____.
    a. 24 watts
    b. 600 watts
    c. 6,000 watts
    d. ¼₄ watt

ANSWERS TO REVIEW QUESTIONS

Answer	**Section**
1. c	Terms/2.1.0
2. b	Terms/2.1.0
3. b	2.1.0
4. d	3.0.0
5. a	4.2.0
6. a	6.1.0
7. a	6.1.0
8. b	7.3.0
9. d	8.0.0
10. b	8.1.0

The NCCER makes every effort to keep these manuals up-to-date and free of technical errors. We appreciate your help in this process. If you have an idea for improving this manual, or if you find an error, a typographical mistake, or an inaccuracy in the NCCER's Craft Training Manuals, please write us, using this form or a photocopy. Be sure to include the exact module number, page number, a description of the problem, and the correction, if possible. Your input will be brought to the attention of the Technical Review Committee. Thank you for your assistance.

Instructors – If you found that additional materials were necessary in order to teach this module effectively, please let us know so that we may include them in the Equipment/Materials list in the Instructor's Guide.

Write: Curriculum Development and Revision Department
National Center for Construction Education and Research
P.O. Box 141104
Gainesville, FL 32614-1104
Fax: 352-334-0932

Craft _____ Module Name _____

Copyright Date _____ Module Number _____ Page Number(s) _____

Description of Problem _____

_____

_____

_____

(Optional) Correction of Problem _____

_____

_____

(Optional) Your Name and Address _____

_____

_____

_____

Electrical Theory II
Module 26105

Electrical Trainee Task Module 26105

NATIONAL
CENTER FOR
CONSTRUCTION
EDUCATION AND
RESEARCH

ELECTRICAL THEORY II

OBJECTIVES

Upon completion of this module, the trainee will be able to:

1. Explain the basic characteristics of a series circuit.
2. Explain the basic characteristics of a parallel circuit.
3. Explain the basic characteristics of a series-parallel circuit.
4. Calculate, using Kirchoff's Voltage Law, the voltage drop in series, parallel, and series-parallel circuits.
5. Calculate, using Kirchoff's Current Law, the total current in parallel and series-parallel circuits.
6. Find the total amount of resistance in a series circuit.
7. Find the total amount of resistance in a parallel circuit.
8. Find the total amount of resistance in a series-parallel circuit.

Prerequisites

Successful completion of the following Task Modules is required before beginning study of this Task Module: Core Curricula; Electrical Level 1, Modules 26101 through 26104.

Required Trainee Materials

1. Trainee Task Module
2. Copy of the latest edition of the *National Electrical Code*
3. Appropriate Personal Protective Equipment

Note: The designations "National Electrical Code," "NE Code," and "NEC," where used in this document, refer to the *National Electrical Code®*, which is a registered trademark of the National Fire Protection Association, Quincy, MA. *All National Electrical Code (NEC) references in this module refer to the 1999 edition of the NEC.*

Course Map

This course map shows all of the task modules in the first level of the Electrical curricula. The suggested training order begins at the bottom and proceeds up. Skill levels increase as a trainee advances on the course map. The training order may be adjusted by the local Training Program Sponsor.

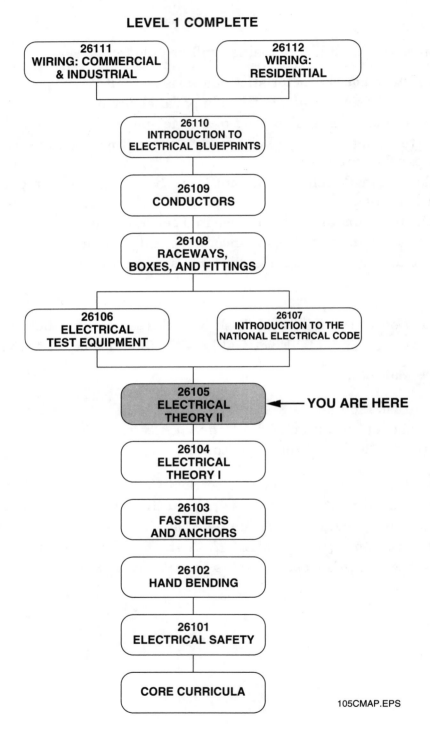

LEVEL 1 COMPLETE

26111
WIRING: COMMERCIAL & INDUSTRIAL

26112
WIRING: RESIDENTIAL

26110
INTRODUCTION TO ELECTRICAL BLUEPRINTS

26109
CONDUCTORS

26108
RACEWAYS, BOXES, AND FITTINGS

26106
ELECTRICAL TEST EQUIPMENT

26107
INTRODUCTION TO THE NATIONAL ELECTRICAL CODE

26105
ELECTRICAL THEORY II ◄— YOU ARE HERE

26104
ELECTRICAL THEORY I

26103
FASTENERS AND ANCHORS

26102
HAND BENDING

26101
ELECTRICAL SAFETY

CORE CURRICULA

105CMAP.EPS

TABLE OF CONTENTS

Trade Terms Introduced In This Module

Kirchoff's Current Law (KCL): The statement that the total amount of current flowing through a parallel circuit is equal to the sum of the amounts of current flowing through each current path.

Kirchoff's Voltage Law (KVL): The statement that the sum of all the voltage drops in a circuit is equal to the source voltage of the circuit.

Parallel circuits: Circuits containing two or more parallel paths through which current can flow.

Series circuits: Circuits with only one path for current flow.

Series-parallel circuits: Circuits that contain both series and parallel current paths.

1.0.0   INTRODUCTION

Ohm's Law was explained in the last module. This fundamental concept is now going to be used to analyze more complex **series circuits**, **parallel circuits**, and **series-parallel circuits**. This module will explain how to calculate resistance, current, and voltage in these complex circuits. Ohm's Law will be used to develop a new law for voltage and current determination. This law is called *Kirchoff's Law* and will become the new foundation for analyzing circuits.

2.0.0   RESISTIVE CIRCUITS

2.1.0   RESISTANCES IN SERIES

A series circuit is a circuit in which there is only one path for current flow. Resistance is measured in ohms (Ω). In the series circuit shown in *Figure 1*, the current (I) is the same in all parts of the circuit. This means that the current flowing through R_1 is the same as the current flowing through R_2 and R_3, and it is also the same as the current supplied by the battery.

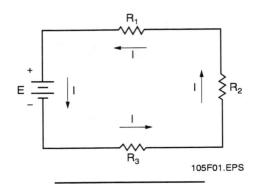

105F01.EPS

Figure 1. Series Circuit

When resistances are connected in series as in this example, the total resistance in the circuit is equal to the sum of the resistances of all the parts of the circuit:

$$R_T = R_1 + R_2 + R_3$$

Where:

R_T = total resistance

$R_1 + R_2 + R_3$ = resistances in series

Example 1:

The circuit shown in *Figure 2(A)* has 50Ω, 75Ω, and 100Ω resistors in series. Find the total resistance of the circuit.

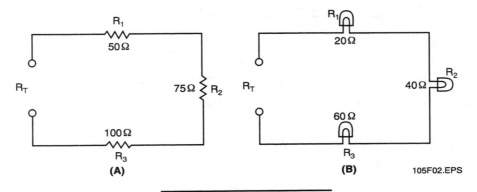

Figure 2. Total Resistance

Add the values of the three resistors in series:

$$R_T = R_1 + R_2 + R_2 = 50 + 75 + 100 = 225\Omega$$

Example 2:

The circuit shown in *Figure 2(B)* has three lamps connected in series with the resistances shown. Find the total resistance of the circuit.

Add the values of the three lamp resistances in series:

$$R_T = R_1 + R_2 + R_3 = 20 + 40 + 60 = 120\Omega$$

2.2.0 RESISTANCES IN PARALLEL

The total resistance in a parallel resistive circuit is given by the formula:

$$R_T = \frac{1}{\frac{1}{R_1} + \frac{1}{R_2} + \frac{1}{R_3} + \frac{1}{R_n}}$$

Where:

> R_T = total resistance in parallel
> R_1, R_2, R_3, and R_n = branch resistances

Example 1:

Find the total resistance of the 2Ω, 4Ω, and 8Ω resistors in parallel shown in *Figure 3*.

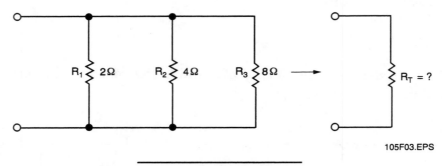

105F03.EPS

Figure 3. Parallel Branch

Write the formula for the three resistances in parallel:

$$R_T = \cfrac{1}{\cfrac{1}{R_1} + \cfrac{1}{R_2} + \cfrac{1}{R_3}}$$

Substitute the resistance values:

$$R_T = \cfrac{1}{\cfrac{1}{2} + \cfrac{1}{4} + \cfrac{1}{8}}$$

$$R_T = \frac{1}{0.5 + 0.25 + 0.125}$$

$$R_T = \frac{1}{0.875}$$

$$R_T = 1.14Ω$$

Note that when resistances are connected in parallel, the total resistance is always less than the resistance of any single branch.

In this case:

> $R_T = 1.14Ω < R_1 = 2Ω$, $R_2 = 4Ω$, and $R_3 = 8Ω$

Example 2:

Add a fourth parallel resistor of 2Ω to the circuit in *Figure 3*. What is the new total resistance, and what is the net effect of adding another resistance in parallel?

Write the formula for four resistances in parallel:

$$R_T = \cfrac{1}{\cfrac{1}{R_1} + \cfrac{1}{R_2} + \cfrac{1}{R_3} + \cfrac{1}{R_4}}$$

Substitute values:

$$R_T = \cfrac{1}{\cfrac{1}{2} + \cfrac{1}{4} + \cfrac{1}{8} + \cfrac{1}{2}}$$

$$R_T = \cfrac{1}{0.5 + 0.25 + 0.125 + 0.5}$$

$$R_T = \cfrac{1}{1.375}$$

$$R_T = 0.73\Omega$$

The net effect of adding another resistance in parallel is a reduction of the total resistance from 1.14Ω to 0.73Ω.

2.2.1 Simplified Formulas

The total resistance of *equal* resistors in parallel is equal to the resistance of one resistor divided by the number of resistors:

$$R_T = \frac{R}{N}$$

Where:

    R_T = total resistance of equal resistors in parallel
    R = resistance of one of the equal resistors
    N = number of equal resistors

If two resistors with the same resistance are connected in parallel, the equivalent resistance is half of that value, as shown in *Figure 4*.

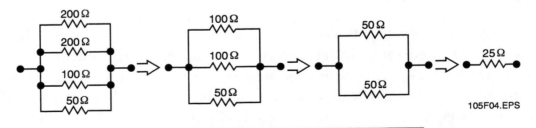

Figure 4. Equal Resistances In A Parallel Circuit

The two 200Ω resistors in parallel are the equivalent of one 100Ω resistor; the two 100Ω resistors are the equivalent of one 50Ω resistor; and the two 50Ω resistors are the equivalent of one 25Ω resistor.

When any two *unequal* resistors are in parallel, it is often easier to calculate the total resistance by multiplying the two resistances and then dividing the product by the sum of the resistances:

$$R_T = \frac{R_1 \times R_2}{R_1 + R_2}$$

Where:

R_T = total resistance of unequal resistors in parallel

R_1, R_2 = two unequal resistors in parallel

Example 1:

Find the total resistance of a 6Ω (R_1) resistor and an 18Ω (R_2) resistor in parallel:

$$R_T = \frac{R_1 \times R_2}{R_1 + R_2} = \frac{6 \times 18}{6 + 18} = \frac{108}{24} = 4.5\Omega$$

Example 2:

Find the total resistance of a 100Ω (R_1) resistor and a 150Ω (R_2) resistor in parallel:

$$R_T = \frac{R_1 \times R_2}{R_1 + R_2} = \frac{100 \times 150}{100 + 150} = \frac{15,000}{250} = 60\Omega$$

2.3.0   SERIES-PARALLEL CIRCUITS

To find current, voltage, and resistance in series circuits and parallel circuits is fairly easy. When working with either type, use only the rules that apply to that type. In a series-parallel circuit, some parts of the circuit are series connected and other parts are parallel connected. Thus, in some parts the rules for series circuits apply, and in other parts, the rules for parallel circuits apply. To analyze or solve a problem involving a series-parallel circuit, it is necessary to recognize which parts of the circuit are series connected and which parts are parallel connected. This is obvious if the circuit is simple. Many times, however, the circuit must be redrawn, putting it into a form that is easier to recognize.

In a series circuit, the current is the same at all points. In a parallel circuit, there are one or more points where the current divides and flows in separate branches. In a series-parallel circuit, there are both separate branches and series loads. The easiest way to find out whether a circuit is a series, parallel, or series-parallel circuit is to start at the negative terminal of the power source and trace the path of current through the circuit back to the positive terminal of the power source. If the current does not divide anywhere, it is a series circuit. If the current divides into separate branches, but there are no series loads, it is a parallel circuit. If the current divides into separate branches and there are also series loads, it is a series-parallel circuit. *Figure 5* shows electric lamps connected in series, parallel, and series-parallel circuits.

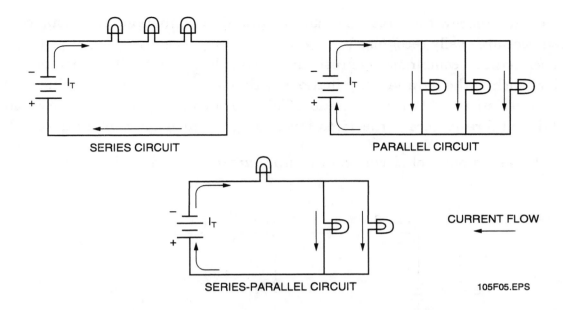

SERIES CIRCUIT

PARALLEL CIRCUIT

SERIES-PARALLEL CIRCUIT

CURRENT FLOW

105F05.EPS

Figure 5. Series, Parallel, And Series-Parallel Circuits

After determining that a circuit is series-parallel, redraw the circuit so that the branches and the series loads are more easily recognized. This is especially helpful when computing the total resistance of the circuit. *Figure 6* shows resistors connected in a series-parallel circuit and the equivalent circuit redrawn to simplify it.

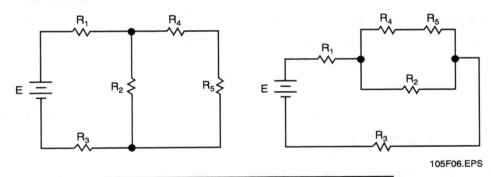

105F06.EPS

Figure 6. Redrawing A Series-Parallel Circuit

2.3.1 Reducing Series-Parallel Circuits

Very often, all that is known about a series-parallel circuit is the applied voltage and the values of the individual resistances. To find the voltage drop across any of the loads or the current in any of the branches, the total circuit current must usually be known. But to find the total current, the total resistance of the circuit must be known. To find the total resistance, reduce the circuit to its simplest form, which is usually one resistance that forms a series circuit with the voltage source. This simple series circuit has the equivalent resistance of the series-parallel circuit it was derived from, and also has the same total current. There are four basic steps in reducing a series-parallel circuit:

- If necessary, redraw the circuit so that all parallel combinations of resistances and series resistances are easily recognized.
- For each parallel combination of resistances, calculate its effective resistance.
- Replace each of the parallel combinations with one resistance whose value is equal to the effective resistance of that combination. This provides a circuit with all series loads.
- Find the total resistance of this circuit by adding the resistances of all the series loads.

Examine the series-parallel circuit shown in *Figure 7* and reduce it to an equivalent series circuit.

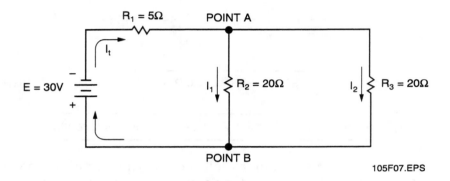

Figure 7. Reducing A Series-Parallel Circuit

In this circuit, resistors R_2 and R_3 are connected in parallel, but resistor R_1 is in series with both the battery and the parallel combination of R_2 and R_3. The current I_T leaving the negative terminal of the voltage source travels through resistor R_1 before it is divided at the junction of resistors R_1, R_2, and R_3 (Point A) to go through the two branches formed by resistors R_2 and R_3.

Given the information in *Figure 7*, calculate the resistance of R_2 and R_3 in parallel and the total resistance of the circuit, R_T.

The total resistance of the circuit is the sum of R_1 and the equivalent resistance of R_2 and R_3 in parallel. To find R_T, first find the resistance of R_2 and R_3 in parallel. Because the two resistances have the same value of 20Ω, the resulting equivalent resistance is 10Ω. Therefore, the total resistance (R_T) is 15Ω (5Ω + 10Ω).

2.4.0 APPLYING OHM'S LAW

2.4.1 Voltage And Current In Series Circuits

In resistive circuits, unknown circuit parameters can be found by using Ohm's Law and the techniques for determining equivalent resistance. Ohm's Law may be applied to an entire series circuit or to the individual parts of the circuit. When it is used on a particular part of a circuit, the voltage across that part is equal to the current in that part multiplied by the resistance of that part.

For example, given the information in *Figure 8*, calculate the total resistance (R_T) and the total current (I_T).

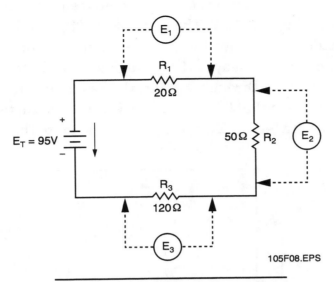

Figure 8. Calculating Voltage Drops

To find R_T:

$$R_T = R_1 + R_2 + R_3$$
$$R_T = 20 + 50 + 120$$
$$R_T = 190\Omega$$

To find I_T using Ohm's Law:

$$I_T = \frac{E_T}{R_T}$$
$$I_T = \frac{95}{190}$$
$$I_T = 0.5A$$

Find the voltage across each resistor. In a series circuit, the current is the same; that is, $I = 0.5A$ through each resistor:

$$E_1 = IR_1 = 0.5(20) = 10V$$
$$E_2 = IR_2 = 0.5(50) = 25V$$
$$E_3 = IR_3 = 0.5(120) = 60V$$

The voltages E_1, E_2, and E_3 found for *Figure 8* are known as *voltage drops* or *IR drops*. Their effect is to reduce the voltage that is available to be applied across the rest of the components in the circuit. The sum of the voltage drops in any series circuit is always equal to the voltage that is applied to the circuit. The total voltage (E_T) is the same as the applied voltage and can be verified in this example ($E_T = 10 + 25 + 60$ or 95V).

2.4.2   Voltage And Current In Parallel Circuits

A parallel circuit is a circuit in which two or more components are connected across the same voltage source, as illustrated in *Figure 9*. The resistors R_1, R_2, and R_3 are in parallel with each other and with the battery. Each parallel path is then a branch with its own individual current. When the total current I_T leaves the voltage source E, part I_1 of the current I_T will flow through R_1, part I_2 will flow through R_2, and the remainder I_3 will flow through R_3. The branch currents I_1, I_2, and I_3 can be different. However, if a voltmeter is connected across R_1, R_2, and R_3, the respective voltages E_1, E_2, and E_3 will be equal to the source voltage E.

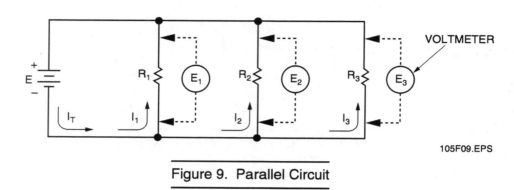

105F09.EPS

Figure 9.   Parallel Circuit

The total current I_T is equal to the sum of all branch currents.

This formula applies for any number of parallel branches whether the resistances are equal or unequal.

Using Ohm's Law, each branch current equals the applied voltage divided by the resistance between the two points where the voltage is applied. Hence, for each branch in *Figure 9* we have the following equations:

Branch 1: $I_1 = \dfrac{E_1}{R_1} = \dfrac{E}{R_1}$

Branch 2: $I_2 = \dfrac{E_2}{R_2} = \dfrac{E}{R_2}$

Branch 3: $I_3 = \dfrac{E_3}{R_3} = \dfrac{E}{R_3}$

With the same applied voltage, any branch that has less resistance allows more current through it than a branch with higher resistance.

Example 1:

The two branches R_1 and R_2, shown in *Figure 10(A)*, across a 110V power line draw a total line current of 20A. Branch R_1 takes 12A. What is the current I_2 in branch R_2?

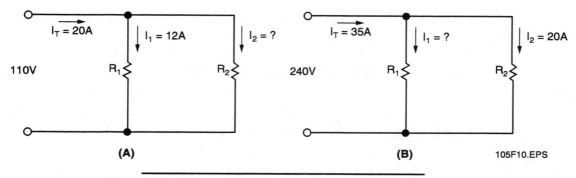

Figure 10. Solving For An Unknown Current

Transpose to find I_2 and then substitute given values:

$I_T = I_1 + I_2$

$I_2 = I_T - I_1$

$I_2 = 20 - 12 = 8A$

Example 2:

As shown in *Figure 10(B),* the two branches R_1 and R_2 across a 240V power line draw a total line current of 35A. Branch R_2 takes 20A. What is the current I_1 in branch R_1?

Transpose to find I_1 and then substitute given values:

$I_T = I_1 + I_2$

$I_1 = I_T - I_2$

$I_1 = 35 - 20 = 15A$

2.4.3   Voltage And Current In Series-Parallel Circuits

Series-parallel circuits combine the elements and characteristics of both the series and parallel configurations. By properly applying the equations and methods previously discussed, the values of individual components of the circuit can be determined. *Figure 11* shows a simple series-parallel circuit with a 1.5V battery.

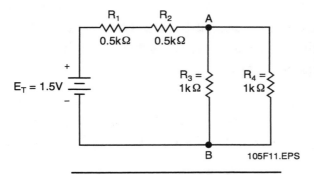

Figure 11. Series-Parallel Circuit

The current and voltage associated with each component can be determined by first simplifying the circuit to find the total current, and then working across the individual components.

This circuit can be broken into two components: the series resistances R_1 and R_2, and the parallel resistances R_3 and R_4.

R_1 and R_2 can be added together to form the equivalent series resistance R_{1+2}:

$$R_{1+2} = R_1 + R_2$$
$$R_{1+2} = 0.5k\Omega + 0.5k\Omega$$
$$R_{1+2} = 1k\Omega$$

R_3 and R_4 can be totaled using either the general reciprocal formula or, since there are two resistances in parallel, the product over sum method. Both methods are shown below.

$$R_{3+4} = \cfrac{1}{\cfrac{1}{R_3} + \cfrac{1}{R_4}} = \cfrac{1}{\cfrac{1}{1k\Omega} + \cfrac{1}{1k\Omega}} = \cfrac{1}{\cfrac{2}{1,000\Omega}} = \frac{1}{.002} = 500\Omega$$

$$R_{3+4} = \frac{R_3 \times R_4}{R_3 + R_4} = \frac{1k\Omega \times 1k\Omega}{1k\Omega + 1k\Omega} = \frac{1,000,000\Omega}{2,000\Omega} = 500\Omega$$

The equivalent circuit containing the R_{1+2} resistance of $1k\Omega$ and the R_{3+4} resistance of 500Ω is shown in *Figure 12*.

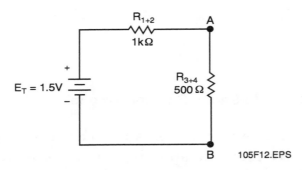

Figure 12. Simplified Series-Parallel Circuit

Using the Ohm's Law relationship that total current equals voltage divided by circuit resistance, the circuit current can be determined. First, however, total circuit resistance must be found. Since the simplified circuit consists of two resistances in series, they are simply added together to obtain total resistance.

$$R_T = R_{1+2} + R_{3+4}$$
$$R_T = 1k\Omega + 500\Omega$$
$$R_T = 1.5k\Omega$$

Applying this to the current/voltage equation:

$$I_T = \frac{E_T}{R_T}$$

$$I_T = \frac{1.5V}{1.5k\Omega}$$

$$I_T = 1mA \text{ or } 0.001A$$

Now that the total current is known, voltage drops across individual components can be determined:

$$E_{R1} = I_T R_1 = 1mA \times 0.5k\Omega = 0.5V$$

$$E_{R2} = I_T R_2 = 1mA \times 0.5k\Omega = 0.5V$$

Since the total voltage equals the sum of all voltage drops, the voltage drop from A to B can be determined by subtraction:

$$E_T = E_{R1} + E_{R2} + E_{A+B}$$

$$E_T - E_{R1} - E_{R2} = E_{A+B}$$

$$1.5V - 0.5V - 0.5V = E_{A+B} = 0.5V$$

Since R_3 and R_4 are in parallel, some of the total current must pass through each resistor. R_3 and R_4 are equal, so the same current should flow through each branch. Using the relationship:

$$I = \frac{E}{R}$$

$$I_{R3} = \frac{E_{R3}}{R_3} \qquad\qquad I_{R4} = \frac{E_{R4}}{R_4}$$

$$I_{R3} = \frac{0.5V}{1k\Omega} \qquad\qquad I_{R4} = \frac{0.5V}{1k\Omega}$$

$$I_{R3} = 0.5mA \qquad\qquad I_{R4} = 0.5mA$$

$$0.5mA + 0.5mA = 1mA$$

Therefore, the total current for the circuit passes through R_1 and R_2 and is evenly divided between R_3 and R_4.

3.0.0   KIRCHOFF'S LAWS

Kirchoff's Laws provide a simple, practical method of solving for unknown parameters in a circuit.

3.1.0   KIRCHOFF'S CURRENT LAW

In its most general form, **Kirchoff's Current Law**, which is also called *Kirchoff's First Law*, states that at any point in a circuit, the total current entering that point must equal the total current leaving that point. For parallel circuits, this implies that the current in a parallel circuit is equal to the sum of the currents in each branch.

When using Kirchoff's Laws to solve circuits, it is necessary to adopt conventions that determine the algebraic signs for current and voltage terms. A convenient system for current is to consider all current flowing into a branch point as positive, and all current directed away from that point as negative.

As an example, in *Figure 13,* the currents can be written as:

$$I_A + I_B - I_C = 0$$

or

$$5A + 3A - 8A = 0$$

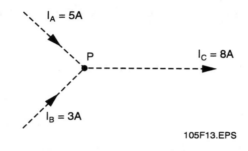

105F13.EPS

Figure 13. Kirchoff's Current Law

Currents I_A and I_B are positive terms because these currents flow into P, but I_C, directed out of P, is negative.

For a circuit application, refer to Point C at the top of the diagram in *Figure 14*. The 6A I_T into Point C divides into the 2A I_3 and 4A I_{4+5}, both directed out. Note that I_{4+5} is the current through R_4 and R_5. The algebraic equation is:

$$I_T - I_3 - I_{4+5} = 0$$

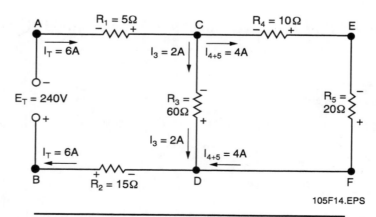

105F14.EPS

Figure 14. Application Of Kirchoff's Current Law

ELECTRICAL — TRAINEE TASK MODULE 26105

Substituting the values for each current:

$$6A - 2A - 4A = 0$$

For the opposite direction, refer to Point D at the bottom of *Figure 14*. Here, the branch currents into Point D combine to equal the mainline current I_T returning to the voltage source. Now, I_T is directed out from Point D, with I_3 and I_{4+5} directed in. The algebraic equation is:

$$I_3 + I_{4+5} - I_T = 0$$
$$2A + 4A - 6A = 0$$

Note that at either Point C or Point D, the sum of the 2A and 4A branch currents must equal the 6A total line current. Therefore, Kirchoff's Current Law can also be stated as:

$$I_{IN} = I_{OUT}$$

For *Figure 14*, the equations for current can be written as shown below.

At Point C:

$$6A = 2A + 4A$$

At Point D:

$$2A + 4A = 6A$$

Kirchoff's Current Law is really the basis for the practical rule in parallel circuits that the total line current must equal the sum of the branch currents.

3.2.0   KIRCHOFF'S VOLTAGE LAW

Kirchoff's Voltage Law states that the algebraic sum of the voltages around any closed path is zero.

Referring to *Figure 15,* the sum of the voltage drops around the circuit must equal the voltage applied to the circuit:

$$E_A = E_1 + E_2 + E_3$$

Where:

E_A = voltage applied to the circuit
E_1, E_2, and E_3 = voltage drops in the circuit

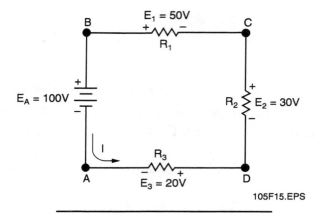

Figure 15. Kirchoff's Voltage Law

Another way of stating this law is that the algebraic sum of the voltage rises and voltage drops must be equal to zero. A voltage source is considered a voltage rise; a voltage across a resistor is a voltage drop. (For convenience in labeling, letter subscripts are shown for voltage sources and numerical subscripts are used for voltage drops.) This form of the law can be written by transposing the right members to the left side:

Voltage applied – sum of voltage drops = 0

Substitute letters:

$$E_A - E_1 - E_2 - E_3 = 0$$
or
$$E_A - (E_1 + E_2 + E_3) = 0$$

3.3.0   LOOP EQUATIONS

Any closed path is called a *loop*. A loop equation specifies the voltages around the loop. Refer to *Figure 16*.

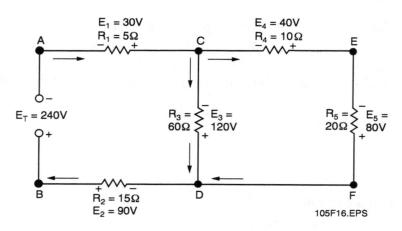

Figure 16. Loop Equation

ELECTRICAL — TRAINEE TASK MODULE 26105

Consider the inside loop A, C, D, B, A, including the voltage drops E_1, E_3, and E_2, and the source E_T. In a clockwise direction, starting at Point A, the algebraic sum of the voltages is:

$$- E_1 - E_3 - E_2 + E_T = 0$$

or

$$- 30V - 120V - 90V + 240V = 0$$

Voltages E_1, E_3, and E_2 have a negative value, because there is a decrease in voltage seen across each of the resistors in a clockwise direction. However, the source E_T is a positive term because an increase in voltage is seen in that same direction.

For the opposite direction, going counterclockwise in the same loop from Point A, E_T is negative while E_1, E_2, and E_3 have positive values. Therefore:

$$- E_T + E_2 + E_3 + E_1 = 0$$

or

$$- 240V + 90V + 120V + 30V = 0$$

When the negative term is transposed, the equation becomes:

$$240V = 90V + 120V + 30V$$

In this form, the loop equation shows that Kirchoff's Voltage Law is really the basis for the practical rule in series circuits that the sum of the voltage drops must equal the applied voltage.

For example, determine the voltage E_B for the circuit shown in *Figure 17*. The direction of the current flow is shown by the arrow. First mark the polarity of the voltage drops across the resistors and trace the circuit in the direction of the current flow starting at Point A. Then write the voltage equation around the circuit:

$$- E_3 - E_B - E_2 - E_1 + E_A = 0$$

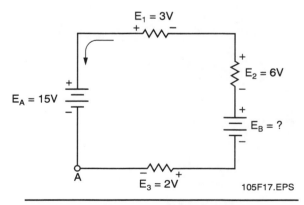

Figure 17. Applying Kirchoff's Voltage Law

Solve for E_B:

$E_B = E_A - E_3 - E_2 - E_1$

$E_B = 15V - 2V - 6V - 3V$

$E_B = 4V$

Since E_B was found to be positive, the assumed direction of current is in fact the actual direction of current.

In its most general form, Kirchoff's Voltage Law (also called *Kirchoff's Second Law*) states that the algebraic sum of all the potential differences in a closed loop is equal to zero. A closed loop means any completely closed path consisting of wire, resistors, batteries, or other components. For series circuits, this implies that the sum of the voltage drops around the circuit is equal to the applied voltage. For parallel circuits, this implies that the voltage drops across all branches are equal.

SUMMARY

The relationships among current, voltage, resistance, and power in Ohm's Law are the same for both DC series and DC parallel circuits. Understanding and being able to apply these concepts is necessary for effective circuit analysis and troubleshooting. DC series-parallel circuits also have these fundamental relationships. Since DC series-parallel circuits are a combination of simple series and parallel circuits, Kirchoff's Voltage and Current Laws will apply. Calculating I, E, R, and P for series-parallel circuits is no more difficult than calculating these values for simple series or parallel circuits. However, for series-parallel circuits, these calculations require more careful circuit analysis in order to use Ohm's Law correctly.

References

For advanced study of topics covered in this task module, the following books are suggested:

Electronics Fundamentals, Latest Edition, Prentice Hall, New York, NY.
Principles of Electric Circuits, Latest Edition, Prentice Hall, New York, NY.

REVIEW QUESTIONS

1. The formula for calculating the total resistance in a series circuit with three resistors is _____.

   a. $R_T = R_1 + R_2 + R_3$
   b. $R_T = R_1 - R_2 - R_3$
   c. $R_T = R_1 \times R_2 \times R_3$
   d. $R_T = \dfrac{1}{\dfrac{1}{R_1} + \dfrac{1}{R_2} + \dfrac{1}{R_3}}$

2. The formula for calculating the total resistance in a parallel circuit with three resistors is _____.

   a. $R_T = R_1 + R_2 + R_3$
   b. $R_T = R_1 - R_2 - R_3$
   c. $R_T = R_1 \times R_2 \times R_3$
   d. $R_T = \dfrac{1}{\dfrac{1}{R_1} + \dfrac{1}{R_2} + \dfrac{1}{R_3}}$

3. The total resistance in *Figure 18* is _____.
   a. $1{,}035\Omega$
   b. 129Ω
   c. 100Ω
   d. 157Ω

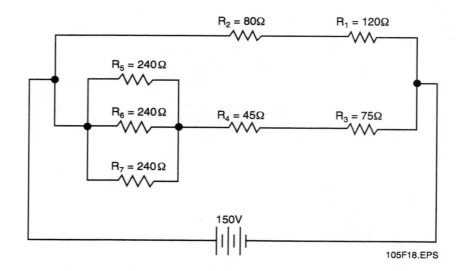

105F18.EPS

Figure 18. Series-Parallel Circuit

4. Find the total resistance in a series circuit with three resistances of 10Ω, 20Ω, and 30Ω.
   a. 15Ω
   b. 1Ω
   c. 20Ω
   d. 60Ω

5. In a parallel circuit, the voltage across each path is equal to the _____.
   a. total circuit resistance times path current
   b. source voltage minus path voltage
   c. path resistance times total current
   d. source voltage

6. The value for total current in *Figure 19* is _____ amps.
   a. 1.25
   b. 2.50
   c. 5
   d. 10

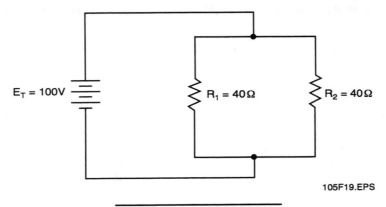

105F19.EPS

Figure 19. Parallel Circuit

7. A resistor of 32Ω is in parallel with a resistor of 36Ω and a 54Ω resistor is in series with the pair. When 350V is applied to the combination, the current through the 54Ω resistor is _____ amps.
   a. 2.87
   b. 3.26
   c. 5.86
   d. 4.94

8. A 242Ω resistor is in parallel with a 180Ω resistor and a 420Ω resistor is in series with the combination. A current of 22mA flows through the 242Ω resistor. The current through the 180Ω resistor is _____ mA.
   a. 29.6
   b. 40.2
   c. 19.8
   d. 36.4

9. Two 24Ω resistors are in parallel, and a 42Ω resistor is in series with the combination. When 78V is applied to the three resistors, the voltage drop across the 42Ω resistor is _____ volts.
   a. 55.8
   b. 60.5
   c. 65.3
   d. 49.8

10. A 19Ω resistor is in parallel with an 18Ω resistor, and in series with the two is a 14Ω resistor. When 70V is applied to this combination, the current through the 19Ω resistor is _____ amps.
    a. 2.06
    b. 1.03
    c. 1.46
    d. 2.37

notes

ANSWERS TO REVIEW QUESTIONS

Answer	Section
1. a	2.1.0
2. d	2.2.0
3. c	2.3.1
4. d	2.1.0
5. d	2.4.2
6. c	2.4.2
7. d	2.4.3
8. a	2.4.3
9. b	3.3.0
10. c	3.3.0

NCCER CRAFT TRAINING USER UPDATES

The NCCER makes every effort to keep these manuals up-to-date and free of technical errors. We appreciate your help in this process. If you have an idea for improving this manual, or if you find an error, a typographical mistake, or an inaccuracy in the NCCER's Craft Training Manuals, please write us, using this form or a photocopy. Be sure to include the exact module number, page number, a description of the problem, and the correction, if possible. Your input will be brought to the attention of the Technical Review Committee. Thank you for your assistance.

Instructors – If you found that additional materials were necessary in order to teach this module effectively, please let us know so that we may include them in the Equipment/Materials list in the Instructor's Guide.

Write: Curriculum Development and Revision Department
              National Center for Construction Education and Research
              P.O. Box 141104
              Gainesville, FL 32614-1104
Fax: 352-334-0932

Craft   Module Name
_____

Copyright Date   Module Number   Page Number(s)
_____

Description of Problem
_____

_____

_____

_____

(Optional) Correction of Problem
_____

_____

_____

(Optional) Your Name and Address
_____

_____

_____

notes

Electrical Test Equipment
Module 26106

**NATIONAL
CENTER FOR
CONSTRUCTION
EDUCATION AND
RESEARCH**

ELECTRICAL TEST EQUIPMENT

OBJECTIVES

Upon completion of this module, the trainee will be able to:

1. Explain the operation of and describe the following pieces of test equipment:
    *   Ammeter
    *   Voltmeter
    *   Ohmmeter
    *   Volt-ohm-milliammeter (VOM)
    *   Wattmeter
    *   Megohmmeter
    *   Frequency meter
    *   Power factor meter
    *   Continuity tester
    *   Voltage tester
    *   Recording instruments
    *   Cable-length meters
2. Explain how to read and convert from one scale to another using the above test equipment.
3. Explain the importance of proper meter polarity.
4. Define frequency and explain the use of a frequency meter.
5. Explain the difference between digital and analog meters.

Prerequisites

Successful completion of the following Task Modules is required before beginning study of this Task Module: Core Curricula; Electrical Level 1, Modules 26101 through 26105.

Required Trainee Materials

1. Trainee Task Module
2. Copy of the latest edition of the *National Electrical Code*
3. Appropriate Personal Protective Equipment

Note: The designations "National Electrical Code," "NE Code," and "NEC," where used in this document, refer to the *National Electrical Code®*, which is a registered trademark of the National Fire Protection Association, Quincy, MA. *All National Electrical Code (NEC) references in this module refer to the 1999 edition of the NEC.*

Course Map

This course map shows all of the task modules in the first level of the Electrical curricula. The suggested training order begins at the bottom and proceeds up. Skill levels increase as a trainee advances on the course map. The training order may be adjusted by the local Training Program Sponsor.

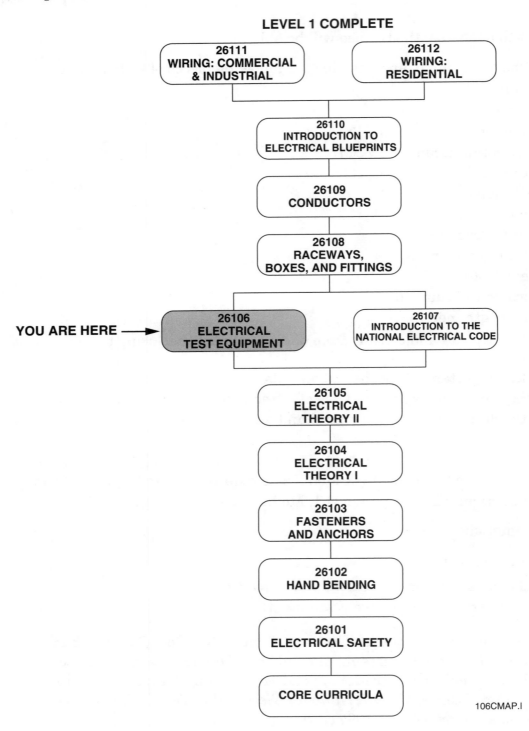

LEVEL 1 COMPLETE

26111
WIRING: COMMERCIAL & INDUSTRIAL

26112
WIRING: RESIDENTIAL

26110
INTRODUCTION TO ELECTRICAL BLUEPRINTS

26109
CONDUCTORS

26108
RACEWAYS, BOXES, AND FITTINGS

YOU ARE HERE →

26106
ELECTRICAL TEST EQUIPMENT

26107
INTRODUCTION TO THE NATIONAL ELECTRICAL CODE

26105
ELECTRICAL THEORY II

26104
ELECTRICAL THEORY I

26103
FASTENERS AND ANCHORS

26102
HAND BENDING

26101
ELECTRICAL SAFETY

CORE CURRICULA

106CMAP.I

TABLE OF CONTENTS

TABLE OF CONTENTS (Continued)

Terms Introduced In This Module

Coil: A number of turns of wire, especially in spiral form, used for electromagnetic effects or for providing electrical resistance.

Continuity: An uninterrupted electrical path for current flow.

d'Arsonval meter movement: A meter movement that uses a permanent magnet and moving coil arrangement to move a pointer across a scale.

Decibel: A unit used to express a relative difference in power between electric signals.

Frequency: The number of cycles completed each second by a given AC voltage; usually expressed in hertz; one hertz = one cycle per second.

1.0.0   INTRODUCTION

The use of electronic test instruments and meters generally involves these three applications:

* Verifying proper operation of instruments and associated equipment
* Calibrating electronic instruments and associated equipment
* Troubleshooting electrical/electronic circuits and equipment

For these applications, specific test equipment is selected to analyze circuits and to determine specific characteristics of discrete components.

The test equipment an electrician chooses for a specific task depends on the type of measurement and the level of accuracy required. Additional factors that may influence selection include:

* Whether the test equipment is portable
* The amount of information that the test equipment provides
* The likelihood that the test equipment may damage the circuit or component being tested (some test equipment can generate enough voltage or current to damage an instrument or electronic circuit)

This module will focus on some of the test equipment that you will be required to use in your job as an electrician. The intent is to familiarize you with the use and operation of such equipment and to provide you with practical experience involving that equipment. Upon completion of this module, you should be able to select the appropriate test equipment and effectively use that equipment to perform an assigned task.

2.0.0 METERS

The functioning of conventional electrical measuring instruments is based upon electro-mechanical principles. Their mechanical components usually work on direct current (DC). Mechanical **frequency** meters are an exception. A meter that measures alternating current (AC) has a built-in rectifier to change the AC to DC and resistors to correct for the various ranges.

Today, many meters are solid-state digital systems; they are superior because they have no moving parts. These meters will work in any position, unlike mechanical meters, which must remain in one position to be read accurately.

2.1.0 D'ARSONVAL METER MOVEMENT

In 1882, a Frenchman named Arsene d'Arsonval invented the galvanometer. This meter used a stationary permanent magnet and a moving **coil** *(Figure 1)* to indicate current flow on a calibrated scale. The early galvanometer was very accurate but could only measure very small currents. Over the following years, many improvements were made that extended the range of the meter and increased its ruggedness. The **d'Arsonval meter movement** is the most commonly used meter movement today.

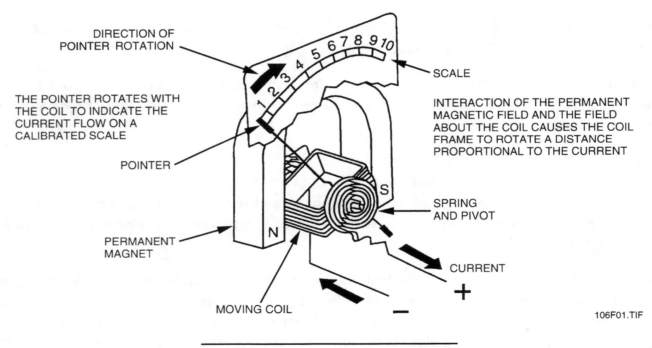

Figure 1. Moving-Coil Meter Movement

A moving-coil meter movement operates on the electromagnetic principle. In its simplest form, the moving-coil meter uses a coil of very fine wire wound on a light aluminum frame. A permanent magnet surrounds the coil. The aluminum frame is mounted on pivots to allow it and the coil to rotate freely between the poles of the permanent magnet. When current flows

ELECTRICAL — TRAINEE TASK MODULE 26106

through the coil, it becomes magnetized, and the polarity of the coil is such that it is repelled by the field of the permanent magnet. This will cause the coil frame to rotate on its pivots, and the distance it rotates is determined by the amount of current that flows through the coil. By attaching a pointer to the coil frame and adding a calibrated scale, the amount of current flowing through the meter can be measured.

The d'Arsonval meter movement uses this same principle of operation *(Figure 2)*. As the current flow increases, the magnetic field around the coil increases, the amount of coil rotation increases, and the pointer swings farther across the meter scale.

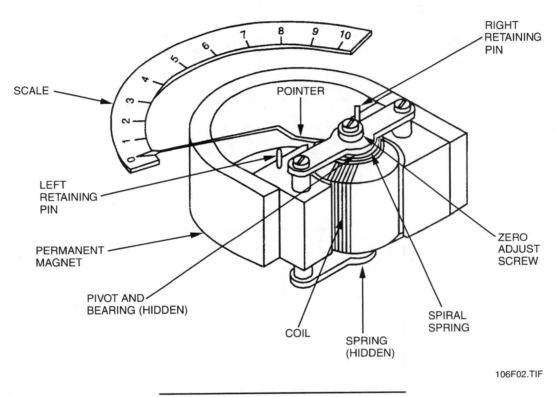

106F02.TIF

Figure 2. d'Arsonval Meter Movement

3.0.0   AMMETER

The ammeter is used to measure current. Most models will measure only small amounts of current. The typical range is in microamperes, µA (0.000001 amperes) or milliamperes, mA (0.001 amperes). Very few ammeters can measure more than 10mA. To increase the range to the ampere level, a shunt is used. To measure above 10mA, a shunt with an extremely low resistance is placed in series with the load, and the meter is connected across the shunt to measure the resulting voltage drop proportional to current flow. A shunt has a very large wattage rating in order to carry a large current.

The meter is connected in parallel with the shunt *(Figure 3)*. Shunts located inside the meter case (internal shunts) are generally used to measure values up to 30 amps; shunts located away from the meter case (external shunts) with leads going to the meter are generally used to measure values greater than 30 amps. Above 30 amps of current, the heat generated could damage the meter if an internal shunt were used. The use of a shunt allows the ammeter to derive current in amps by actually measuring the voltage drop across the shunt. Ammeter connections are shown in *Figure 4.*

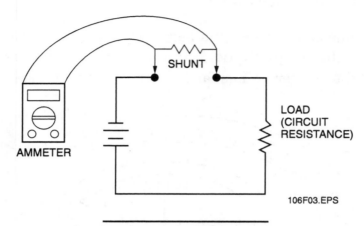

Figure 3. Ammeter Shunt

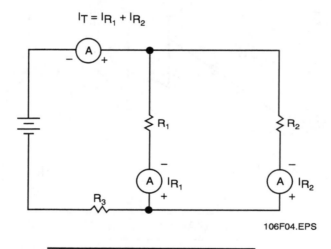

Figure 4. Ammeter Connections

Never connect an ammeter in parallel with a load. Because of the low resistance in the ammeter, this will cause a short circuit, probable damage to the meter and/or circuit, and personal injury. When connecting an ammeter in a DC circuit, you must observe proper polarity. In other words, you must connect the negative terminal of the meter to the negative or low potential point in the circuit, and connect the positive terminal of the meter to the positive or high potential point in the circuit (*Figure 5*). Current must flow through the meter from minus (−) to plus (+). If you connect the meter with the polarities reversed, the meter coil will move in the opposite direction, and the pointer might strike the left retaining pin. You will not obtain a current reading, and you might bend the pointer of the meter.

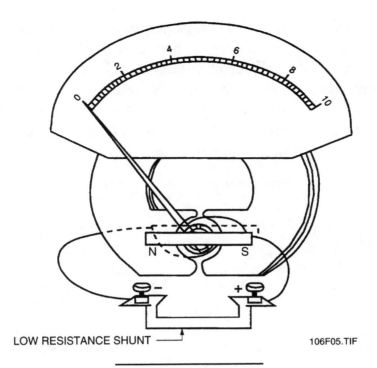

LOW RESISTANCE SHUNT

106F05.TIF

Figure 5. DC Ammeter

It is not very practical to use an ammeter that has only one range; therefore, a multirange ammeter needs to be discussed. A multirange ammeter is one containing a basic meter movement and several shunts that can be connected across the meter movement. See *Figure 6*.

The range of this 1mA meter movement has been extended to measure 0-10mA, 0-100mA, and 0-1A by using multiple shunts.

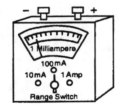

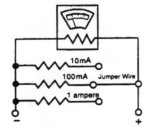

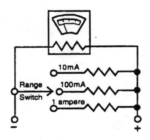

A range switch provides the simplest way of setting the meter to the desired range.

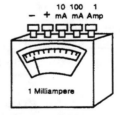

When separate terminals are used to select the desired range, a jumper must be connected from the positive terminal to connect the shunt across the meter movement.

106F06.TIF

Figure 6. Multirange Ammeter

A range switch is normally used to select the particular shunt for the desired current range. Sometimes, however, separate terminals for each range are mounted on the meter case. Some multirange ammeters have only one set of values on the scale, even though they measure several different current ranges. For example, if the scale is calibrated in values from 0 to 1 milliamp (mA), and the range switch is in the 1mA position, read the current directly. However, if the range switch is in the 10mA position, multiply the scale reading by 10 to find the amount of current flowing through the circuit. See *Figure 7*.

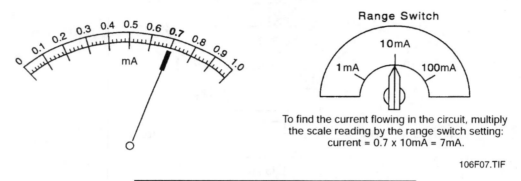

Some multirange current meters have only one set of values marked on the scale.

To find the current flowing in the circuit, multiply the scale reading by the range switch setting:
current = 0.7 x 10mA = 7mA.

106F07.TIF

Figure 7. Multirange Single-Scale Ammeter

Other current meters have a separate set of values on the calibrated scale that correspond to the different positions of the range switch. In this case, be sure that you read the set of values that correspond to the position of the range switch *(Figure 8)*.

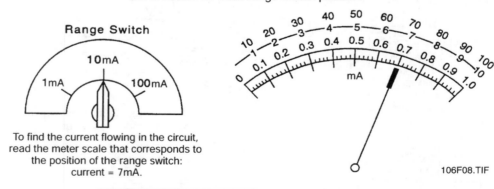

Some multirange current meters have a set of values for each range switch position.

To find the current flowing in the circuit, read the meter scale that corresponds to the position of the range switch:
current = 7mA.

106F08.TIF

Figure 8. Multirange Multiscale Ammeter

When measuring AC current at levels greater than one ampere, a clamp-on ammeter is often used. This meter also measures DC. The clamp-on ammeter may have a mechanical movement or a digital readout. These meters clamp over the wire and do not break the insulation. This type of ammeter senses current flow by measuring the magnetic field surrounding the conductor. When using this type of ammeter, be sure to measure only one conductor at a time. This type of meter will often measure up to 1,000 amperes at 600 volts. A digital clamp-on ammeter is shown in *Figure 9*.

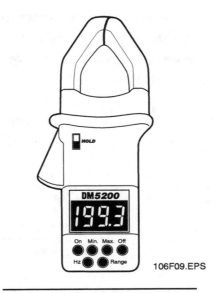

Figure 9. Clamp-On Ammeter

4.0.0 VOLTMETER

The basic current meter movement already covered, whether AC or DC, can also be used to measure voltage (electromotive force or emf). The meter coil has a fixed resistance, and, therefore, when current flows through the coil, a voltage drop will be developed across this resistance. According to Ohm's Law, the voltage drop will be directly proportional to the amount of current flowing through the coil. Also, the amount of current flowing through the coil is directly proportional to the amount of voltage applied to it. Therefore, by calibrating the meter scale in units of voltage instead of current, the voltage in a circuit can be measured.

Since a basic current meter movement has a low coil resistance and low current-handling capabilities, its use as a voltmeter is very limited. In fact, the maximum voltage that could be measured with a one milliamp meter movement is one volt *(Figure 10)*.

Since the voltage across the meter coil resistance is proportional to the current flowing through the coil, the 1mA current meter movement can measure voltage directly by calibrating the meter scale in the units of voltage that produce the current through the coil.

$$E = I_R R_M = 0.001 \times 1000 = 1 \text{ volt}$$

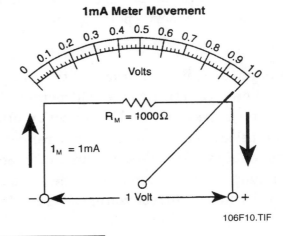

Figure 10. 1mA Meter Movement

The voltage range of a meter movement can be extended by adding a resistor, called a *multiplier resistor*, in series. The value of this resistor must be such that, when added to the meter coil resistance, the total resistance limits the current to the full-scale current rating of the meter for any applied voltage *(Figure 11)*.

By connecting a multiplier resistor in series with the meter resistance, the range of a basic meter movement can be extended to measure voltages higher than the $I_M R_M$ voltage drop across the meter coil.

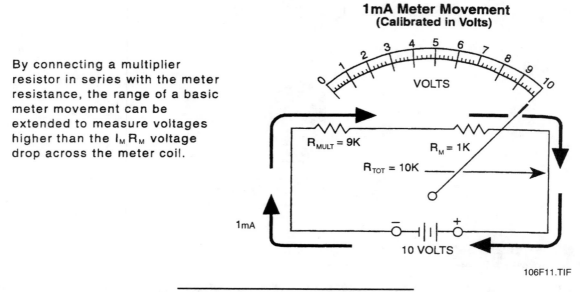

Figure 11. Adding Multiplier Resistor

Voltmeters must be in parallel with the circuit component being measured. On the higher ranges, the amount of current flowing through the meter is much lower due to its very high total resistance. However, an inaccurate reading will result if the voltmeter is placed in series rather than in parallel with a circuit component. When connecting a DC voltmeter, always observe the proper polarity. The negative lead of the meter must be connected to the negative or low potential end of the component, and the positive lead to the positive or high potential end of the component. As was the case when using an ammeter, if you connect a voltmeter to the component with opposing polarities, the meter coil will move to the left, and the pointer could be bent. In an AC circuit, the voltage constantly reverses polarity, so there is no need to observe polarity when connecting the voltmeter to a component in an AC circuit.

As with ammeters, it is impractical to have a voltmeter that will only measure one range of voltages; therefore, multirange voltmeters also need to be discussed. To make a voltmeter capable of measuring multiple ranges, an electrician needs several multiplier resistors that are switch-selectable for the different ranges desired. Reading the scale of a voltmeter is as simple as reading the scale of an ammeter. Some multirange voltmeters have only one range of values marked on the scale, and the scale reading must be multiplied by the range switch setting to obtain the correct voltage *(Figure 12)*.

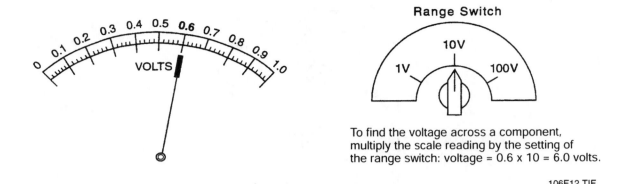

Figure 12. Multirange Single-Scale Voltmeter

Other voltmeters have different ranges on the scale for each setting of the range switch. In this case, be sure that you read the set of values that correspond to the position of the range switch *(Figure 13)*.

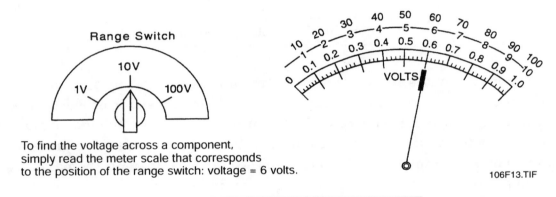

Figure 13. Multirange Multiscale Voltmeter

Some digital voltmeters are autoranging. These types of voltmeters do not have a range switch. The internal construction of the meter itself will select the proper resistance for the current being detected. However, when using a voltmeter that is *not* autoranging, always start with the highest voltage setting and work down until the indication reads somewhere between half-scale and three-quarter scale. This will give a more accurate reading and will prevent damage to the voltmeter.

5.0.0 OHMMETER

An ohmmeter is a device that measures the resistance of a circuit or component. It can also be used to locate open circuits or shorted circuits. Basically, an ohmmeter consists of a DC current meter movement, a low-voltage DC power source (usually a battery), and current-limiting resistors, all of which are connected in series *(Figure 14)*.

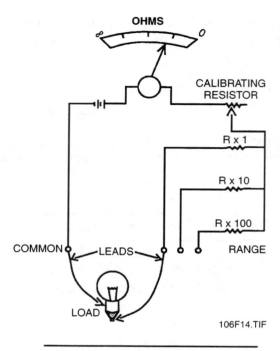

Figure 14. Ohmmeter Schematic

This combination of devices allows the meter to calculate resistance by deriving it using Ohm's Law. Before measuring the resistance of an unknown resistor or electrical circuit, the test leads are shorted together. This causes current to flow through the meter movement and, the pointer deflects to the right. This means that there is zero resistance present across the input terminals of the ohmmeter, and when zero resistance is present, the pointer will deflect full scale. Therefore, full-scale deflection of an ohmmeter indicates zero resistance. Most ohmmeters have a zero adjustment knob. This is used to correct for the fact that as the batteries of the meter age, their output voltage decreases. As the voltage drops, the current through the circuit decreases, and the meter will no longer deflect full scale. By correcting for this change before each use, the internal resistance of the meter, along with the resistance of the leads, is lowered and nulled to deflect the pointer full scale.

After the ohmmeter is adjusted to zero, it is ready to be connected in a circuit to measure resistance. The circuit must be verified as deenergized by using a voltmeter prior to taking a reading with an ohmmeter. If the circuit were energized, its voltage could cause a damaging current to flow through the meter. This can cause damage to the meter and/or circuit as well as personal injury.

When making resistance measurements in circuits, each component in the circuit can be tested individually by removing the component from the circuit and connecting the ohmmeter leads across it. Actually, the component does not have to be totally removed from the circuit. Usually, the part can be effectively isolated from the rest of the circuit by disconnecting one of its leads from the circuit. However, this method can still be somewhat time consuming. Some manufacturers provide charts that list the resistance readings that should be obtained from various test points to a reference point in the equipment. There are usually many parts of the circuit between the test point and the reference point, so if you get an abnormal reading, you must begin checking smaller groups of components, or individual components, to isolate the defective one. If resistance charts are not available, be very careful to ensure that other components are not in parallel with the component being tested.

6.0.0 VOLT-OHM-MILLIAMMETER

The volt-ohm-milliammeter (VOM), also known as the *multimeter*, is a multipurpose instrument. It is a combination of the three previous meters discussed: the milliammeter, the voltmeter, and the ohmmeter. One common analog multimeter is shown in *Figure 15*. There are many different models of this basic multimeter. To prevent having to discuss each and every meter, one version will be explained here. Any controls or functions on your meter that are not covered here should be reviewed in the applicable owner's manual.

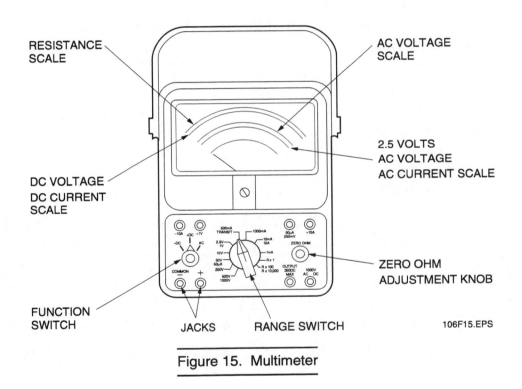

Figure 15. Multimeter

The typical volt-ohm-milliammeter is a rugged, accurate, compact, easy-to-use instrument. The instrument can be used to make accurate measurements of DC and AC voltages, current, resistance, **decibels,** and output voltage. The output voltage function is used for measuring the AC component of a mixture of AC and DC voltages. This occurs primarily in amplifier circuits.

This meter has the following features: a 0-1 volt DC range, 0-500 volt DC and AC ranges, a TRANSIT position on the range switch, rubber plug bumpers on the bottom of the case to reduce sliding, and an externally-accessible battery and fuse compartment.

6.1.0   SPECIFICATIONS

The specifications of the example multimeter are:

- DC voltage:
  - Sensitivity: 20KΩ per volt
  - Accuracy: 1¾% of full scale
- AC voltage:
  - Sensitivity: 5KΩ per volt
  - Accuracy: 3% of full scale
- DC current:
  - 250mV to 400mV drop
  - Accuracy: 1¾% of full scale
- Resistance:
  - Accuracy: 1.75° of arc
  - Nominal open circuit voltage 1.5V (9V on the 10KΩ ohm range)
- Nominal short circuit current:
  - 1Ω range: 1.25mA
  - 100Ω range: 1.25mA
  - 10KΩ range: 75mA
- Meter frequency response:
  - Up to 100kHz

6.2.0   OVERLOAD PROTECTION

In the example multimeter, a 1A, 250V fuse is provided to protect the circuits on the ohmmeter ranges. It also protects the milliampere ranges from excessive overloads. If the instrument fails to indicate, the fuse may be burned out. The fuse is mounted in a holder in the battery and fuse compartment. A spare fuse is located in a well between the + terminal of the D cell and the side of the case. Access to the compartment is obtained by loosening the single captivating screw on the compartment cover. To replace a burned-out fuse, remove it from the holder and replace it with a fuse of the exact same type. When removing the fuse from its holder, first remove the battery.

In addition to the fuse, a varistor protects the indicating instrument circuit. The varistor limits the current through the moving coil in case of overload.

The fuse and varistor will prevent serious damage to the meter in most cases of accidental overload. However, no overload protection system is completely foolproof, and misapplication on high-voltage circuits can damage the instrument. Care and caution should always be exercised to protect both you and the VOM.

Beside the actual steps used in making a measurement, some other points to consider while using a multimeter are:

- Keep the instrument in a horizontal position when storing and away from the edge of a workbench, shelf, or other area where it may be knocked off and damaged.
- Avoid rapid or extreme temperature changes. For example, do not leave the meter in your truck during hot or cold weather. Rapid and extreme temperature changes will advance the aging of the meter components and adversely affect meter life and accuracy.
- Avoid overloading the measuring circuits of the instrument. Develop a habit of checking the range position before connecting the test leads to a circuit. Even slight overloads can damage the meter. Even though it may not be noticeable in blown fuses or a bent needle, damage has been done. Slight overloads will advance the aging of components, again causing changes in meter life and accuracy.
- Place the range switch in the TRANSIT position when the instrument is not in use or when it is being moved. This reduces the swinging of the pointer when the meter is carried. Every meter does not have a TRANSIT position, but if the meter does, it should be used. Random, uncontrolled swings of the meter movement may damage the movement, bend the needle, or reduce its accuracy.
- If the meter has not been used for a long period of time, rotate the function and range switches in both directions to wipe the switch contacts. Most switch contacts are plated with copper or silver. Over a period of time, these materials will oxidize (tarnish). This will create a high resistance through the switch, causing a large inaccuracy. Rotating through the switch positions will clean the tarnish off and provide good electrical contact.

6.3.0 MAKING MEASUREMENTS

CAUTION: When using the meter as a millivoltmeter, care must be taken to prevent damage to the indicating instrument from excessive voltage. Before using the 250mV range, first use the 1.0V range to determine that the voltage measured is no greater than 250mV (or 0.25VDC).

6.3.1 Measuring DC Voltage, 0-250 Millivolts

Step 1 Set the function switch to +DC.

Step 2 Plug the black test lead into the – (COMMON) jack and the red test lead into the +50µA/250mV jack.

Step 3 Set the range switch to 50µA (dual position with 50V).

Step 4 Connect the black test lead to the negative side of the circuit being measured and the red test lead to the positive side of the circuit.

Step 5 Read the voltage on the black scale marked DC and use the figures marked 0-250. Read directly in millivolts.

6.3.2 Measuring DC Voltage, 0-1 Volt

Step 1 Set the function switch to –DC. Plug the black test lead into the – (COMMON) jack and the red test lead into the +1V jack.

Step 2 Set the range switch to 1V (dual position with 2.5V).

Step 3 Connect the black test lead to the negative side of the circuit being measured and the red test lead to the positive side of the circuit.

Step 4 Read the voltage on the black scale marked DC. Use the figures marked 0-10 and divide the reading by 10.

6.3.3 Measuring DC Voltage, 0-2.5 Through 0-500 Volts

WARNING! Be extremely careful when working with higher voltages. Do not touch the instrument test leads while power is on in the circuit being measured.

Step 1 Set the function switch to +DC.

Step 2 Set the range switch to one of the five voltage range positions marked 2.5V, 10V, 50V, 250V, or 500V. When in doubt as to the voltage present, always use the highest voltage range as a protection to the instrument. If the voltage is within a lower range, the switch may be set for the lower range to obtain a more accurate reading.

Step 3 Plug the black test lead into the – (COMMON) jack and the red test lead into the + jack.

Step 4 Connect the black test lead to the negative side of the circuit being measured and the red test lead to the positive side of the circuit.

Step 5 Read the voltage on the black scale marked DC. For the 2.5V range, use the 0-250 figures and divide by 100. For the 10V, 50V, and the 250V ranges, read the figures directly. For the 500V range, use the 0-50 figures and multiply by 10.

6.3.4 Measuring DC Voltage, 0-1,000 Volts

WARNING! Be extremely careful when working with higher voltages. Do not touch the instrument test leads while power is on in the circuit being measured.

Step 1 Set the function switch to +DC.

Step 2 Set the range switch to 1,000V (dual position with 500V).

Step 3 Plug the black test lead into the − (COMMON) jack and the red test lead into the 1,000V jack.

Step 4 Be sure power is off in the circuit being measured and all capacitors have been discharged. Connect the black test lead to the negative side of the circuit being measured and the red test lead to the positive side of the circuit.

Step 5 Turn on the power in the circuit being measured.

Step 6 Read the voltage using the 0-10 figures on the black scale marked DC. Multiply the reading by 100.

6.3.5 Measuring AC Voltage, 0-2.5 Through 0-500 Volts

WARNING!   Be extremely careful when working with higher voltages.
Do not touch the instrument or test leads while power is
on in the circuit being measured.

CAUTION: When measuring line voltage such as from a 120V, 240V, or
480V source, be sure that the range switch is set to the proper
voltage position.

Step 1 Set the function switch to AC.

Step 2 Set the range switch to one of the five voltage range positions marked 2.5V, 10V, 50V, 250V, or 500V. When in doubt as to the actual voltage present, always use the highest voltage range as a protection to the instrument. If the voltage is within a lower range, the switch may be set for the lower range to obtain a more accurate reading.

Step 3 Plug the black test lead into the − (COMMON) jack and the red test lead into the + jack.

Step 4 Connect the test leads across the voltage source (in parallel with the circuit).

Step 5 Turn on the power in the circuit being measured.

Step 6 For the 2.5V range, read the value directly on the AC scale marked 2.5V. For the 10V, 50V, and 250V ranges, read the red scale marked AC and use the black figures immediately above the scale. For the 500V range, read the red scale marked AC and use the 0-50 figures. Multiply the reading by 10.

6.3.6 Measuring AC Voltage, 0-1,000 Volts

WARNING!   Be extremely careful when working with higher voltages. Do not touch the instrument or test leads while power is on in the circuit being measured.

Step 1   Set the function switch to AC.

Step 2   Set the range switch to 1,000V (dual position with 500V).

Step 3   Plug the black test lead into the – (COMMON) jack and the red test lead into the 1,000V jack.

Step 4   Be sure the power is off in the circuit being measured and that all capacitors have been discharged. Connect the test leads to the circuit.

Step 5   Turn on the power in the circuit being measured.

Step 6   Read the voltage on the red scale marked AC. Use the 0-10 figures and multiply by 100.

6.3.7 Measuring Output Voltage

It is often desired to measure the AC component of an output voltage where both AC and DC voltage levels exist. This occurs primarily in amplifier circuits. The meter has a 0.1µF, 400V capacitor in series with the OUTPUT jack. The capacitor blocks the DC component of the current in the test circuit but allows the AC or desired component to pass on to the indicating instrument circuit. The blocking capacitor may alter the AC response at low frequencies but is usually ignored at audio frequencies.

CAUTION: When using OUTPUT, do not apply it to a circuit whose DC voltage component exceeds the 400V rating of the blocking capacitor.

Step 1   Set the function switch to AC.

Step 2   Plug the black test lead into the – (COMMON) jack and the red test lead into the OUTPUT jack.

Step 3   Set the range switch to one of the range positions marked 2.5V, 10V, 50V, or 250V.

Step 4   Connect the test leads across the circuit being measured, with the black test lead attached to the ground side.

Step 5   Turn on the power in the test circuit. Read the output voltage on the appropriate AC voltage scale. For the 2.5V range, read the value directly on the AC scale marked 2.5V. For the 10V, 50V, or 250V ranges, use the red scale marked AC and read the black figures immediately above the scale.

6.3.8 Measuring Decibels

For some applications, mockup audio frequency voltages are measured in terms of decibels. The decibel scale (dB) at the bottom of the dial is marked from −20 to +10.

Step 1 To measure decibels, read the dB scale in accordance with instructions for measuring AC. For example, when the range switch is set to the 2.5V position, the dB scale is read directly.

Step 2 The dB readings on the scale are referenced to a 0dB power level of .001W across 600Ω, or .775VAC across 600Ω.

Step 3 For the 10V range, read the dB scale and add +12dB to the reading. For the 50V range, read the dB scale and add +26dB to the reading. For the 250V range, read the dB scale and add +40dB to the reading.

Step 4 If the 0dB reference level is .006W across 500Ω, subtract +7dB from the reading.

6.4.0 DIRECT CURRENT MEASUREMENTS

6.4.1 Voltage Drop

The voltage drop across the meter on all milliampere current ranges is approximately 250mV measured at the jacks. An exception is the 0-500mA range with a drop of approximately 400mV. This voltage drop will not affect current measurements. In some transistor circuits, however, it may be necessary to compensate for the added voltage drop when making measurements.

6.4.2 Measuring Direct Current, 0-50 Microamperes

CAUTION: Never connect the test leads directly across voltage when the meter is used as a current-indicating instrument. Always connect the instrument in series with the load across the voltage source.

Step 1 Set the function switch to +DC.

Step 2 Plug the black test lead into the − (COMMON) jack and the red test lead into the +50μA/250mV jack.

Step 3 Set the range switch to 50μA (dual position with 50V).

Step 4 Open the circuit in which the current is being measured. Connect the instrument in series with the circuit. Connect the red test lead to the positive side and the black test lead to the negative side.

Step 5 Read the current on the black DC scale. Use the 0-50 figures to read directly in microamperes.

Note: In all direct current measurements, be certain the power to the circuit being tested has been turned off before disconnecting test leads and restoring circuit **continuity**.

6.4.3 Measuring Direct Current, 0-1 Through 0-500 Milliamperes

Step 1 Set the function switch to +DC.

Step 2 Plug the black test lead into the – (COMMON) jack and the red test lead into the + jack.

Step 3 Set the range switch to one of the four range positions (1mA, 10mA, 100mA, or 500mA).

Step 4 Open the circuit in which the current is being measured. Connect the VOM in series with the circuit. Connect the red test lead to the positive side and the black test lead to the negative side of the part of the circuit you are measuring.

Step 5 Read the current in milliamperes on the black DC scale. For the 1mA range, use the 0-10 figures and divide by 10. For the 10mA range, use the 0-10 figures and multiply by 10. For the 500mA range, use the 0-50 figures and multiply by 10.

6.4.4 Measuring Direct Current, 0-10 Amperes

Step 1 Plug the black test lead into the –10A jack and the red test lead into the +10A jack.

Step 2 Set the range switch to 10A (dual position with 10mA).

Step 3 Open the circuit in which the current is being measured. Connect the instrument in series with the circuit. Connect the red test lead to the positive side and the black test lead to the negative side.

Note: The function switch has no effect on polarity for the 10A range.

Step 4 Read the current on the black DC scale. Use the 0-10 figures to read directly in amperes.

CAUTION: When using the 10A range, never remove a test lead from its panel jack while current is flowing through the circuit. Otherwise, damage may occur to the plug and jack.

6.4.5 Zero Ohm Adjustment

When resistance is measured, the VOM batteries furnish power for the circuit. Since batteries are subject to variation in voltage and internal resistance, the instrument must be adjusted to zero prior to measuring a resistance.

Step 1 Set the range switch to the desired ohms range.

Step 2   Plug the black test lead into the – (COMMON) jack and the red test lead into the + jack.

Step 3   Connect the ends of the test leads to short the VOM resistance circuit.

Step 4   Rotate the ZERO OHM control until the pointer indicates zero ohms. If the pointer cannot be adjusted to zero, one or both of the batteries must be replaced.

Step 5   Disconnect the ends of the test leads and connect them to the component being measured.

6.4.6   Measuring Resistance

CAUTION: Before measuring resistance, be sure power is off to the circuit being tested. Disconnect the component from the circuit before measuring its resistance.

Step 1   Set the range switch to one of the resistance range positions:

- Use R x 1 for resistance readings from 0 to 200 ohms.
- Use R x 100 for resistance readings from 200 to 20,000 ohms.
- Use R x 10,000 for resistance readings above 20,000 ohms.

Step 2   Set the function switch to either the –DC or +DC position. The operation is the same in either position.

Step 3   Adjust the ZERO OHM control for each resistance range.

- Observe the reading on the OHMS scale at the top of the dial. Note that the OHMS scale reads from right to left for increasing values of resistance.
- To determine the actual resistance value, multiply the reading by the factor at the switch position (K on the OHMS scale equals one thousand).

Step 4   If there is a forward and backward resistance such as in diodes, the resistance should be relatively low in one direction (for forward polarity) and higher in the opposite direction.

CAUTION: Check that the OHMS range being used will not damage any of the semiconductors.

Step 5   If the purpose of the resistance measurement is to check a semiconductor in or out of a circuit (forward and reverse bias resistance measurements), check the following prior to making the measurement:

- The polarity of the voltage at the input jacks is identical to the input jack markings. Therefore, be certain that the polarity of the test leads is correct for the application.
- Ensure that the range selected will not damage the semiconductor (use R x 100 or below).

- Refer to the meter specifications and review the limits of the semiconductor according to the manufacturer's ratings.
  - If the semiconductor is a silicon diode or conventional silicon transistor, no precautions are normally required.
  - If the semiconductor material is germanium, check the ratings of the device and refer to its specifications.

Step 6   Rotate the function switch between the two DC positions to reverse polarity. This will determine if there is a difference between the resistance in the two directions.

Step 7   The resistance of such diodes will measure differently from one resistance range to another on the same VOM with the function switch in a given position. For example, a diode that measures 80Ω on the R x 1 range may measure 300Ω on the R x 100 range. The difference in values is a result of the diode characteristic and does not indicate any fault in the VOM.

7.0.0   DIGITAL METERS

Digital meters have revolutionized the test equipment world. Improved accuracy is very easily attainable, more functions can be incorporated into one meter, and both autoranging and automatic polarity indication can be used. Technically, digital multimeters are classified as electronic multimeters; however, digital multimeters do not use a meter movement. Instead, a digital meter's input circuit converts a current into a digital signal, which is then processed by electronic circuits and displayed numerically on the meter face.

A major limitation with many meters that use meter movements is that the scale reading must be estimated if the meter pointer falls between scale divisions. Digital multimeters eliminate the need to estimate these readings by displaying the reading as a numerical display.

With digital meters, technicians must revise the way the indications are viewed. For example, if a technician were reading the AC voltage on a normal wall outlet with an analog voltmeter, any indication within the range of 120VAC would be considered acceptable. But, when read with a digital meter, the technician might think something was wrong if the meter showed a reading of 114.53VAC. Bear in mind that the digital meter is very precise in its reading, sometimes more precise than is called for, or is usable. Also, be aware that the indicated parameter may change with the range used. This is primarily due to the change in accuracy and where the meter is rounding off.

There are many types of digital multimeters. Some are bench-type multimeters, while others are designed to be handheld. Most types of digital multimeters have an input impedance of 10 megohms and above. They are very sensitive to small changes in current and are therefore very accurate.

An example of a digital meter is shown in *Figure 16*. The internal operation of this meter is basically the same as other digital meters. The following paragraphs discuss the operation and use of this particular meter. For specific instructions, always refer to the owner's manual supplied with your meter.

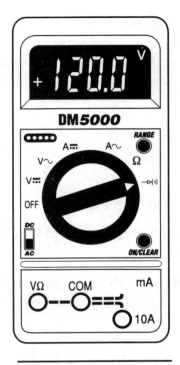

106F16.EPS

Figure 16. Digital Meter

7.1.0 FEATURES

The example meter offers the following features:

- *Autorange/manual range modes* – The meter features autoranging for all measurement ranges. Press the RANGE button to enter manual range mode. A flashing symbol may be used to show that you are in the manual range mode. Press the RANGE button as required to select the desired range. To switch back to auto range, press the ON/CLEAR button once (clear mode) or select another function.

- *Automatic off* – The example meter turns itself off after one hour of non-use. The current draw while the meter is turned off does not affect battery life. If the meter turns itself off while a parameter is being monitored, press the ON/CLEAR button to turn it on again. To protect against electrical damage, the meter also turns itself off if a test lead is inserted into the 10A jack while the meter is in any mode other than A --- or A ~.

- *Dangerous voltage indication* – The meter shows the symbol for any range over 20V. In the autoranging mode, the meter also beeps when it changes to any range over 20V.

- *Overload display* – The meter displays OL and a rapidly flashing decimal point (position determined by range) when the input is too large to display.

- *Audible acknowledgment* – The meter acknowledges each press of a button or actuation of the selector switch with a beep.

7.2.0 OPERATION

This section will discuss the use of various controls and explain how measurements should be taken.

7.2.1 Dual Function ON/CLEAR Button

Press the ON/CLEAR button to turn the meter on. Operation begins in the autorange mode, and the range for maximum resolution is selected automatically. Press the ON/CLEAR button again to turn the meter off.

7.2.2 Measuring Voltage

Step 1   Select V --- or V ~.

Step 2   Connect the test leads as shown in *Figure 17.*

Step 3   Observe the voltage reading on the display. Depending on the range, the meter displays units in mV or V.

To avoid shock hazard or meter damage, do not apply more than 1,500VDC or 1,000VAC to the meter input or between any input jack and earth ground.

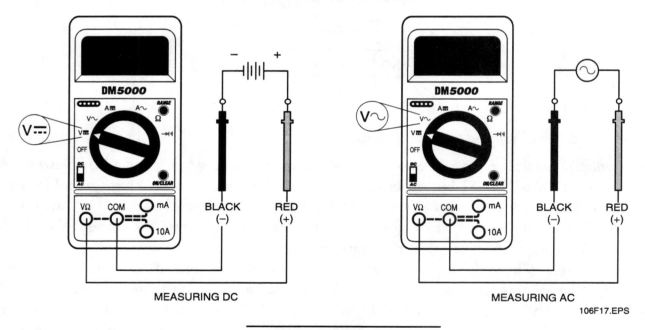

MEASURING DC

MEASURING AC

106F17.EPS

Figure 17. Measuring Voltage

7.2.3 Measuring Current

Step 1   Select A --- or A ~.

Step 2   Insert the meter in series with the circuit with the red lead connected to either:

- The mA jack for input up to 200 milliamps
- The 10A jack for input up to 20 amps

Step 3   Make hookups as shown in *Figures 18* and *19*.

Note: The meter shuts itself off if a test lead is inserted into the 10A jack when the meter is in any function other than A --- or A~.

Step 4   Observe the current reading.

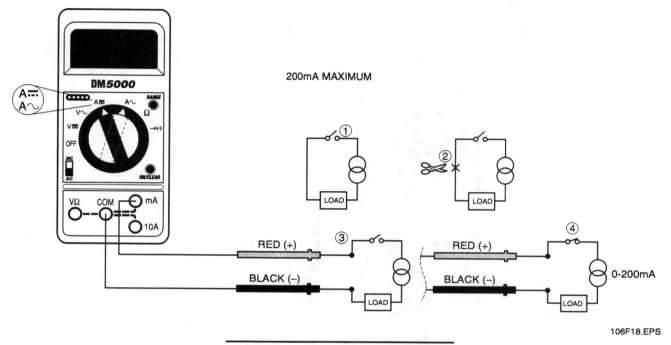

Figure 18. Measuring Current (mA)

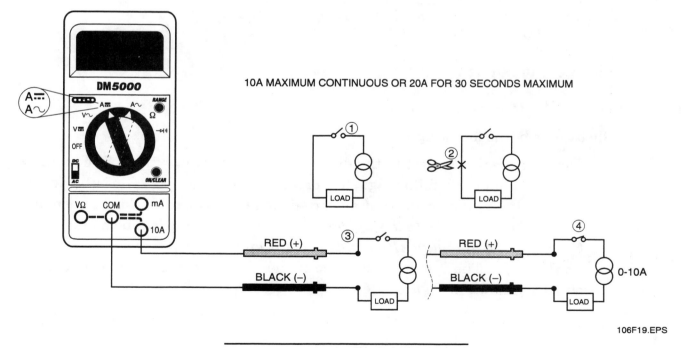

Figure 19. Measuring Current (Amps)

7.2.4 Measuring Resistance

When measuring resistance, any voltage present will cause an incorrect reading. For this reason, the capacitors in a circuit in which resistance measurements are about to be taken should first be discharged.

Step 1   Select Ω (ohms).

Step 2   Connect the test leads as shown in *Figure 20*.

Step 3   Observe the resistance reading.

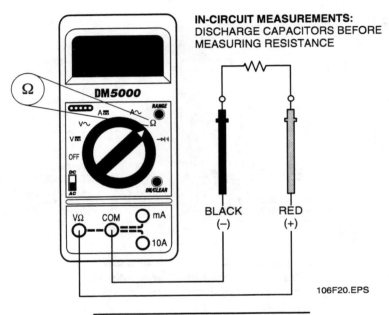

Figure 20. Measuring Resistance

7.2.5 Diode/Continuity Test

Step 1   Select ⊲⊢⋙ (diode/continuity).

Step 2   Choose one of the following:

- *Forward bias* – Connect to the diode as shown in *Figure 21(A)*. The meter will display one of the following: the forward voltage drop (V_F) of a good diode (<0.7V), a very low reading for a shorted diode (<0.3V), or OL for an open diode.
- *Reverse bias or open circuit* – Reverse the leads to the diode. The meter displays OL, as shown in *Figure 21(B)*. It does not beep.
- *Continuity* – The meter beeps once if the circuit resistance is less than 150 ohms, as shown in *Figure 21(C)*.

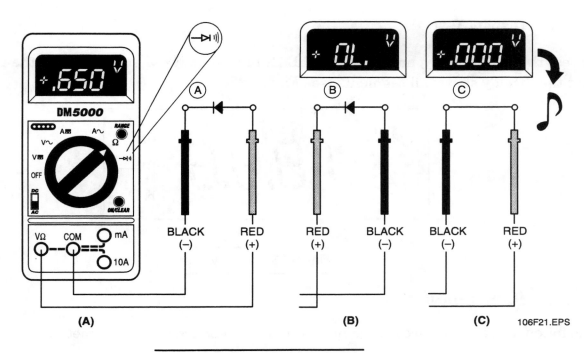

Figure 21. Diode/Continuity Test

7.2.6 Transistor Junction Test

Test transistors in the same manner as diodes by checking the two diode junctions formed between the base and emitter, and the base and collector of the transistor. *Figure 22* shows the orientation of these effective diode junctions for PNP and NPN transistors. Also check between the collector and emitter to determine if a short is present.

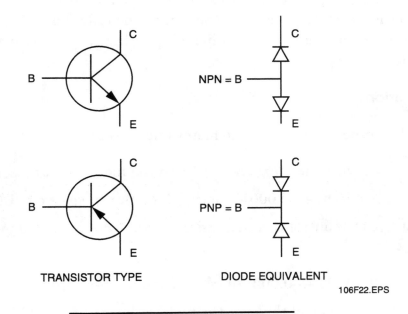

TRANSISTOR TYPE   DIODE EQUIVALENT

106F22.EPS

Figure 22. Transistor Junction Test

7.2.7 Display Test

To test the LCD display, hold the ON/CLEAR button down when turning on the meter. Verify that the display shows all segments (see *Figure 23*).

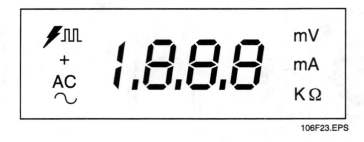

106F23.EPS

Figure 23. Display Test

7.3.0 MAINTENANCE

The following sections discuss the necessary maintenance for a multimeter.

7.3.1 Battery Replacement

Replace the battery as soon as the meter's decimal point starts blinking during normal use; this indicates that <100 hours of battery life remain. Remove the case back and replace the battery with the same or equivalent 9V alkaline battery.

7.3.2 Fuse Replacement

The meter uses two input protection fuses for the mA and 10A inputs. Remove the case back to gain access to the fuses. Replace with the same type only. The large fuse should be readily available. A spare for the smaller fuse is included in the case. If necessary, this fuse must be reordered from the factory.

7.3.3 Calibration

Have a qualified technician calibrate the meter once a year.

Step 1   Remove the case back (*Figure 24*). Turn the meter ON and select V ---.

Step 2   Apply +1.900VDC +/–0.0001V to the V-n input (negative to COM).

Step 3   Adjust the DC control through the hole in the circuit board for a display of +1.900V.

Step 4   Select V.

Step 5   Apply +1.900VAC +/–0.002VAC @ 60Hz to the V-n input.

Step 6   Adjust the AC control through the hole in the circuit board for a display of +1.900V.

Step 7   Reassemble the meter.

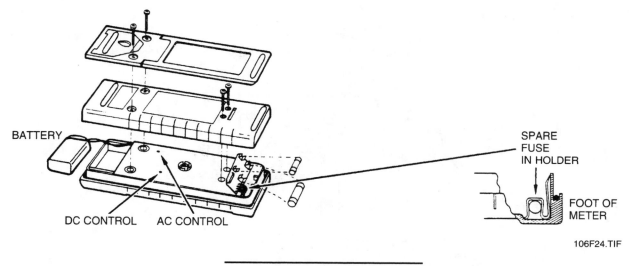

BATTERY

SPARE
FUSE
IN HOLDER

FOOT OF
METER

DC CONTROL   AC CONTROL

106F24.TIF

Figure 24. Meter Maintenance

8.0.0   WATTMETER

Rather than performing two measurements and then calculating power, a power-measuring meter called a *wattmeter* can be connected into a circuit to measure power. The power can be read directly from the scale of this meter. Not only does a wattmeter simplify power measurements, but it has two other advantages:

First, voltage and current in an AC circuit are not always in phase; current sometimes either leads or lags the voltage (this is known as the *power factor*). When this happens, multiplying the voltage times the current results in apparent power, not true power. Therefore, in an AC circuit, measuring the voltage and current and then multiplying them can often result in an incorrect value of power dissipation by the circuit. However, the wattmeter takes the power factor into account and always indicates true power.

Second, voltmeters and ammeters consume power. The amount consumed depends on the levels of the voltage and the current in the circuit, and it cannot be accurately predicted. Therefore, very accurate power measurements cannot be made by measuring voltage and current and then calculating power. However, some wattmeters compensate for their own power losses so that only the power dissipated in the circuit is measured. If the wattmeter is not compensated, the power that is dissipated is sometimes marked on the meter, or it can easily be determined so that a very accurate measurement can be made. Typically, the accuracy of a wattmeter is within one percent.

The basic wattmeter consists of two stationary coils connected in series and one movable coil (*Figure 25*). The moving coil, wound with many turns of fine wire, has a high resistance.

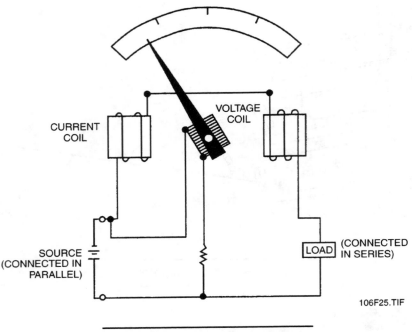

Figure 25. Wattmeter Schematic

The stationary coils, wound with a few turns of a larger wire, have a low resistance. The interaction of the magnetic fields around the different coils will cause the movable coil and its pointer to rotate in proportion to the voltage across the load and the current through the load. Thus, the meter indicates E times I, or power.

The two circuits in the wattmeter will be damaged if too much current passes through them. This fact is of special importance because the reading on the meter does not tell the user that the coils are being overheated. If an ammeter or voltmeter is overloaded, the pointer will indicate beyond full-scale deflection. In a wattmeter, both the current and potential (voltage) circuits may be carrying such an overload that their insulation is burning, and yet the pointer may only be partway up the scale. A low-power factor circuit will give a low reading on the wattmeter even when the current and potential circuits are loaded to their maximum safe limits.

9.0.0 MEGOHMMETER (MEGGER)

An ordinary ohmmeter cannot be used for measuring resistances of several million ohms, such as those found in conductor insulation or between motor or transformer windings, and so on. To adequately test these types of very high resistances, it is necessary to use a much higher potential than is furnished by the battery of an ohmmeter. For this purpose, a megger is used. There are three types of meggers: hand, battery, and electric.

The megger is similar to a moving-coil meter except that it has two windings (coils). See *Figure 26*. Coil A is in series with resistor R_2 across the output of the generator.

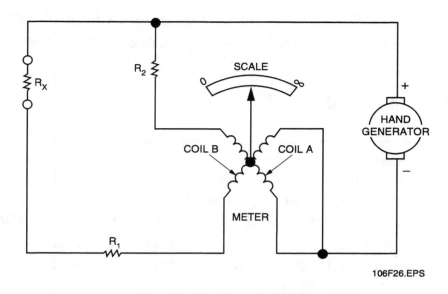

Figure 26. Megger Schematic

This coil is wound so that it causes the pointer to move toward the high-resistance end of the scale when the generator is in operation. Winding B is in series with R_1 and R_X (the unknown resistance to be measured). This winding is wound so that it causes the pointer to move toward the low or zero-resistance end of the scale when the generator is in operation.

When an extremely high resistance appears across the input terminals of the megger, the current through coil A causes the pointer to read infinity. Conversely, when a relatively low resistance appears across the input terminals, the current through coil B causes the pointer to deflect toward zero. The pointer stops at a point on the scale determined by the current through coil B, which is controlled by R_X.

Digital meggers *(Figure 27)* use the same operational principles. Instead of having a scaled meter movement, these meters give the value of resistance in a digital readout display. The digital readout makes reading the measurement much easier and helps to eliminate errors.

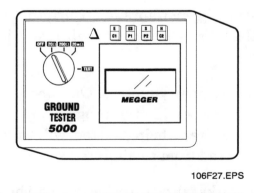

Figure 27. Digital Readout Megger

To avoid excessive test voltages, most hand meggers are equipped with friction clutches. When the generator is cranked faster than its rated speed, the clutch slips, and the generator speed and output voltage are maintained at their rated values.

9.1.0 SAFETY PRECAUTIONS

WARNING!   When a megger is used, the generator voltage is present on the test leads. This voltage could be hazardous to you or the equipment you are testing. *NEVER TOUCH THE TEST LEADS WHILE THE TESTER IS BEING USED.* Isolate the item you are testing from the circuit before using the megger. Protect all parts of the test subject from contact by others.

When using a megger, you could be injured or cause damage to the equipment being worked on if the following minimum safety precautions are not observed:

- Use meggers on high-resistance measurements *only* (such as insulation measurements or to check two separate conductors in a cable).
- *Never* touch the test leads while the handle is being cranked.
- Deenergize and verify the deenergization of the circuit completely before connecting the meter.
- Disconnect the item being checked from other circuitry, if possible, *before* using the meter.
- After the test, ground the tested circuit to discharge any energy that may be left in the circuit.

10.0.0 FREQUENCY METER

Frequency is the number of cycles completed each second by a given AC voltage, and it is usually expressed in hertz (one hertz = one cycle per second). The frequency meter is used in AC power-producing devices such as generators to ensure that the correct frequency is being produced. Failure to produce the correct frequency will result in excess heat and component damage.

There are two common types of frequency meters. One operates with a set of reeds having natural vibration frequencies that respond in the range being tested. The reed with a natural frequency closest to that of the current being tested will vibrate most strongly when the meter operates. The frequency is read from a calibrated scale.

A moving-disk frequency meter works with two coils, one of which is a magnetizing coil whose current varies inversely with the frequency. A disk with a pointer mounted between the coils turns in the direction determined by the stronger coil. Solid-state frequency meters are also available.

10.1.0 PHASE ROTATION TESTER

A phase rotation tester is an indicator that allows you to see the phase rotation of incoming current.

10.2.0 INFRARED SENSING DEVICE

An infrared sensing device is an optical device that measures the infrared heat emitted from an object.

10.3.0 CABLE-LENGTH METER

A cable-length meter measures the length and condition of a cable by sending a signal down the cable and then reading the signal that is reflected back

10.4.0 HARMONIC TEST SET

A harmonic is a sine wave whose frequency is a whole number multiple of the original base frequency. For example, a standard 60Hz sine wave may have second and third harmonics at 120Hz and 180Hz, respectively. Harmonics may be caused by various circuit loads such as fluorescent lights and by certain three-phase transformer connections.

The results of harmonics are heating in wiring and a voltage or current that cannot be detected by most digital meters.

11.0.0 POWER FACTOR METER

The power factor is the ratio of true (actual) power to apparent power. The power factor of a circuit or piece of equipment may be found by using an ammeter, wattmeter, and voltmeter. To calculate the power factor, the wattmeter reading (true power) is divided by the product of the ammeter and voltmeter readings (apparent power or EI). The ideal power factor is one.

$$\text{Power Factor} = \frac{\text{True Power (Wattmeter Reading)}}{\text{Apparent Power (EI)}}$$

It is not necessary to calculate these readings if a power factor meter is available to read the power factor directly (*Figure 28*). This meter indicates the equivalent of pure resistance or unit power factor, which is a one-to-one ratio.

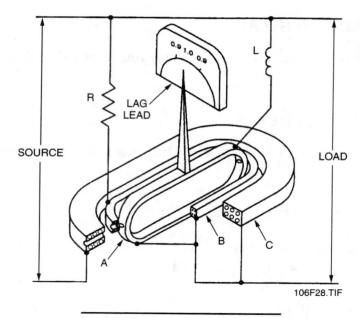

Figure 28. Power Factor Meter

CONTINUITY TESTER

12.0.0 CONTINUITY TESTER

An ohmmeter can be used to test continuity, but this means carrying an expensive, often bulky test instrument with you. Pocket-type continuity testers are just as reliable and much more compact and portable.

There are two types of continuity testers: audio and visual. These are specialized devices for identifying conductors in a conduit by checking continuity.

12.1.0 AUDIO CONTINUITY TESTER

This type of tester is used to *ring out* wires in conduit runs. At one end of the conduit run, select one wire, strip off a little insulation, and connect that wire to the conduit. At the other end of the conduit run, clip one lead of the tester to the conduit. Touch the other lead to one wire at a time until the audible alarm sounds, which indicates continuity (a closed circuit). Then, put matching tags on the wire. Continue this procedure, one wire at a time, to identify the other wires.

12.2.0 VISUAL CONTINUITY TESTER

This type of tester is useful if you are working in an area where background noise might make it hard to hear the audio tester. The procedure is the same. When the proper wire is tested, the light will come on.

13.0.0 VOLTAGE TESTER

A voltage tester is a simple aid that determines whether there is a potential difference between two points. It *does not* calculate the value of that difference. If the actual value of the potential difference is needed, use a voltmeter.

13.1.0 WIGGY®

Pocket-type voltage testers are inexpensive and portable. They can easily be carried in your tool pouch, eliminating the need to carry a delicate voltmeter on the job. Simple neon testers are becoming quite popular. Another type is known as a *Wiggy®* (see *Figure 29*).

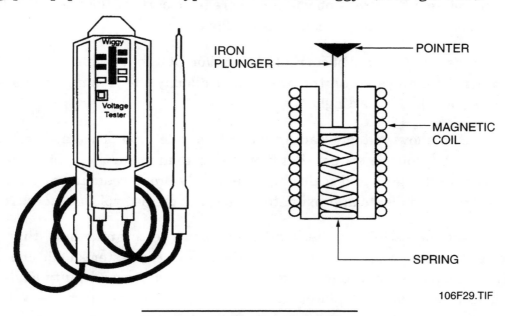

106F29.TIF

Figure 29. Wiggy® Voltage Tester

13.1.1 Principle Of Operation

The operation of a Wiggy® is fairly simple. The basic component is a solenoid. When a current flows through the coil, it will produce a magnetic field which will pull the plunger down against a spring. The spring will limit how far the plunger can be drawn into the cylinder. As the current increases, the plunger will move farther. The amount of current depends upon the potential difference applied across the coil. A pointer on the plunger indicates the potential difference on the scale.

CAUTION: This is an approximation only. If the actual value of the voltage needs to be known, use a voltmeter.

The scale on the tester has voltage indications for AC on one side of the pointer and DC on the other side. The AC scale indicates 120, 240, 480, and 600 volts. The DC scale indicates 125, 240, and 600 volts.

CAUTION: To avoid damage to the tester, do not use a Wiggy® above its stated range.

To use a voltage tester, the probes are placed across a possible source of voltage. If there is voltage present, current flows through the coil inside the tester, creating a magnetic field. The magnetized coil pulls the indicator along the scale until it reaches a point corresponding to the approximate voltage. If there is no voltage, there is no current flow, so the indicator on the scale does not move, and no voltage reading is displayed.

The range of voltage and the type of current (AC and/or DC) that a voltage tester is capable of measuring are usually indicated on the scales that display the reading. The scales for each type of current are marked accordingly.

The methods used to show voltage readings vary from one type of voltage tester to another. For example, there are voltage testers that have lights to indicate the approximate amount of voltage registered. Both lights and scales have relatively broad readout ranges because these voltage testers can only indicate approximate values, not precise voltage measurements.

A voltage tester should always be checked before each use to make sure that it is in good condition and is operating correctly. The external check of the tester should include a careful inspection of the insulation on the leads for cracks or frayed areas. Faulty leads constitute a safety hazard, so they must be replaced. As a check to make sure that the voltage tester is operating correctly, the probes of the tester are connected to a power source that is known to be energized. The voltage indicated on the tester should match the voltage of the power source. If there is no indication, the voltage tester is not operating correctly, and it must be repaired or replaced. It must also be repaired or replaced if it indicates a voltage different from the known voltage of the source.

It is essential to check a voltage tester before use. A faulty voltage tester can be dangerous to the electrician using it and to other personnel. For example, damaged insulation or a cracked casing could expose the electrician to electrical shock. Also, a faulty voltage tester might indicate that power is off when it is really on. This would create a serious safety hazard for personnel involved in equipment repair. The face plate or the scale on the front of the tester should be checked before a voltage tester is used to be sure that the tester can handle the amount of voltage that the power source may contain. Care should be taken when placing the probes of the tester across the power source. A voltage tester is designed to take a quick reading. If the probes are left in the circuit too long, the tester will burn out. A voltage tester should never be connected for more than a few seconds at a time.

CAUTION: Do not use a voltage tester in hazardous locations.

Voltage testers are used to make sure that power is available when it is needed and to make sure that power has been cut off when it should have been. In a troubleshooting situation, it might be necessary to verify that power is available in order to be sure that lack of power is not the problem. For example, if there were a problem with a power tool, such as a drill, a voltage tester might be used to make sure that power is available to run the drill. A voltage tester might also be used to verify that there is power available to a three-phase motor that will not start.

For safety reasons, it is always necessary to make sure that the power is turned off before working on any electrical equipment. A voltage tester can be used for such a test.

Keep the following in mind when using a voltage tester:

- Check the tester before each use.
- Handle the tester as if it were a calibrated instrument.
- Use good safety practices when operating the tester.
- Do not use circuits expected to be above the scale on the tester.
- Do not use if damage is indicated to the tester.
- Do not use in classified, hazardous areas or in high-frequency circuits.

14.0.0   RECORDING INSTRUMENTS

The term *recording instrument* describes many instruments that make a permanent record of measured quantities over a period of time. Recording instruments can be divided into three general groups:

- Instruments that record electrical quantities, including potential difference, current, power, resistance, and frequency.
- Instruments that record nonelectrical quantities by electrical means (such as a temperature recorder that uses a potentiometer system to record thermocouple output).
- Instruments that record nonelectrical quantities by mechanical means (such as a temperature recorder that uses a bimetallic element to move a pen across an advancing strip of paper).

It is often necessary to know the conditions that exist in an electrical circuit over a period of time to determine such things as peak loads, voltage fluctuations, etc. It may be neither practical nor economical to assign a worker to watch an indicating instrument and record its readings. An automatic recording instrument can be connected to take continuous readings, and the record can be collected for review and analysis.

Recording instruments are basically the same as the indicating meters already covered, but they have recording mechanisms attached to them. They are generally made of the same parts, use the same electrical mechanisms, and are connected in the same way. The only basic difference is the permanent record.

14.1.0   STRIP-CHART RECORDERS

Strip-chart recorders are the most widely-used recording instruments for electrical measurement. Their name comes from the fact that the record is made on a strip of paper, usually four to six inches wide and perhaps up to 60 feet long. These can be used to record either voltage or current. A recording voltmeter is shown in *Figure 30*.

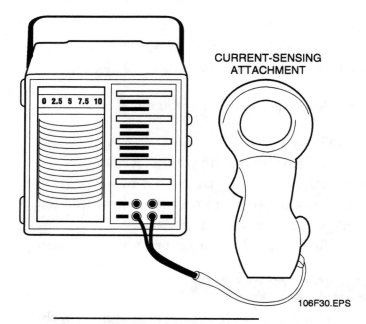

Figure 30. Recording Voltmeter

Strip-chart recorders offer several advantages in electrical measurement. The long charts allow the recording to cover a considerable length of time with little attention, and strip-chart recorders can be operated at a relatively high speed to provide very detailed records.

14.1.1   Typical Strip-Chart Recorder

A typical strip-chart recorder includes the following parts:

- The frame, which supports the other parts
- The moving system, which consists of the parts that move as a direct result of variations in the quantity being measured
- The graphic record, which is the line traced on the chart
- The chart carriage, which consists of the clock mechanism, timing gears, timing drum, chart spool, rewinding mechanism, and writing table
- A scale on the paper strip, which makes it possible to read the values of the quantity being measured
- The inking system, which consists of a special pen with an ink reservoir
- A case, which provides convenience in installation and removal, as well as in replacement of the chart paper

The moving parts used in recording instruments are basically the same as those used in the indicating instrument used to measure the same electrical quantity. However, the parts of a recording instrument are larger and require more power because of the added friction of the pen on the chart paper.

The paper chart is graduated in two directions. In one direction, the graduations correspond to the indicating scale of the instrument. In the other, they indicate time (seconds, minutes, hours, or days depending on the clock mechanism). The time graduations are uniformly spaced because the chart moves at a constant speed.

The chart carriage includes the entire mechanism for handling the chart paper. At the top, it holds the roll of paper. Just above the roll, and directly beneath the pen, is the writing table. The table is a metal plate that provides a flat surface under the chart paper at all points on the arc of the pen's motion. This prevents the pen from puncturing or tearing the paper.

The heart of the chart carriage is the clock mechanism, which drives the timing drum. The mechanism may be driven by an electric clock. Some of these mechanisms also contain a motor-wound spring clock or a battery-powered clock for a backup in the event of a loss of power.

The timing drum is connected to the clock through a gear unit, which determines the speed of the drum. Drive pins around the timing drum fit into holes at the edges of the chart paper.

The inking system usually includes an inkwell containing enough ink for a considerable time.

15.0.0 SAFETY

In the interest of safety, all test equipment should be inspected and tested before being taken to the job site. A thorough visual inspection, checking for broken meters or knobs, damaged plugs, or frayed cords is important.

Perform an operational check. For example:

- On an ohmmeter, short the probes and ensure that you can zero the meter.
- With an oscilloscope, make sure that you can obtain a trace, and if a test signal is available, connect a test probe and check it.
- A voltmeter can be checked against an AC wall receptacle or a battery.

If a meter has a calibration sticker, check to see if it has been calibrated recently. For precise measurements, a recently calibrated meter is a more reliable instrument.

Every person who works with electronic equipment should be constantly alert to the hazards to which personnel may be exposed, and should also be capable of rendering first aid. The hazards considered in this section are: electric shock, burns, and related hazards.

Safety must be the primary responsibility of all personnel. The installation, maintenance, and operation of electrical equipment enforces a stern safety code. Carelessness on the part of the technician or operator can result in serious injury or death due to electrical shock, falls, burns, flying objects, etc. After an accident has occurred, investigation almost invariably shows that it could have been prevented by the exercise of simple safety precautions and procedures. Each person concerned with electrical equipment is responsible for reading and becoming thoroughly familiar with the safety practices and procedures contained in all safety codes and equipment technical manuals *before* performing work on electrical equipment. It is your personal responsibility to identify and eliminate unsafe conditions and unsafe acts which cause accidents.

You must bear in mind that deenergizing main supply circuits by opening supply switches will not necessarily deenergize all circuits in a given piece of equipment. A source of danger that has often been neglected or ignored, sometimes with tragic results, is the input to electrical equipment from other sources, such as backfeeds. Moreover, the rescue of a victim shocked by the power input from a backfeed is often hampered because of the time required to determine the source of power and isolate it. Therefore, turn off *all* power inputs before working on equipment and tag and lockout, then check with an operating tester (e.g., wattmeter) to be sure that the equipment is safe to work on. Take the time to be safe when working on electrical circuits and equipment. Carefully study the schematics and wiring diagrams of the entire system, noting what circuits must be deenergized in addition to the main power supply. Remember, electrical equipment commonly has more than one source of power. Be certain that all power sources are deenergized before servicing the equipment. Do not service any equipment with the power on unless absolutely necessary. Remember that the 115V power supply voltage is not a low, relatively harmless voltage but is the voltage that has caused more deaths than any other medium.

Safety can never be stressed enough. There are times when your life literally depends on it. The following is a listing of common-sense safety precautions that must be observed at all times:

- Use only one hand when turning power switches on or off. Keep the doors to switch and fuse boxes closed except when working inside or replacing fuses. Use a fuse puller to remove cartridge fuses, after first making certain that the circuit is dead.
- Your company will make the determination as to whether or not you are qualified to work on an electrical circuit.
- Do not work with energized equipment by yourself; have another person (safety observer), qualified in first aid for electrical shock, present at all times. The person stationed nearby should also know which circuits and switches control the equipment, and that person should be given instructions to pull the switch immediately if anything unforeseen happens.

- Always be aware of the nearness of high-voltage lines or circuits. Use rubber gloves where applicable and stand on *approved* rubber matting. Not all rubber mats are good insulators.
- Inform those in charge of operations as to the circuit on which work is being performed.
- Keep clothing, hands, and feet dry. When it is necessary to work in wet or damp locations, use a dry platform and place a rubber mat or other nonconductive material on top of the wood. Use insulated tools and insulated flashlights of the molded type when required to work on exposed parts.
- Do *not* work on energized circuits unless absolutely necessary.
- All power supply switches or cutout switches from which power could possibly be fed must be secured in the OPEN (safety) position and locked and tagged.
- Never short out, tamper with, or block open an interlock switch.
- Keep clear of exposed equipment; when it is absolutely necessary to work on it, use only one hand as much as possible.
- Avoid reaching into enclosures except when absolutely necessary. When reaching into an enclosure, use rubber blankets to prevent accidental contact with the enclosure.
- Do not use bare hands to remove hot vacuum tubes from their sockets. Wear protective gloves or use a tube puller.
- Use a shorting stick to discharge all high-voltage capacitors.
- Make certain that the equipment is properly grounded. Ground all test equipment to the equipment under test.
- Turn off the power before connecting alligator clips to any circuit.
- When measuring circuits over 300V, do not hold the insulated test prods with bare hands.

15.1.0   USE OF HIGH-VOLTAGE PROTECTION EQUIPMENT

Anyone working on or near energized circuitry must use special equipment to provide protection from electrical shock. Protective equipment includes gloves, leather sleeves, rubber blankets, and rubber mats. It should be noted that this electrical protective equipment is in addition to the regular protective equipment normally required for maintenance work. Regular protective equipment typically includes hard hats which are rated for electrical resistance, eye protection, safety shoes, and long sleeves.

Gloves that are approved for protection from electrical shock are made of rubber. A separate leather cover protects the rubber from punctures or other damage. They protect the worker by insulating the hands from electrical shock. Gloves are rated as providing protection from certain amounts of voltage. Whenever an individual is going to be working around exposed conductors, the gloves chosen should be rated for at least as much voltage as the conductors are carrying.

Rubber sleeves are used along with gloves to provide additional protection. The combination of sleeves and gloves protects the hands and arms from electrical shock.

Rubber blankets and floor mats have many uses. Blankets are used to cover energized conductors while work is going on around them. They might be used to cover the energized main buses in a breaker panel before working on a deenergized breaker. Rubber floor mats are used to insulate workers from the ground. If a worker is standing on a rubber mat and then contacts an energized conductor, the current cannot flow through the body to the ground, so the worker will not get shocked.

15.2.0   GENERAL TESTING OF ELECTRICAL EQUIPMENT

As stated earlier, personnel must be familiar with the proper use of available test equipment. Also, personnel must be familiar with and use the local instructions governing the testing of electrical circuits to protect both the person and the equipment.

The next section includes an example of instructions from a local utility and covers work permits, test permits, authorized employees for testing, circuit isolation, and relay, instrument, and meter testing. Each topic is discussed individually.

15.2.1   Example Policy – Work Permits Or Test Permits

A written work permit must be obtained by an authorized technician or supervisor from the operator in charge before testing equipment under the operator's control. A test permit is required to make a dielectric proof test.

The following must be on the work or test permit:

- Designation of the equipment to be tested
- Scope and limits of tests to be made
- Method of isolation and protection provided

If it is necessary to alter the nature or to extend the scope of tests for which a work permit or test permit has been granted, return the permit to the operator and request a new permit before proceeding with the revised tests. When the test is completed, return the work permit or test permit to the operator, and report whether the results of the test were satisfactory or unsatisfactory.

Whenever a work permit or test permit remains in effect beyond the working hours of the person to whom it was issued and a different person will be in charge of the tests, consult the operator for the proper turnover procedure. The person assuming the responsibility for continuance of the work must verify that the items on the work or test permit are correct.

If a work permit or test permit is issued for testing that is to continue over a period of several days, the permit must be returned to the operator at the end of each working day and picked up at the beginning of the next working day.

15.2.2 Example Policy – Authorized Employees For Testing

Only employees authorized to perform operations at a given station shall:

- Open, remove, close, or replace doors or covers on electrical compartments.
- Remove or replace potential transformer fuses.
- Open or close switches.
- Open or close fuse cutouts or disconnecting potheads.
- Make tests for the presence of potential.
- Place or remove blocks and protective tags.
- Apply or remove grounds or short circuits.
- Attach or remove leads to high- or intermediate-voltage conductors.
- Make operating tests.
- Test links, fuses, switches associated with generators, buses, feeders, transformers, etc. normally shall be removed, replaced, or operated only under the direct supervision of an authorized employee when the associated equipment is live or is available for normal operation.

In some cases, when it is necessary for the equipment to remain in service, or in any case when the associated equipment has been removed from service, the operator-in-charge may authorize the person to whom the work permit was issued to remove and replace test links and fuses and to operate the heel and toe switches and potential switches in low-voltage circuits.

15.2.3 Example Policy – Circuit Isolation

All equipment to be tested which is rated at 350V phase-to-ground or higher must be isolated from all sources of supply by either two breaks, or a single break and a ground, unless test equipment is used which can be operated safely with one break. Where possible, two breaks in series are preferred over one break.

If no potential is to be applied to the equipment, the ground may be applied on the equipment side of the single break.

If a test potential is to be applied to the equipment, the ground has to be on the far side of the single break. If a test potential of less than 400V is to be applied to the equipment, the ground may be replaced with an approved protective discharge gap on the equipment to be tested. The protective gaps and test leads are to be connected or disconnected while the equipment is grounded. If disconnect switches are used to provide the breaks, the switches shall be locked open, and the operating motor fuses removed.

On equipment operating at 350V or less, a test potential may be applied with only a single break. An open truck-type or elevating-type circuit breaker withdrawn or lowered to the disconnect position or a swinging-type disconnect switch shall be considered as a single break.

If any of the following are the only type of disconnecting device available, short circuits and grounds must be applied at the nearest available point to the device.

- A disconnect switch with the blade prevented from falling closed by an insulating block placed over the jaw
- A disconnect switch in which the separation of the blade and jaw cannot be verified visually
- An oil circuit breaker or oil switch having no associated disconnecting device
- An oil-filled fusible cutout having no provision for disconnecting the leads

Potential transformers that may become energized from the low-voltage side shall be isolated on that side by opening switches, removing links, removing fuses, or by removing connections temporarily. The low-voltage side of the potential transformer shall also be short circuited and grounded.

Equipment which has been subjected to DC, high-voltage tests shall be grounded to discharge the residual voltage before any person is permitted to touch the test cables or any current-carrying part of the equipment tested. If the equipment tested has considerable capacitance, then it should be grounded for at least 30 minutes after the test potential is removed. Equipment of low capacitance should be grounded for a length of time equivalent to the time that the potential was applied.

15.2.4   Example Policy – Clearances

Adequate clearances are to be maintained between energized and exposed conductors and personnel (refer to **NEC Section 110-26**). Where DC voltages are involved, clearances specified shall be used with specified voltages considered as DC line-to-ground values.

If adequate clearances cannot be maintained from exposed live parts of apparatus in the normal course of free movement within the area during test, then access to that area shall be restricted by fences and barricades. Signs clearly indicating the hazard shall be posted in conspicuous locations. This requirement applies to equipment in service as well as to equipment to which test voltages are applied.

Whenever there is any question of the adequacy of clearance between the specific area in which work is to be done and exposed live parts of adjacent equipment, a field inspection shall be made by management representatives of the group involved before starting the job. The result of this inspection should be to outline the protection necessary to complete the work safely, including watchers where needed.

15.2.5   Electrical Circuit Tests

Testing of electrical circuits can be required at new installations or existing installations where wiring has been modified or new wiring has been interfaced with existing circuits. It is broken down into two sections:

- Electrical circuit tests performed on AC relay protection and metering circuits include:
  - Point-to-point wire checks to verify that the wiring has been installed according to the diagram of connections
  - Current transformer polarity, impedance, continuity, and where applicable, tap progression and intercore coupling tests
  - Voltage transformer polarity, ratio, and self-induced high-potential (hi-pot) tests
  - Insulation resistance tests
- Electrical circuit tests performed on control, power, alarm, and annunciator circuits include:
  - Insulation resistance tests
  - Operation tests

An important piece of information is the minimum distance allowed when working near energized electrical circuits because large voltages can arc across an air gap. Personnel must maintain a distance that is greater than that arc distance. This is especially true when using a *hot stick* to open a disconnect. These distances are listed in *Table 1.*

Note: These clearances apply to qualified personnel.

Voltage Range (Phase-to-Phase) Kilovolts	Minimum Working and Clear Hot stick Distance
2.1 to 15	2 ft. 0 in.
15.2 to 35	2 ft. 4 in.
35.1 to 46	2 ft. 6 in.
46.1 to 72.5	3 ft. 0 in.
72.6 to 121	3 ft. 4 in.
138 to 145	3 ft. 6 in.
161 to 169	3 ft. 8 in.
230 to 242	5 ft. 0 in.
345 to 362	*7 ft. 0 in.
500 to 552	*11 ft. 0 in.
700 to 765	*15 ft. 0 in.
* For voltages above 345kV, the minimum working and clear hot stick distances may be reduced provided that such distances are not less than the shortest distance between the energized part and a grounded surface.	

Table 1. OSHA Table

15.3.0 METER LOADING EFFECTS

It is important to understand how a piece of test equipment will affect the circuit to which it is attached. Even when properly used, a piece of test equipment can put a load on the circuit under testing.

An ammeter becomes a part of the circuit when it is used. Because of this, the meter can sometimes significantly alter the voltages and currents in the circuit. The following example illustrates how this can happen.

In Circuit A on the left in *Figure 31*, the current is calculated to be 10µA. To measure this current using a meter which has an internal resistance of 2,000Ω, the meter would be placed as shown in Circuit B. The 2,000Ω resistance of the meter adds to the 10,000Ω resistance, reducing the current in the circuit to 8.33µA. The presence of the meter, therefore, alters the current flowing in the circuit, resulting in an incorrect reading.

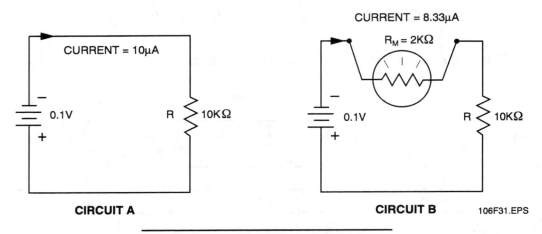

Figure 31. Meter Loading – Example 1

A voltmeter can also alter circuit parameters. The following example illustrates how this can happen.

In Circuit A in *Figure 32*, the voltage across the 40,000Ω resistor is to be measured using a voltmeter with a 1,000Ω internal resistance. We can determine the effect that the meter will have on the circuit when it is connected as shown in Circuit B.

Connecting the voltmeter in parallel with the 40,000Ω resistor, as shown in Circuit B, significantly changes the total resistance in the circuit. The parallel combination of 40,000Ω and 200,000Ω is:

$$R = \frac{40,000 \times 200,000}{40,000 + 200,000} = 33,333\Omega$$

Add the 10KΩ resistance in series for a total circuit resistance of 43,333Ω.

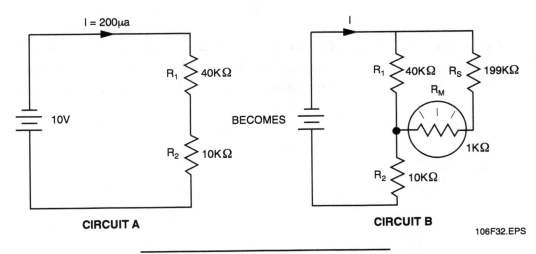

CIRCUIT A   BECOMES   **CIRCUIT B**

106F32.EPS

Figure 32. Meter Loading – Example 2

Therefore, with the voltmeter in the circuit, the current is:

$E \div R = I$ or $10 \div 43{,}333 = 2.31 \times 10^{-4}A$

This current causes a voltage drop across the combination of the meter and the $40{,}000\Omega$ resistor of:

$E = IR$ or $(2.31 \times 10^{-4}) \times 33{,}333 = 7.7V$

The voltmeter will read 7.7V, whereas the voltage drop across the $40{,}000\Omega$ resistor without the voltmeter connected is 8V.

In most cases, the current flowing through the voltmeter movement is negligible compared to the current flowing through the element whose voltage is being measured. When this is the case, the voltmeter has a negligible effect on the circuit.

In most commercial voltmeters, the internal resistance is expressed as the ohms-per-volt rating of the meter. This quantity tells what the internal resistance of the meter is on any particular full-scale setting. In general, the meter's internal resistance is the ohms-per-volt rating times the full-scale voltage. The higher the ohms-per-volt rating, the higher the internal resistance of the meter, and the smaller the effect of the meter on the circuit.

Care should be taken when selecting either an ammeter or a voltmeter in which the meter does not load down the circuit under test. Also, realize that the same circuit under test with a different meter will indicate differently due to the separate loading effects of each meter.

SUMMARY

The use of test equipment is an important part of your job as an electrician. Selecting the proper instrument to be used in a specific application will help you to fully perform your task. You will use the information provided by these instruments to help evaluate the work being accomplished.

References

For advanced study of topics covered in this task module, the following books are suggested:

Electronics Fundamentals, Latest Edition, Prentice Hall, New York, NY.
Principles of Electric Circuits, Latest Edition, Prentice Hall, New York, NY.

REVIEW QUESTIONS

1. An ammeter is used to measure _____.
   a. current
   b. voltage
   c. resistance
   d. insulation value

2. Ammeters use a _____ to measure values higher than 10mA.
   a. multiplier resistor
   b. rectifier bridge
   c. transformer coil
   d. shunt

3. Voltmeters use a _____ to measure values higher than one volt.
   a. multiplier resistor
   b. rectifier bridge
   c. transformer coil
   d. shunt

4. Electromotive force (emf) is measured using a(n) _____.
   a. ammeter
   b. wattmeter
   c. voltmeter
   d. ohmmeter

5. A voltmeter is connected _____ with the circuit being tested.
   a. in parallel
   b. in series
   c. in a sequential configuration
   d. in a looped configuration

6. Which of the following is *not* true regarding an ohmmeter?
   a. It is powered by a DC source, usually a battery.
   b. It contains current-limiting resistors.
   c. It must be used in an energized circuit.
   d. It should be adjusted to zero before each use.

7. Which of the following values can be measured using a VOM?
   a. Decibels
   b. Voltage
   c. Resistance
   d. All of the above

8. When in doubt as to the voltage present, always use the _____ range to protect the VOM.
   a. lowest
   b. highest
   c. middle
   d. it does not matter with today's precision instruments

9. Wattmeters are used to measure _____.
   a. power
   b. voltage
   c. resistance
   d. impedance

10. To check the resistance between motor or transformer windings, use a _____.
    a. standard ohmmeter
    b. wattmeter
    c. megger
    d. power factor meter

11. A Wiggy® is used to _____.
    a. measure wattage
    b. measure impedance
    c. test for potential difference
    d. provide an exact voltage reading

12. Recording instruments can be used to measure _____.
    a. power
    b. current
    c. frequency
    d. all of the above

13. The minimum working distance for a 15kV circuit is _____.
    a. 2 ft. 0 in.
    b. 5 ft. 0 in.
    c. 7 ft. 0 in.
    d. 15 ft. 0 in.

14. Connecting a voltmeter may affect the total _____ in a circuit.
    a. voltage
    b. current
    c. resistance
    d. electromotive force

15. Which of the following is *not* true regarding meter loading?
   a. It may cause different meters to produce different readings.
   b. The higher the ohms-per-volt rating, the smaller the effect of the meter on the circuit.
   c. It may impact both voltage and resistance.
   d. It can be prevented by removing power to the circuit being tested.

ANSWERS TO REVIEW QUESTIONS

Answer		**Section**
1.	a	3.0.0
2.	d	3.0.0
3.	a	4.0.0
4.	c	4.0.0
5.	a	4.0.0
6.	c	5.0.0
7.	d	6.0.0
8.	b	6.3.3
9.	a	8.0.0
10.	c	9.0.0
11.	c	13.1.1
12.	d	14.0.0
13.	a	15.2.5
14.	c	15.3.0
15.	d	15.3.0

NCCER CRAFT TRAINING USER UPDATES

The NCCER makes every effort to keep these manuals up-to-date and free of technical errors. We appreciate your help in this process. If you have an idea for improving this manual, or if you find an error, a typographical mistake, or an inaccuracy in the NCCER's Craft Training Manuals, please write us, using this form or a photocopy. Be sure to include the exact module number, page number, a description of the problem, and the correction, if possible. Your input will be brought to the attention of the Technical Review Committee. Thank you for your assistance.

Instructors – If you found that additional materials were necessary in order to teach this module effectively, please let us know so that we may include them in the Equipment/Materials list in the Instructor's Guide.

Write: Curriculum Development and Revision Department
National Center for Construction Education and Research
P.O. Box 141104
Gainesville, FL 32614-1104

Fax: 352-334-0932

Craft _____ Module Name _____

Copyright Date _____ Module Number _____ Page Number(s) _____

Description of Problem _____

_____

_____

_____

(Optional) Correction of Problem _____

_____

_____

(Optional) Your Name and Address _____

_____

_____

notes

Introduction to the National Electrical Code

Module 26107

Electrical Trainee Task Module 26107

INTRODUCTION TO THE NATIONAL ELECTRICAL CODE®

OBJECTIVES

Upon completion of this module, the trainee will be able to:

1. Explain the purpose and history of the *National Electrical Code* (NEC).
2. Describe the layout of the NEC.
3. Explain how to navigate the NEC.
4. Describe the purpose of the National Electrical Manufacturers' Association (NEMA) and the National Fire Protection Association (NFPA).
5. Explain the role of testing laboratories.

Prerequisites

Successful completion of the following Task Modules is required before beginning study of this Task Module: Core Curricula; Electrical Level 1, Task Modules 26101 through 26106.

Required Trainee Materials

1. Trainee Task Module
2. Copy of the latest edition of the *National Electrical Code*

Note: The designations "National Electrical Code," "NE Code," and "NEC," where used in this document, refer to the *National Electrical Code®*, which is a registered trademark of the National Fire Protection Association, Quincy, MA. *All National Electrical Code (NEC) references in this module refer to the 1999 edition of the NEC.*

Course Map

This course map shows all of the task modules in the first level of the Electrical curricula. The suggested training order begins at the bottom and proceeds up. Skill levels increase as a trainee advances on the course map. The training order may be adjusted by the local Training Program Sponsor.

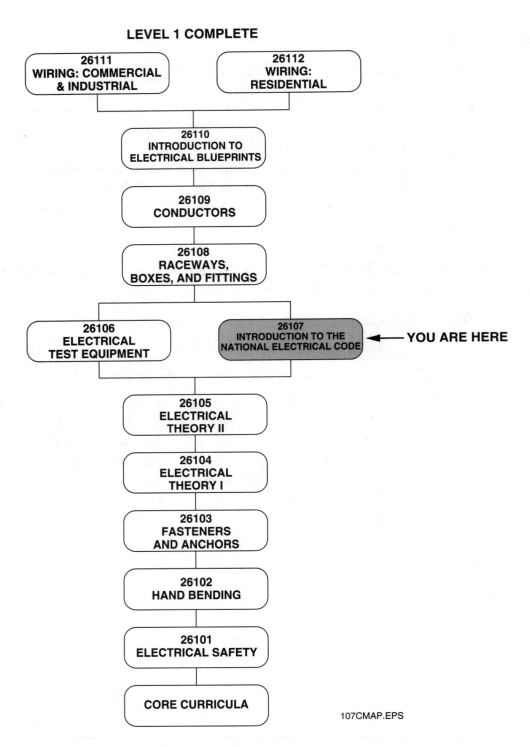

LEVEL 1 COMPLETE

26111
WIRING: COMMERCIAL
& INDUSTRIAL

26112
WIRING:
RESIDENTIAL

26110
INTRODUCTION TO
ELECTRICAL BLUEPRINTS

26109
CONDUCTORS

26108
RACEWAYS,
BOXES, AND FITTINGS

26106
ELECTRICAL
TEST EQUIPMENT

26107
INTRODUCTION TO THE
NATIONAL ELECTRICAL CODE ← YOU ARE HERE

26105
ELECTRICAL
THEORY II

26104
ELECTRICAL
THEORY I

26103
FASTENERS
AND ANCHORS

26102
HAND BENDING

26101
ELECTRICAL SAFETY

CORE CURRICULA

107CMAP.EPS

TABLE OF CONTENTS

Trade Terms Introduced In This Module

Articles: The articles are the main topics of the NEC, beginning with *NEC Article 90, Introduction,* and ending with *NEC Article 820, Community Antenna Television and Radio Distribution Systems.*

Chapters: Nine chapters form the broad structure of the NEC.

Examples: Examples in *NEC Chapter 9* provide methods for performing load calculations for various types of buildings, feeders, and branch circuits.

Exceptions: Exceptions follow the applicable sections of the NEC and allow alternative methods to be used under specific conditions.

Fine Print Note (FPN): Explanatory material that follows specific NEC sections.

National Electrical Manufacturers' Association (NEMA): The association that maintains and improves the quality and reliability of electrical products.

National Fire Protection Association (NFPA): The publishers of the NEC. The NFPA develops standards to minimize the possibility and effects of fire.

Nationally Recognized Testing Laboratories (NRTL): Product safety certification laboratories that are responsible for testing and certifying electrical equipment.

Parts: Certain articles in the NEC are subdivided into parts. Parts have letter designations (e.g., Part A).

Sections: Parts and articles are subdivided into sections. Sections have numeric designations that follow the article number and are preceded by a hyphen (e.g., 501-4).

Tables: Tables are located within chapters to provide more detailed information explaining NEC content.

1.0.0   INTRODUCTION

The *National Electrical Code* (NEC) is published by the **National Fire Protection Association (NFPA).** The NEC is one of the most important tools for the electrician. When used together with the electrical code for your local area, the NEC provides the minimum requirements for the installation of electrical systems. Unless otherwise specified, always use the latest edition of the NEC as your on-the-job reference. It specifies the minimum provisions necessary for protecting people and property from electrical hazards. In some areas, however, local laws may specify different editions of the NEC, so be sure to use the edition specified by your employer. Also, bear in mind that the NEC only specifies *minimum* requirements, so local or job requirements may be more stringent.

2.0.0 PURPOSE AND HISTORY OF THE NEC

The primary purpose of the NEC is the practical safeguarding of persons and property from hazards arising from the use of electricity [**NEC Section 90-1(a)**]. A thorough knowledge of the NEC is one of the first requirements for becoming a trained electrician. The NEC is probably the most widely used and generally accepted code in the world. It has been translated into several languages. It is used as an electrical installation, safety, and reference guide in the United States. Many other parts of the world use it as well. Compliance with NEC standards increases the safety of electrical installations—the reason the NEC is so widely used.

Although the NEC itself states "This Code is not intended as a design specification nor an instruction manual for untrained persons," it does provide a sound basis for the study of electrical installation procedures—under the proper guidance. The NEC has become the standard reference of the electrical construction industry. Anyone involved in electrical work should obtain an up-to-date copy and refer to it frequently.

Whether you are installing a new electrical system or altering an existing one, all electrical work must comply with the current NEC and all local ordinances. Like most laws, the NEC is easier to work with once you understand the language and know where to look for the information you need.

Note: This module is *not* a substitute for the NEC. You need to acquire a copy of the most recent edition and keep it handy at all times. The more you know about the NEC, the better an electrician you will become.

2.1.0 HISTORY

In 1881, The National Association of Fire Engineers met in Richmond, Virginia. From this meeting came the idea to draft the first *National Electrical Code*. The first nationally-recommended electrical code was published by the National Board of Fire Underwriters (now the American Insurance Association) in 1895.

In 1896, the National Electric Light Association (NELA) was working to make the requirements of the fire insurance organizations and electrical utilities fit together. NELA succeeded in promoting a conference which would result in producing a standard national code. The NELA code would serve the interests of the insurance industry, operating concerns, manufacturing, and industry.

The conference produced a set of requirements that was unanimously accepted. In 1897, the first edition of the NEC was published, and the NEC became the first cooperatively-produced national code. The organization which produced the NEC was known as the *National Conference on Standard Electrical Rules*. This group became a permanent organization, and its job was to develop the NEC.

In 1911, the NFPA took over administration and control of the NEC. However, the National Bureau of Fire Underwriters (NBFU) continued to publish the NEC until 1962. During the period from 1911 until now, the NEC has experienced several major changes, as well as regular three-year updates. In 1923, the NEC was rearranged and rewritten, and in 1937, it was editorially revised. In 1959, the NEC was revised to include a numbering system under which each **section** of each **article** was identified with an article/section number.

2.1.1 Who Is Involved?

The creation of a universally-accepted set of rules is an involved and complicated process. Rules made by a committee have the advantage that they usually do not leave out the interest of any of the groups represented at the committee. However, since the rules must represent the interests and requirements of an assortment of groups, they are often quite complicated and wordy.

In 1949, the NFPA reorganized the NEC committee into its present structure. The present structure consists of a Correlating Committee and twenty Code Making Panels (CMPs).

The Correlating Committee consists of ten principal voting members and six alternates. The principal function of the Correlating Committee is to ensure that:

- No conflict of requirements exists
- Correlation has been achieved
- NFPA regulations governing committee projects have been followed
- A practical schedule of revision and publication is established and maintained

Each of the twenty CMPs have members who are experts on particular subjects and have been assigned certain articles to supervise and revise as required. Members of the CMPs represent such special interest groups as trade associations, electrical contractors, electrical designers and engineers, electrical inspectors, electrical manufacturers and suppliers, electrical testing laboratories, and insurance organizations.

Each panel is structured so that not more than one-third of its members are from a single interest group. The members of the NEC Committee create or revise requirements for the NEC through probing, debating, analyzing, weighing, and reviewing new input. Anyone, including you, can submit proposals to amend the NEC. Sample forms for this purpose may be obtained from the Secretary of the Standards Council at NFPA Headquarters. In addition to written proposals, the NFPA also holds meetings to discuss code changes and proposals.

The NFPA membership is drawn from the fields listed above. In addition to publishing the NEC, the duties of the NFPA include the following:

- Developing, publishing, and distributing standards that are intended to minimize the possibility and effects of fire and explosion
- Conducting fire safety education programs for the general public

- Providing information on fire protection, prevention, and suppression
- Compiling annual statistics on causes and occupancies of fires, large-loss fires (over one million dollars), fire deaths, and firefighter casualties
- Providing field service by specialists on electricity, flammable liquids and gases, and marine fire problems
- Conducting research projects that apply statistical methods and operations research to develop computer modes and data management systems

3.0.0 THE LAYOUT OF THE NEC

The NEC begins with a brief history. *Figure 1* shows how the NEC is organized.

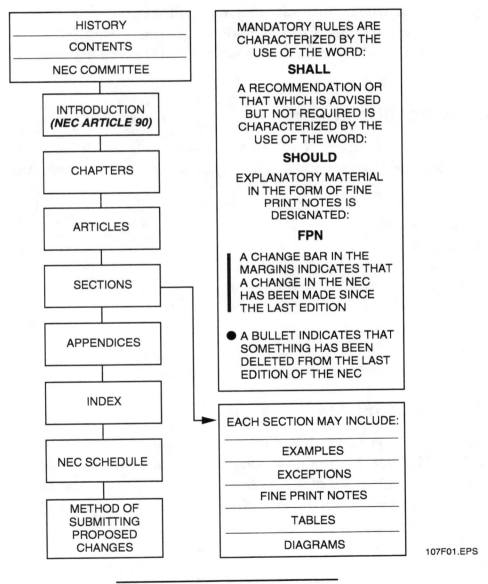

Figure 1. The Layout Of The NEC

3.1.0   TYPES OF RULES

There are two basic types of rules in the NEC: mandatory rules and advisory rules. Here is how to recognize the two types of rules and how they relate to all types of electrical systems:

- *Mandatory rules* – All mandatory rules have the word *shall* in them. The word *shall* means *must*. If a rule is mandatory, you must comply with it.
- *Advisory rules* – All advisory rules have the word *should* in them. The word *should* in this case means *recommended but not necessarily required*. If a rule is advisory, compliance is discretionary. If you want to comply with it, do so, but it is not required by the NEC.

Be alert to local amendments to the NEC. Local ordinances may amend the language of the NEC, changing it from *should* to *shall*. This means that you must do in that county or city what may only be recommended in some other area. The office that issues building permits will either sell you a copy of the code that is enforced in that county or tell you where the code is sold.

3.2.0   NEC INTRODUCTION

The main body of the text begins with an *Introduction*, also entitled **NEC Article 90**. This introduction gives you an overview of the NEC. Items included in this section are:

- Purpose of the NEC
- Scope of the codebook
- Code arrangement
- Code enforcement
- Mandatory rules and explanatory material
- Formal interpretation
- Examination of equipment for safety
- Wiring planning
- Metric units of measurement

3.3.0   THE BODY OF THE NEC

The remainder of the book is organized into nine **chapters**. See *Figure 1* for an overview of the layout. **NEC Chapter 1** contains a complete list of definitions used in the NEC. These definitions are referred to as **NEC Article 100**. **NEC Article 110** gives the general requirements for electrical installations. It is important for you to be familiar with this general information and the definitions.

3.3.1 NEC Definitions

There are many definitions included in **NEC Article 100**. You should become familiar with the definitions. Here are two that you should become especially familiar with:

- *Labeled* – "Equipment or materials to which has been attached a label or other identifying mark of an organization that is acceptable to the authority having jurisdiction and concerned with product evaluation, that maintains periodic inspection of production of labeled equipment or materials, and by whose labeling the manufacturer indicates compliance with appropriate standards or performance in a specified manner."
- *Listed* – "Equipment or materials included in a list published by an organization acceptable to the authority having jurisdiction and concerned with product evaluation, that maintains periodic inspection of production of listed equipment or materials, and whose listing states either that the equipment or material meets appropriate designated standards or has been tested and found suitable for use in a specified manner."

Besides installation rules, you will also have to be concerned with the type and quality of materials that are used in electrical wiring systems. Nationally-recognized testing laboratories are product safety certification laboratories. Underwriters' Laboratories, also called *UL*, is one such laboratory. These laboratories establish and operate product safety certification programs to make sure that items produced under the service are safeguarded against reasonable foreseeable risks. Some of these organizations maintain a worldwide network of field representatives who make unannounced visits to manufacturing facilities to counter-check products bearing their seal of approval. The UL label is shown below (*Figure 2*).

107F02.EPS

Figure 2. Underwriters' Laboratories Label

3.4.0 THE REFERENCE PORTION OF THE NEC

NEC Chapters 2 through 9 each contain numerous articles and sections. Each chapter focuses on a general category of electrical application, such as **NEC Chapter 2**, *Wiring and Protection*. Each article emphasizes a more specific **part** of that category, such as **NEC Article 210**, *Branch Circuits*, **Part A**, *General Provisions*. Each section gives examples of a specific application of the NEC, such as **NEC Section 210-4**, *Multiwire Branch Circuits*.

3.4.1 Organization Of The Chapters

The chapters of the NEC are organized into three major categories:

- *NEC Chapters 1, 2, 3, 4, and 9* – The first four chapters present the rules for the design and installation of electrical wiring systems. *NEC Chapter 9* contains **tables** which specify the properties of conductors and rules for the use of conduit to enclose the conductors. The **examples** in this chapter demonstrate the use of the rules for design given in the first four chapters.
- *NEC Chapters 5, 6, and 7* – These chapters are concerned with special occupancies, equipment, and conditions. Rules in these chapters may modify or amend those in the first four chapters.
- *NEC Chapter 8* – This chapter covers communications systems, such as the telephone and telegraph, as well as radio and television receiving equipment.

3.5.0 TEXT IN THE NEC

When you open the NEC, you will notice several different types of text or printing used. Here is an explanation of each type of text:

- *Bold black letters* – Headings for each NEC application.
- *Exceptions* – **Exceptions** explain the circumstances under which a specific part of the NEC does not apply. Exceptions are written in *italics* under that part of the NEC to which they pertain.
- *Fine print notes* – **Fine print notes (FPNs)** explain something in an application, suggest other sections to read about the application, or provide tips about the application. These are defined in the text by the term (*FPN*) shown in parentheses before a paragraph in smaller print.
- *Diagrams* – These may be included with explanations to give you a picture of what your application may look like.
- *Tables* – Tables are often included when there is more than one possible application of the NEC. You would use a table to look up the specifications of your application.

4.0.0 NAVIGATING THE NEC

To locate information for a particular procedure being performed, use the following steps:

Step 1 Familiarize yourself with *NEC Articles 90, 100, and 110* to gain an understanding of the material covered in the NEC and the definitions used in it.

Step 2 Turn to the *Table of Contents* at the beginning of the NEC.

Step 3 Locate the chapter that focuses on the desired category.

Step 4 Find the article pertaining to your specific application.

Step 5 Turn to the page indicated. Each application will begin with a bold heading.

Note: An index is provided at the end of the NEC. The index lists specific topics and provides a reference to the location of the material within the NEC. The index is helpful when you are looking for a specific topic rather than a general category.

Once you are familiar with **NEC Articles 90, 100, and 110**, you can move on to the rest of the NEC. There are several key sections used often in servicing electrical systems.

- *Wiring design and protection* – **NEC Chapter 2** discusses wiring design and protection, the information electrical technicians need most often. It covers the use and identification of grounded conductors, branch circuits, feeders, calculations, services, overcurrent protection, and grounding. This is essential information for all types of electrical systems. This chapter is also a *how-to* chapter. It explains how to provide proper spacing for conductor supports, how to provide temporary wiring, and how to size the proper grounding conductor or electrode. If you run into a problem related to the design or installation of a conventional electrical system, you can probably find a solution for it in this chapter.
- *Wiring methods and materials* – **NEC Chapter 3** lists the rules on wiring methods and materials. The materials and procedures to use on a particular system depend on the type of building construction, the type of occupancy, the location of the wiring in the building, the type of atmosphere in the building or in the area surrounding the building, mechanical factors, and the relative costs of different wiring methods. The provisions of this chapter apply to all wiring installations except remote control switching (**NEC Article 725**), low-energy power circuits (**NEC Article 725**), signal systems (**NEC Article 725**), and communications systems and conductors (**NEC Article 800**) when these items form an integral part of equipment, such as in motors and motor controllers.

There are four basic wiring methods used in most modern electrical systems. Nearly all wiring methods are a variation of one or more of these methods.

- Sheathed cables of two or more conductors, such as NM cable and BX armored cable (**NEC Articles 330 through 339**)
- Raceway wiring systems, such as rigid and EMT conduit (**NEC Articles 342 through 358**)
- Busways (**NEC Article 364**)
- Cable trays (**NEC Article 318**)

NEC Article 310 gives a complete description of all types of electrical conductors. Electrical conductors come in a wide range of sizes and forms. Be sure to check the working drawings and specifications to see what sizes and types of conductors are required for a specific job. If the conductor type and size are not specified, choose the most appropriate type and size meeting standard NEC requirements.

NEC Articles 318 through 384 give rules for raceways, boxes, cabinets, and raceway fittings. Outlet boxes vary in size and shape, depending on their use, the size of the raceway, the number of conductors entering the box, the type of building construction, and the atmospheric conditions of the area. This chapter should answer most questions on the selection and use of these items.

The NEC does not describe in detail all types and sizes of outlet boxes. However, the manufacturers of outlet boxes provide excellent catalogs showing their products. Collect these catalogs, since these items are essential to your work.

NEC Article 380 covers the switches, pushbuttons, pilot lamps, receptacles, and convenience outlets you will use to control electrical circuits or to connect portable equipment to electric circuits. Again, get the manufacturers' catalogs on these items. They will provide you with detailed descriptions of each of the wiring devices.

NEC Article 384 covers switchboards and panelboards, including their location, installation methods, clearances, grounding, and overcurrent protection.

4.1.0   EQUIPMENT FOR GENERAL USE

NEC Chapter 4 begins with the use and installation of flexible cords and cables, including the trade name, type letter, wire size, number of conductors, conductor insulation, outer covering, and use of each. This chapter also covers fixture wires, again giving the trade name, type letter, and other important details.

NEC Article 410 on lighting fixtures is especially important. It gives installation procedures for fixtures in specific locations. For example, it covers fixtures near combustible material and fixtures in closets. However, the NEC does not describe how many fixtures will be needed in a given area to provide a certain amount of illumination.

NEC Article 430 covers electric motors, including mounting the motor and making electrical connections to it. Motor controls and overload protection are also covered.

NEC Articles 440 through 460 cover air conditioning and heating equipment, transformers, and capacitors.

NEC Article 480 provides requirements related to battery-operated electrical systems. Storage batteries are seldom thought of as part of a conventional electrical system, but they often provide standby emergency lighting service. They may also supply power to security systems that are separate from the main AC electrical system.

4.2.0 SPECIAL OCCUPANCIES

NEC Chapter 5 covers special occupancy areas. These are areas where the sparks generated by electrical equipment may cause an explosion or fire. The hazard may be due to the atmosphere of the area or the presence of a volatile material in the area. Commercial garages, aircraft hangers, and service stations are typical special occupancy locations.

NEC Articles 500 and 501 cover the different types of special occupancy atmospheres where an explosion is possible. The atmospheric groups were established to make it easy to test and approve equipment for various types of uses.

NEC Articles 501-4, 502-4, and 503-3 cover the installation of explosion-proof wiring. An explosion-proof system is designed to prevent the ignition of a surrounding explosive atmosphere when arcing occurs within the electrical system.

There are three main classes of special occupancy location:

- *Class I* – Areas containing flammable gases or vapors in the air. Class I areas include paint spray booths, dyeing plants where hazardous liquids are used, and gas generator rooms (*NEC Article 501*).
- *Class II* – Areas where combustible dust is present, such as grain-handling and storage plants, dust and stock collector areas, and sugar-pulverizing plants (*NEC Article 502*). These are areas where, under normal operating conditions, there may be enough combustible dust in the air to produce explosive or ignitable mixtures.
- *Class III* – Areas that are hazardous because of the presence of easily ignitable fibers or other particles in the air, although not in large enough quantities to produce ignitable mixtures (*NEC Article 503*). Class III locations include cotton mills, rayon mills, and clothing manufacturing plants.

NEC Articles 511 and 514 regulate garages and similar locations where volatile or flammable liquids are used. While these areas are not always considered critically hazardous locations, there may be enough danger to require special precautions in the electrical installation. In these areas, the NEC requires that volatile gases be confined to an area not more than four feet above the floor. So in most cases, conventional raceway systems are permitted above this level. If the area is judged to be critically hazardous, explosion-proof wiring (including seal-offs) may be required.

NEC Article 520 regulates theaters and similar occupancies where fire and panic can cause hazards to life and property. Drive-in theaters do not present the same hazards as enclosed auditoriums, but the projection rooms and adjacent areas must be properly ventilated and wired for the protection of operating personnel and others using the area.

NEC Chapter 5 also covers residential storage garages, aircraft hangars, service stations, bulk storage plants, health care facilities, and mobile homes and parks.

4.3.0 SPECIAL EQUIPMENT

Residential electrical workers will seldom need to refer to the articles in **NEC Chapter 6**, but this chapter is of great concern to commercial and industrial electrical workers.

NEC Article 600 covers electric signs and outline lighting. **NEC Article 610** applies to cranes and hoists. **NEC Article 620** covers the majority of the electrical work involved in the installation and operation of elevators, dumbwaiters, escalators, and moving walks. The manufacturer is responsible for most of this work. The electrician usually just furnishes a feeder terminating in a disconnect means in the bottom of the elevator shaft. The electrician may also be responsible for a lighting circuit to a junction box midway in the elevator shaft for connecting the elevator cage lighting cable and exhaust fans. The articles in this chapter list most of the requirements for these installations.

NEC Article 630 regulates electric welding equipment. It is normally treated as a piece of industrial power equipment requiring a special power outlet, but there are special conditions that apply to the circuits supplying welding equipment. These are outlined in detail in this chapter.

NEC Article 640 covers wiring for sound recording and similar equipment. This type of equipment normally requires low-voltage wiring. Special outlet boxes or cabinets are usually provided with the equipment, but some items may be mounted in or on standard outlet boxes. Some sound recording systems require direct current. It is supplied from rectifying equipment, batteries, or motor generators. Low-voltage alternating current comes from relatively small transformers connected on the primary side to a 120V circuit within the building.

Other items covered in **NEC Chapter 6** include X-ray equipment (**NEC Article 660**), induction and dielectric heat-generating equipment (**NEC Article 665**), and machine tools (**NEC Article 670**).

If you ever have work that involves **NEC Chapter 6**, study the chapter before work begins. That can save a lot of installation time. Another way to cut down on labor hours and prevent installation error is to acquire a set of rough-in drawings of the equipment being installed. It is easy to install the wrong outlet box or to install the right box in the wrong place. Having a set of rough-in drawings can prevent these simple but costly errors.

4.4.0 SPECIAL CONDITIONS

In most commercial buildings, the NEC and local ordinances require a means of lighting public rooms, halls, stairways, and entrances. There must be enough light to allow the occupants to exit from the building if the general building lighting is interrupted. Exit doors must be clearly indicated by illuminated exit signs.

NEC Chapter 7 covers the installation of emergency systems. These circuits should be arranged so that they can automatically transfer to an alternate source of current, usually storage batteries or gasoline-driven generators. As an alternative in some types of occupancies, you can connect them to the supply side of the main service, so disconnecting the main service switch would not disconnect the emergency circuits. This chapter also covers a variety of other equipment, systems, and conditions that are not easily categorized elsewhere in the NEC.

NEC Chapter 8 is a special category for wiring associated with electronic communications systems including telephone and telegraph, radio and TV, fire and burglar alarms, and community antenna systems.

4.5.0 EXAMPLES OF NAVIGATING THE NEC

4.5.1 Installing Type SE Cable

Suppose you are installing Type SE (service-entrance) cable on the side of a home. You know that this cable must be secured, but you are not sure of the spacing between cable clamps. To find out this information, use the following procedure:

Step 1 Look in the NEC *Table of Contents* and follow down the list until you find an appropriate category. (Or, you can use the index at the end of the book.)

Step 2 *NEC Article 230* will probably catch your eye first, so turn to the page where it begins.

Step 3 Scan down through the section numbers until you come to *NEC Section 230-51*, *Mounting Supports*. Upon reading this section, you will find in paragraph *(a)* *Service-Entrance Cables* that "Service-entrance cable shall be supported by straps or other approved means within 12 inches (305mm) of every service head, gooseneck, or connection to a raceway or enclosure, and at intervals not exceeding 30 inches (762mm)."

After reading this section, you will know that a cable strap is required within 12 inches of the service head and within 12 inches of the meter base. Furthermore, the cable must be secured in between these two termination points at intervals not exceeding 30 inches.

4.5.2 Installing Track Lighting

Assume that you are installing track lighting in a residential occupancy. The owners want the track located behind the curtain of their sliding glass patio doors. To determine if this is a NEC violation, follow these steps:

Step 1 Look in the NEC *Table of Contents* and find the chapter that contains information about the general application you are working on. *NEC Chapter 4*, *Equipment for General Use*, covers track lighting.

Step 2   Now look for the article that fits the specific category you are working on. In this case, **NEC Article 410** covers lighting fixtures, lampholders, lamps, and receptacles.

Step 3   Next locate the section within **NEC Article 410** that deals with the specific application. For this example, refer to *Part R, Lighting Track*.

Step 4   Turn to the page listed.

Step 5   Read **NEC Section 410-100,** *Definitions*, to become familiar with track lighting. Continue down the page with **NEC Section 410-101** and read the information contained therein. Note that paragraph *(c) Locations Not Permitted* under **NEC Section 410-101** states the following: "Lighting track shall not be installed (1) where likely to be subjected to physical damage; (2) in wet or damp locations; (3) where subject to corrosive vapors; (4) in storage battery rooms; (5) in hazardous (classified) locations; (6) where concealed; (7) where extended through walls or partitions; (8) less than 5 feet (1.52m) above the finished floor except where protected from physical damage or track operating at less than 30 volts RMS open-circuit voltage."

Step 6   Read **NEC Section 410-101(c)** carefully. Do you see any conditions that would violate any NEC requirements if the track lighting is installed in the area specified? In checking these items, you will probably note condition (6), "where concealed." Since the track lighting is to be installed behind a curtain, this sounds like a NEC violation. We need to check further.

Step 7   You need the NEC definition of concealed. Therefore, turn to **NEC Article 100,** *Definitions* and find the main term *concealed*. It reads: "Concealed: Rendered inaccessible by the structure or finish of the building . . ."

Step 8   Although the track lighting may be out of sight if the curtain is drawn, it will still be readily accessible for maintenance. Consequently, the track lighting is really not concealed according to the NEC definition.

When using the NEC to determine electrical installation requirements, keep in mind that you will nearly always have to refer to more than one section. Sometimes the NEC itself refers the reader to other articles and sections. In some cases, the user will have to be familiar enough with the NEC to know what other sections pertain to the installation at hand. It can be a confusing situation, but time and experience in using the NEC will make it much easier. A pictorial road map of some NEC topics is shown in *Figures 3* and *4*.

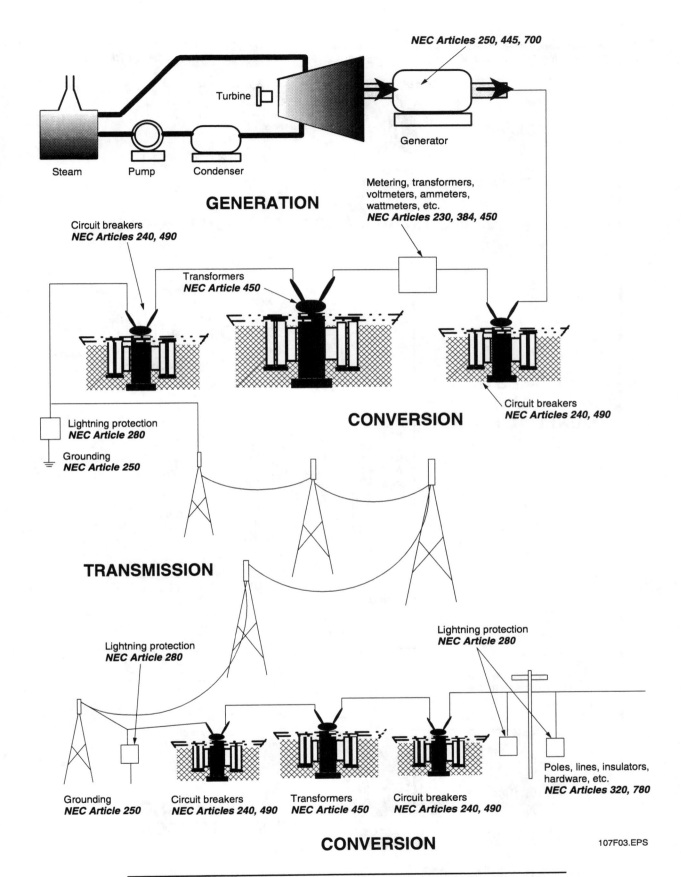

Figure 3. NEC References For Power Generation And Transmission

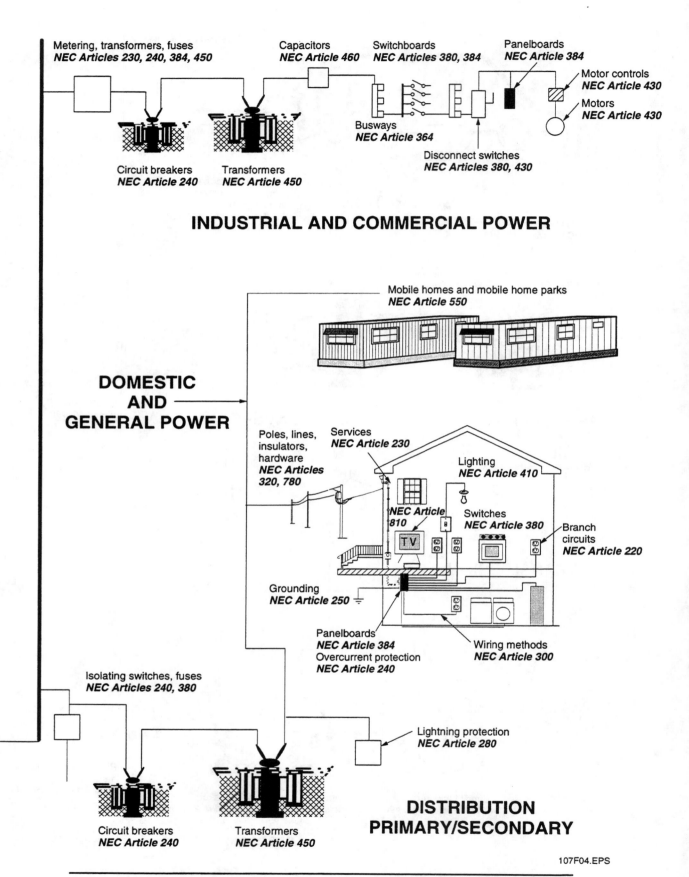

Metering, transformers, fuses
NEC Articles 230, 240, 384, 450

Capacitors
NEC Article 460

Switchboards
NEC Articles 380, 384

Panelboards
NEC Article 384

Motor controls
NEC Article 430

Motors
NEC Article 430

Busways
NEC Article 364

Disconnect switches
NEC Articles 380, 430

Circuit breakers
NEC Article 240

Transformers
NEC Article 450

INDUSTRIAL AND COMMERCIAL POWER

Mobile homes and mobile home parks
NEC Article 550

DOMESTIC AND GENERAL POWER

Poles, lines, insulators, hardware
NEC Articles 320, 780

Services
NEC Article 230

Lighting
NEC Article 410

NEC Article 810

Switches
NEC Article 380

Branch circuits
NEC Article 220

Grounding
NEC Article 250

Panelboards
NEC Article 384
Overcurrent protection
NEC Article 240

Wiring methods
NEC Article 300

Isolating switches, fuses
NEC Articles 240, 380

Lightning protection
NEC Article 280

Circuit breakers
NEC Article 240

Transformers
NEC Article 450

DISTRIBUTION PRIMARY/SECONDARY

107F04.EPS

Figure 4. NEC References For Industrial, Commercial, And Residential Power

5.0.0 OTHER ORGANIZATIONS

5.1.0 THE ROLE OF TESTING LABORATORIES

As mentioned earlier, testing laboratories are an integral part of the development of the NEC. The NFPA and other organizations provide testing laboratories to conduct research into electrical equipment and its safety. These laboratories perform extensive testing of new products to make sure they are built to NEC standards for electrical and fire safety. These organizations receive statistics and reports from agencies all over the United States concerning electrical shocks and fires and their causes. Upon seeing trends developing concerning the association of certain equipment and dangerous situations or circumstances, this equipment will be specifically targeted for research. All the reports from these laboratories are used in the generation of changes or revisions to the NEC.

5.2.0 NATIONALLY RECOGNIZED TESTING LABORATORIES

Nationally Recognized Testing Laboratories (NRTL) are product safety certification laboratories. They establish and operate product safety certification programs to make sure that items produced under the service are safeguarded against reasonably foreseeable risks. NRTLs maintain a worldwide network of field representatives who make unannounced visits to factories to check products bearing their safety marks.

5.3.0 NATIONAL ELECTRICAL MANUFACTURERS' ASSOCIATION

The **National Electrical Manufacturers' Association (NEMA)** was founded in 1926. It is made up of companies that manufacture equipment used for generation, transmission, distribution, control, and utilization of electric power. The objectives of NEMA are to maintain and improve the quality and reliability of products; to ensure safety standards in the manufacture and use of products; and to develop product standards covering such matters as naming, ratings, performance, testing, and dimensions. NEMA participates in developing the NEC and the *National Electrical Safety Code* and advocates their acceptance by state and local authorities.

SUMMARY

The NEC specifies the minimum provisions necessary for protecting people and property from hazards arising from the use of electricity and electrical equipment. As an electrician, you must be aware of how to use and apply the NEC on your job. Using the NEC will help you to safely install and maintain the electrical equipment and systems you come into contact with.

References

For advanced study of topics covered in this task module, the following book is suggested:

National Electrical Code Handbook, Latest Edition, National Fire Protection Association, Quincy, MA.

REVIEW QUESTIONS

1. The NEC provides the _____ requirements for the installation of electrical systems.
    a. minimum
    b. most stringent
    c. design specification
    d. complete

2. Which of the following groups is usually *not* represented on the Code Making Panels?
    a. Trade associations
    b. Electrical inspectors
    c. Insurance organizations
    d. Government lobbyists

3. What is the difference between a mandatory rule and an advisory rule?
    a. You must comply with mandatory rules.
    b. Mandatory rules include the word *shall*.
    c. It is recommended that you comply with advisory rules.
    d. All of the above.

4. What is covered in *NEC Article 110?*
    a. Branch circuits
    b. Definitions
    c. General requirements for electrical installations
    d. Wiring design and protection

5. Which NEC chapters cover the design and installation of electrical systems?
    a. 1, 2, and 7
    b. 1, 2, 3, 4, and 9
    c. 6, 7, and 9
    d. 5, 6, 7, and 9

6. Which of the following covers busways?
    a. *NEC Articles 330 through 339*
    b. *NEC Articles 342 through 358*
    c. *NEC Article 364*
    d. *NEC Article 318*

7. Which article of the NEC provides installation procedures for lighting fixtures in specific locations?
    a. *NEC Article 410*
    b. *NEC Article 501*
    c. *NEC Article 364*
    d. *NEC Article 460*

8. Which article of the NEC covers theaters?
   a. *NEC Article 339*
   b. *NEC Article 110*
   c. *NEC Article 430*
   d. *NEC Article 520*

9. *NEC Article 600* covers _____.
   a. track lighting
   b. electric signs and outline lighting
   c. X-ray equipment
   d. emergency lighting systems

10. Which chapter of the NEC covers devices such as radios, televisions, and telephones?
    a. *NEC Chapter 8*
    b. *NEC Chapter 7*
    c. *NEC Chapter 6*
    d. *NEC Chapter 5*

notes

ANSWERS TO REVIEW QUESTIONS

Answer		**Section**
1.	a	1.0.0
2.	d	2.1.1
3.	d	3.1.0
4.	c	3.3.0
5.	b	3.4.1
6.	c	4.0.0
7.	a	4.1.0
8.	d	4.2.0
9.	b	4.3.0
10.	a	3.4.1/4.4.0

The NCCER makes every effort to keep these manuals up-to-date and free of technical errors. We appreciate your help in this process. If you have an idea for improving this manual, or if you find an error, a typographical mistake, or an inaccuracy in the NCCER's Craft Training Manuals, please write us, using this form or a photocopy. Be sure to include the exact module number, page number, a description of the problem, and the correction, if possible. Your input will be brought to the attention of the Technical Review Committee. Thank you for your assistance.

Instructors – If you found that additional materials were necessary in order to teach this module effectively, please let us know so that we may include them in the Equipment/Materials list in the Instructor's Guide.

Write: Curriculum Development and Revision Department
National Center for Construction Education and Research
P.O. Box 141104
Gainesville, FL 32614-1104
Fax: 352-334-0932

Craft _____ Module Name _____

Copyright Date _____ Module Number _____ Page Number(s) _____

Description of Problem _____

_____

_____

_____

(Optional) Correction of Problem _____

_____

_____

(Optional) Your Name and Address _____

_____

_____

Raceways, Boxes, and Fittings
Module 26108

Electrical Trainee Task Module 26108

NATIONAL
CENTER FOR
CONSTRUCTION
EDUCATION AND
RESEARCH

RACEWAYS, BOXES, AND FITTINGS

OBJECTIVES

Upon completion of this module, the trainee will be able to:

1. Describe various types of cable trays and raceways.
2. Identify and select various types and sizes of raceways.
3. Identify and select various types and sizes of cable trays.
4. Identify and select various types of raceway fittings.
5. Identify various methods used to install raceways.
6. Demonstrate knowledge of NEC raceway requirements.
7. Describe procedures for installing raceways and boxes on masonry surfaces.
8. Describe procedures for installing raceways and boxes on concrete surfaces.
9. Describe procedures for installing raceways and boxes in a metal stud environment.
10. Describe procedures for installing raceways and boxes in a wood frame environment.
11. Describe procedures for installing raceways and boxes on drywall surfaces.
12. Recognize safety precautions that must be followed when working with boxes and raceways.

Prerequisites

Successful completion of the following Task Modules is required before beginning study of this Task Module: Core Curricula; Electrical Level 1, Modules 26101 through 26107.

Required Trainee Materials

1. Trainee Task Module
2. Copy of the latest edition of the *National Electrical Code*
3. Appropriate Personal Protective Equipment
4. *OSHA Electrical Safety Guidelines* (pocket guide)

Note: The designations "National Electrical Code," "NE Code," and "NEC," where used in this document, refer to the *National Electrical Code®*, which is a registered trademark of the National Fire Protection Association, Quincy, MA. *All National Electrical Code (NEC) references in this module refer to the 1999 edition of the NEC.*

Course Map

This course map shows all of the task modules in the first level of the Electrical curricula. The suggested training order begins at the bottom and proceeds up. Skill levels increase as a trainee advances on the course map. The training order may be adjusted by the local Training Program Sponsor.

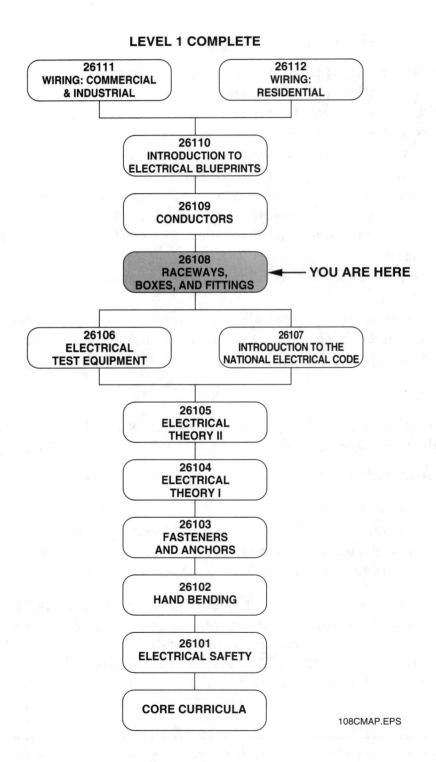

LEVEL 1 COMPLETE

26111
WIRING: COMMERCIAL
& INDUSTRIAL

26112
WIRING:
RESIDENTIAL

26110
INTRODUCTION TO
ELECTRICAL BLUEPRINTS

26109
CONDUCTORS

26108
RACEWAYS,
BOXES, AND FITTINGS ◀── YOU ARE HERE

26106
ELECTRICAL
TEST EQUIPMENT

26107
INTRODUCTION TO THE
NATIONAL ELECTRICAL CODE

26105
ELECTRICAL
THEORY II

26104
ELECTRICAL
THEORY I

26103
FASTENERS
AND ANCHORS

26102
HAND BENDING

26101
ELECTRICAL SAFETY

CORE CURRICULA

108CMAP.EPS

TABLE OF CONTENTS

TABLE OF CONTENTS (Continued)

Trade Terms Introduced In This Module

Accessible: Able to be reached, as for service or repair.

Approved: Meeting the requirements of an appropriate regulatory agency.

Bonding wire: A wire used to make a continuous grounding path between equipment and ground.

Cable trays: Rigid structures used to support electrical conductors.

Conductors: Wires or cables used to carry electrical current.

Conduit: A round raceway, similar to pipe, that houses conductors.

Exposed location: Not permanently closed in by the structure or finish of a building; able to be installed or removed without damage to the structure.

Kick: A bend in a piece of conduit, usually less than 45°, made to change the direction of the conduit.

Raceways: Enclosed channels designed expressly for holding wires, cables, or busbars, with additional functions as permitted in the NEC.

Splice: Connection of two or more conductors.

Tap: Intermediate point on a main circuit where another wire is connected to supply electrical current to another circuit.

Trough: A long, narrow box used to house electrical connections that could be exposed to the environment.

Underwriters' Laboratories (UL): An agency that evaluates and approves electrical components and equipment.

Wireways: Steel troughs designed to carry electrical wire and cable.

1.0.0 INTRODUCTION

Electrical **raceways** present challenges and new requirements involving proper installation techniques, general understanding of raceway systems, and applications of the NEC to raceway systems. Acquiring quality installation skills for raceway systems requires practice, knowledge, and training.

A presentation of the various types of raceway systems and fittings, basic raceway installation skills, and NEC requirements applicable to raceway systems are included in this module. This module also covers raceway supports and environmental considerations for raceway systems, as well as general raceway information.

Along with the study of this module, the following NEC Articles should be referenced:

- **NEC Article 250** – *Grounding*
- **NEC Article 318** – *Cable Trays*
- **NEC Article 345** – *Intermediate Metal Conduit*
- **NEC Article 346** – *Rigid Metal Conduit*
- **NEC Article 347** – *Rigid Nonmetallic Conduit*
- **NEC Article 348** – *Electrical Metallic Tubing*
- **NEC Article 350** – *Flexible Metal Conduit*
- **NEC Article 351** – *Liquid-Tight Flexible Metal Conduit and Flexible Nonmetallic Conduit*
- **NEC Article 362** – *Wireways*

Note: Mandatory rules in the NEC are characterized by the use of the word *shall*. Explanatory material is in the form of Fine Print Notes (FPNs). When referencing specific sections of the NEC, always check to see if any exceptions apply.

2.0.0 RACEWAYS

Raceway is a general term referring to a wide range of circular and rectangular enclosed channels used to house electrical wiring. Raceways can be metallic or nonmetallic and come in different shapes. Depending on the particular purpose for which they are intended, raceways include enclosures such as underfloor raceways, flexible metal **conduit**, **wireways**, surface metal raceways, and surface nonmetallic raceways.

3.0.0 CONDUIT

Conduit is a raceway with a circular cross section, similar to pipe, that contains wires or cables. Conduit is used to provide protection for **conductors** and route them from one place to another. In addition, conduit makes it easier to replace or add wires to existing structures. Metal conduit also provides a permanent electrical path to ground. This equipment should be listed per the NEC.

3.1.0 CONDUIT AS A GROUND PATH

For safety reasons, most equipment that receives electrical power and has a metallic frame is grounded. In order to ground the equipment, an electrical connection must be made to connect the metal frame of the electrically-powered equipment to the grounding point at the service-entrance equipment. This is usually done in one or both of the following ways:

- The frame of the equipment is connected to a wire (equipment grounding conductor) which is directly connected to the ground point at the grounding terminal.
- The frame of the equipment is connected (bonded) to a metal conduit or other type of raceway system, which provides an uninterrupted and low impedance circuit to the ground point at the service-entrance equipment. The metal raceway or conduit acts as the equipment grounding conductor.

Note: According to *NEC Section 250-96*, metal raceways, **cable trays**, cable armor, cable sheath, enclosures, frames, fittings, and other metal noncurrent-carrying parts that are to serve as grounding conductors with or without the use of supplementary equipment grounding conductors shall be effectively bonded where necessary to ensure electrical continuity and the capacity to safely conduct any fault current likely to be imposed on them. The purpose of the equipment grounding conductor is to provide a low resistance path to ground for all equipment which receives power. This is done so that if an ungrounded conductor comes in contact with the frame of a piece of equipment, the circuit overcurrent device immediately acts to open the circuit. It also reduces the voltage to ground that would be present on the faulted equipment if a person came in contact with the equipment frame.

3.2.0 TYPES OF CONDUIT

There are many types of conduit used in the construction industry. The size of conduit which is to be used is determined by engineering specifications, local codes, and the NEC. Refer to *NEC Chapter 9, Tables 1 through 8 and Appendix C* for conduit fill with various conductors. There are several common types of conduit to examine.

3.2.1 Electrical Metallic Tubing

Electrical metallic tubing (EMT) is the lightest duty tubing available for enclosing and protecting electrical wiring. EMT is widely used for residential, commercial, and industrial wiring systems. It is lightweight, easily bent and/or cut to shape, and is the least costly type of metallic conduit. Because the wall thickness of EMT is less than that of rigid conduit, it is often referred to as *thinwall conduit*. A comparison of inside and outside diameters of EMT to rigid metal conduit and intermediate metal conduit (IMC) is shown in *Figure 1*.

NEC Section 348-4 permits the installation of EMT for either exposed or concealed work where it will not be subject to severe physical damage during installation or after construction. Installation of EMT is permitted in wet locations such as outdoors or indoors in dairies, laundries, and canneries using waterproof fittings.

Note: Refer to *NEC Section 348-5* for restrictions that apply to the use of EMT.

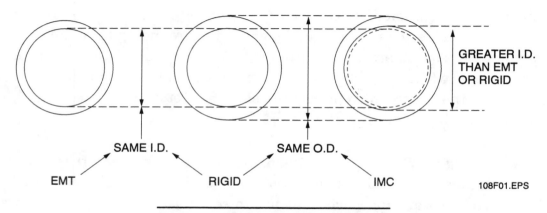

Figure 1. EMT Conduit Comparison

EMT shall not be used (1) where, during installation or afterward, it will be subject to severe physical damage; (2) where protected from corrosion solely by enamel; (3) in cinder concrete or cinder fill where subject to permanent moisture unless protected on all sides by a layer of non-cinder concrete at least two inches thick or unless the tubing is at least 18 inches under the fill; (4) in any hazardous (classified) locations except as permitted by **NEC Sections 502-4, 503-3, and 504-20**; or (5) for the support of fixtures or other equipment.

In a wet area, EMT and other conduit must be installed to prevent water from entering the conduit system. In locations where walls are subject to regular wash-down (see **NEC Section 300-6**), the entire conduit system must be installed to provide a ¼-inch air space between it and the wall or supporting surface. The entire conduit system is considered to include conduit, boxes, and fittings. To ensure resistance to corrosion caused by wet environments, EMT is galvanized. The term *galvanized* is used to describe the procedure in which the interior and exterior of the conduit is coated with a corrosion-resistant zinc compound.

EMT, being a good conductor of electricity, may be used as an equipment grounding conductor. In order to qualify as an equipment grounding conductor [see **NEC Section 250-68(b)**], the conduit system must be tightly connected at each joint and provide a continuous grounding path from each electrical load to the service equipment. The connectors used in an EMT system ensure electrical and mechanical continuity throughout the system (see **NEC Sections 250-96, 300-10, and 348-10**).

Note: Support requirements for EMT are also covered in **NEC Section 348-13**. The types of supports will be discussed later in this training module.

EMT fittings are manufactured in two basic types. One type of fitting is the compression coupling. (See *Figure 2*).

Because EMT is too thin for threads, special fittings must be used. For wet or damp locations, compression fittings such as those shown in *Figure 2* are used. These fittings contain a compression ring made of metal that forms a watertight seal.

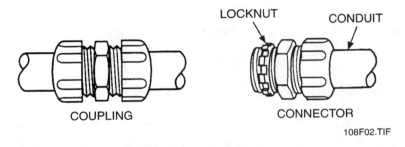

Figure 2. Compression Fittings

When EMT compression couplings are used, they must be securely tightened, and when installed in masonry concrete, they must be of the concrete-tight type. If installed in a wet location, they must be the raintight type. Refer to **NEC Article 348**.

EMT fittings for dry locations can be either the setscrew type or the indenting type. To use the setscrew type, the ends of the EMT are inserted into the sleeve and the setscrews are tightened to make the connection. Various types of setscrew couplings are shown in *Figure 3*.

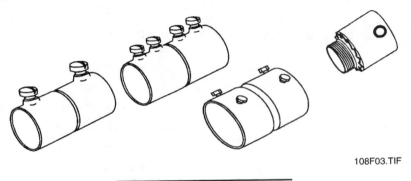

Figure 3. Setscrew Fittings

EMT sizes of 2½ inches and larger have the same outside diameter as corresponding sizes of galvanized rigid steel conduit (GRC). GRC threadless connectors may be used to connect EMT.

Note: EMT connectors, although they are the same size as GRC threadless connectors, may not be used to connect GRC.

Both setscrew and compression couplings are available in die-cast or steel construction. Steel couplings are stronger than die-cast couplings and have superior quality.

Support requirements for EMT are presented in **NEC Section 348-13**. As with most other metal conduit, EMT must be supported at least every 10 feet and within 3 feet of each outlet box, junction box, cabinet, fitting, or terminating end of the conduit. An exception to **NEC Section 348-13** allows the fastening of unbroken lengths of EMT to be increased to a distance of 5 feet (1.52m) where structural members do not readily permit fastening within 3 feet (914mm).

Electrical nonmetallic tubing (ENT) is also available. It provides an economical alternative to EMT, but it can only be used behind a 15-minute barrier inside non-fire-rated walls in certain applications.

3.2.2   Rigid Metal Conduit

Rigid metal conduit is conduit that is constructed of metal having sufficient thickness to permit the cutting of pipe threads at each end. Rigid metal conduit provides the best physical protection for conductors of any of the various types of conduit. Rigid metal conduit is supplied in 10-foot lengths including a threaded coupling on one end.

Note: Specific information on rigid metal conduit may be found in *NEC Article 346*.

Rigid metal conduit may be fabricated from steel or aluminum. Rigid metal steel conduit may be galvanized, or enamel-coated inside and out. Because of its threaded fittings, rigid metal conduit provides an excellent equipment grounding conductor as defined in *NEC Section 250-118(2)*. A piece of rigid metal conduit is shown in *Figure 4*. The support requirements for rigid metal conduit are presented in *NEC Section 346-12*.

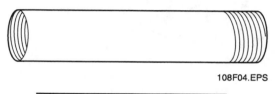

108F04.EPS

Figure 4. Rigid Metal Conduit

3.2.3   Galvanized Rigid Steel Conduit

GRC is mostly used in industrial applications. GRC is heavier than EMT and IMC. It is more difficult to cut and bend, requires threading of each end, and has a higher purchase price than EMT and IMC. As a result, the cost of installing GRC is generally higher than the cost of installing EMT and IMC.

3.2.4   Plastic-Coated GRC

Plastic-coated GRC has a thin coating of PVC over the GRC. This combination is useful when an environment calls for the ruggedness of GRC along with the corrosion resistance of PVC. Typical installations where plastic-coated GRC may be required are:

• Chemical plants
• Food plants
• Refineries
• Fertilizer plants

- Paper mills
- Wastewater treatment plants

Plastic-coated GRC requires special threading and bending techniques.

3.2.5   Aluminum Conduit

Aluminum conduit has several characteristics that distinguish it from steel conduit. Because it has better resistance to wet environments and some chemical environments, aluminum conduit generally requires less maintenance in installations such as sewage treatment plants.

Direct burial of aluminum conduit results in a self-stopping chemical reaction on the conduit surface which forms a coating on the conduit. This coating acts to prevent further corrosion, increasing the life of the installation.

Note: Caution must be exercised to avoid burial of aluminum conduit in soil or concrete that contains calcium chloride. Calcium chloride may interfere with the corrosion resistance of aluminum conduit. Calcium chloride and similar materials are often added to concrete to speed concrete setting. It is important to determine if chlorides are to be used in the concrete prior to installing aluminum conduit. If chlorides are to be used, aluminum conduit must be avoided. Check with local authorities regarding this type of usage.

Since aluminum conduit is lighter than steel conduit, there are some installation advantages to using aluminum. For example, a 10-foot section of 3-inch aluminum conduit weighs about 23 pounds, compared to the 68-pound weight of its steel counterpart.

3.2.6   Black Enamel Steel Conduit

Rigid black enamel steel conduit (often called *black conduit*) is steel conduit that is coated with a black enamel. In the past, this type of conduit was used exclusively for indoor wiring. Black enamel steel conduit is no longer manufactured for sale in the United States. It is mentioned only because it may still be found in existing installations.

3.2.7   Intermediate Metal Conduit

Intermediate metal conduit (IMC) has a wall thickness that is less than rigid metal conduit but greater than that of EMT. The weight of IMC is approximately ⅔ that of rigid metal conduit. Because of its lower purchase price, lighter weight, and thinner walls, IMC installations are generally less expensive than comparable rigid metal conduit installations. However, IMC installations still have high strength ratings.

Note: Additional information on intermediate metal conduit may be found in *NEC Article 345*.

The outside diameter of a given size of IMC is the same as that of the comparable size of rigid metal conduit. Therefore, rigid metal conduit fittings may be used with IMC. Since the threads on IMC and rigid metal conduit are the same size, no special threading tools are needed to thread IMC. Some electricians feel that threading IMC is more difficult than threading rigid metal conduit because IMC is somewhat harder.

The internal diameter of a given size of IMC is somewhat larger than the internal diameter of the same size of rigid metal conduit because of the difference in wall thickness. Bending IMC is considered easier than bending rigid metal conduit because of the reduced wall thickness. However, bending is sometimes complicated by kinking which may be caused by the increased hardness of IMC.

The NEC requires that intermediate metal conduit be identified along its length at 5-foot intervals with the letters *IMC*. **NEC Sections 110-21 and 345-16(c)** describe this marking requirement.

Intermediate metal conduit, like rigid metal conduit, is permitted to act as an equipment grounding conductor, as defined in **NEC Section 250-118(3)**.

The use of IMC may be restricted in some jurisdictions. It is important to investigate the requirements of each jurisdiction before selecting any materials.

3.2.8   Rigid Nonmetallic Conduit

The most common type of rigid nonmetallic conduit is manufactured from polyvinyl chloride (PVC). Because PVC conduit is noncorrosive, chemically inert, and non-aging, it is often used for installation in wet or corrosive environments. Corrosion problems found with steel and aluminum rigid metal conduit do not occur with PVC. However, PVC conduit may deteriorate under some conditions, such as extreme sunlight.

All PVC conduit is marked according to standards established by the National Electrical Manufacturers' Association (NEMA) or **Underwriters' Laboratories (UL)**. A section of PVC conduit is shown in *Figure 5*.

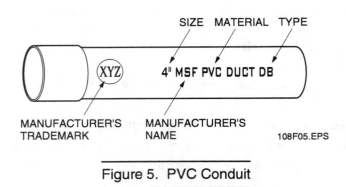

Figure 5. PVC Conduit

Since PVC conduit is lighter than steel or aluminum rigid conduit, IMC, or EMT, it is considered easier to handle. PVC conduit can usually be installed much faster than other types of conduit because the joints are made up with cement and require no threading.

PVC conduit contains no metal. This characteristic reduces the voltage drop of conductors carrying alternating current in PVC compared to identical conductors in steel conduit.

Because PVC is nonconducting, it cannot be used as an equipment grounding conductor. An equipment grounding conductor sized in accordance with **NEC Table 250-122** must be pulled in each PVC conductor run (except for underground service-entrance conductors).

PVC is available in lengths up to 20 feet. However, some jurisdictions require it to be cut to 10-foot lengths prior to installation. PVC expands and contracts significantly when exposed to changing temperatures. This expansion and contraction characteristic is worse in long runs. To avoid damage to PVC conduit caused by temperature changes, expansion couplings such as the one shown in *Figure 6* are used. The inside of the coupling is sealed with one or more O-rings. This type of coupling may allow up to six inches of movement. Check the requirements of the local jurisdiction prior to installing PVC.

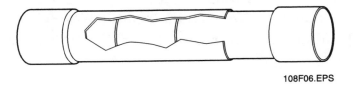

108F06.EPS

Figure 6. PVC Expansion Coupling

PVC conduit is manufactured in two types:

*   *Type EB* – Thin wall for underground use only when encased in concrete. Also referred to as *Type I*.
*   *Type DB* – Thick wall for underground use without encasement in concrete. Also referred to as *Type II*.

Type DB is available in two wall thicknesses, Schedule 40 and Schedule 80.
*   Schedule 40 is heavy wall for direct burial in the earth and above-ground installations.
*   Schedule 80 is extra heavy wall for direct burial in the earth, above-ground installations for general applications, and installations where the conduit is subject to physical damage.

PVC conduit is affected by higher-than-usual ambient temperatures. Support requirements for PVC are found in **NEC Table 347-8**. As with other conduit, it must be supported within three feet of each termination, but the maximum spacing between supports depends upon the size of the conduit. Some of the regulations for the maximum spacing of supports are:

- ½- to 1-inch conduit: every 3 feet
- 1¼- to 2-inch conduit: every 5 feet
- 2½- to 3-inch conduit: every 6 feet
- 3½- to 5-inch conduit: every 7 feet
- 6-inch conduit: every 8 feet

3.2.9 Liquid-Tight Flexible Nonmetallic Conduit

Liquid-tight flexible nonmetallic conduit (LFNC) was developed as a raceway for industrial equipment where flexibility was required and protection of conductors from liquids was also necessary. This is covered by *NEC Article 351, Part B*. Usage of LFNC has been expanded from industrial applications to outside and direct burial usage where listed.

Several varieties of LFNC have been introduced. The first product (LFNC-A) was commonly referred to as *hose*. It consisted of an inner and outer layer of neoprene with a nylon reinforcing web between the layers. A second-generation product (LFNC-B) consisted of a smooth wall, flexible PVC with a rigid PVC integral reinforcement rod. The third product (LFNC-C) was a nylon corrugated shape without any integral reinforcements. These three permitted LFNC raceway designs must be flame resistant with fittings **approved** for installation of electrical conductors. Nonmetallic connectors are listed for use and some liquid-tight metallic flexible conduit connectors are dual-listed for both metallic and nonmetallic liquid-tight flexible conduit.

LFNC is sunlight-resistant and suitable for use at conduit temperatures of 80°C dry and 60°C wet. It is available in ⅜-inch through 4-inch sizes. *NEC Section 351-23(b)* states that LFNC cannot be used where subject to physical damage or in lengths longer than six feet, except where properly secured, where flexibility is required, or as permitted by *NEC Section 351-23(a)(5)*. Also, it cannot be used to contain conductors in excess of 600 volts nominal except as permitted by *NEC Section 600-31*.

Liquid-tight flexible metal conduit is a raceway of circular cross section having an outer liquid-tight, nonmetallic, sunlight-resistant jacket over an inner flexible metal core with associated couplings and connectors.

Flex connectors are used to connect flexible conduit to boxes or equipment. They are available in straight, 45°, and 90° configurations (*Figure 7*).

STRAIGHT CONNECTOR

45° CONNECTOR

90° CONNECTOR

108F07.TIF

Figure 7. Flex Connectors

3.2.10 Flexible Metal Conduit

Flexible metal conduit, sometimes called *flex*, may be used for many kinds of wiring systems. Flexible metal conduit is made from a single strip of steel or aluminum, wound and interlocked. It is typically available in sizes from ⅜ inch to 4 inches in diameter. An illustration of flexible metal conduit is shown in *Figure 8*.

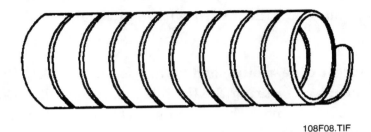

108F08.TIF

Figure 8. Flexible Metal Conduit

Flexible metal conduit is often used to connect equipment or machines that vibrate or move slightly during operation. Also, final connection to equipment having an electrical connection point that is marginally **accessible** is often accomplished with flexible metal conduit.

Flexible metal conduit is easily bent, but the minimum bending radius is the same as for other types of conduit. It should not be bent more than the equivalent of four quarter bends (360° total) between pull points (e.g., conduit bodies and boxes). It can be connected to boxes with a flexible conduit connector and to rigid conduit or EMT by using a combination coupling.

Note: When using a combination coupling, be sure the flexible conduit is pushed as far as possible into the coupling. This covers the end and protects the conductors from damage.

Two types of combination couplings are shown in *Figure 9*.

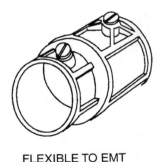
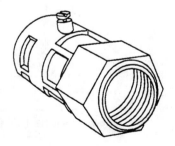

FLEXIBLE TO EMT

FLEXIBLE TO RIGID

108F09.TIF

Figure 9. Combination Couplings

Flexible metal conduit is generally available in two types: nonliquid-tight and liquid-tight. **NEC Articles 350 and 351** cover the uses of flexible metal conduit.

Liquid-tight flexible metal conduit has an outer covering of liquid-tight, sunlight-resistant flexible material that acts as a moisture seal. It is intended for use in wet locations. It is used primarily for equipment and motor connections when movement of the equipment is likely to occur. The number of bends, size, and support requirements for liquid-tight conduit are the same as for all flexible conduit. Fittings used with liquid-tight conduit must also be of the liquid-tight type.

Support requirements for flexible metal conduit are found in **NEC Sections 350-18 and 351-8.** Straps or other means of securing the flexible metal conduit must be spaced every 4½ feet and within 12 inches of each end. (This spacing is closer together than for rigid conduit.) However, at terminals where flexibility is necessary, lengths of up to 36 inches without support are permitted. Failure to provide proper support for flexible conduit can make pulling conductors difficult.

4.0.0 METAL CONDUIT FITTINGS

A large variety of conduit fittings are available to do electrical work. Manufacturers design and construct fittings to permit a multitude of applications. The type of conduit fitting used in a particular application depends upon the size and type of conduit, the type of fitting needed for the application, the location of the fitting, and the installation method. The requirements and proper applications of boxes and fittings (conduit bodies) are found in **NEC Section 300-15.** Some of the more common types of fittings are examined in the following sections.

4.1.0 COUPLINGS

Couplings are sleeve-like fittings that are typically threaded inside to join two male threaded pieces of rigid conduit or IMC. A piece of conduit with a coupling is shown in *Figure 10*.

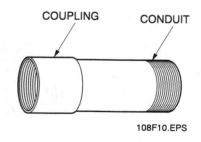

COUPLING   CONDUIT

108F10.EPS

Figure 10. Conduit And Coupling

Other types of couplings may be used depending upon the location and type of conduit. Several types are shown in *Figure 11*.

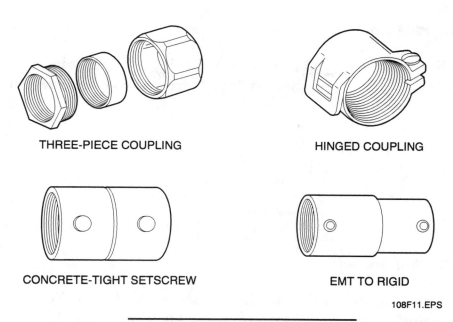

THREE-PIECE COUPLING

HINGED COUPLING

CONCRETE-TIGHT SETSCREW

EMT TO RIGID

108F11.EPS

Figure 11. Metal Conduit Couplings

4.2.0 CONDUIT BODIES

Conduit bodies are a separate portion of a conduit or tubing system that provide access through a removable cover(s) to the interior of the system at a junction of two or more sections of the system or at a terminal point of the system. They are usually cast and are significantly higher in cost than the stamped steel boxes permitted with EMT. However, there are situations in which conduit bodies are preferable, such as in outdoor locations, for appearance's sake in an **exposed location**, or to change types or sizes of raceways. Also, conduit bodies do not have to be supported, as do stamped steel boxes. They are also used when elbows or bends would not be appropriate.

NEC Section 370-16(c) states that conduit bodies cannot contain **splices**, **taps**, or devices unless they are durably and legibly marked by the manufacturer with their cubic inch capacity. The maximum number of conductors permitted in a conduit body is found using ***NEC Table 370-16(b)***. (See *Table 1*.)

Size of Conductor	Free Space Within Box for Each Conductor
No. 18	1.5 cubic inches
No. 16	1.75 cubic inches
No. 14	2.0 cubic inches
No. 12	2.25 cubic inches
No. 10	2.5 cubic inches
No. 8	3.0 cubic inches
No. 6	5.0 cubic inches

Table 1. Volume Required Per Conductor

4.2.1　Type C Conduit Bodies

Type C conduit bodies may be used to provide a pull point in a long conduit run or a conduit run that has bends totaling more than 360°. A Type C conduit body is shown in *Figure 12*.

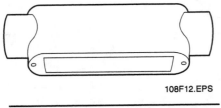

108F12.EPS

Figure 12. Type C Conduit Body

4.2.2　Type L Conduit Bodies

When referring to conduit bodies, the letter *L* represents an *elbow*. A Type L conduit body is used as a pulling point for conduit that requires a 90° change in direction. The cover is removed, then the wire is pulled out, coiled on the ground or floor, reinserted into the other conduit body's opening, and pulled. The cover and its associated gasket are then replaced. Type L conduit bodies are available with the cover on the back (Type LB), on the sides (Type LL or LR), or on both sides (Type LRL). Several Type L conduit bodies are shown in *Figure 13*.

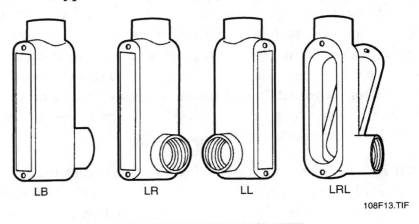

LB　　　　　LR　　　　　LL　　　　　LRL

108F13.TIF

Figure 13. Type L Conduit Bodies

Note:　　The cover and gasket must be ordered separately. Do not assume that these parts come with conduit bodies when they are ordered.

To identify Type L conduit bodies, use the following method:

Step 1　　Hold the body like a pistol.

Step 2　　Locate the opening on the body:

- If the opening is to the left, it is a Type LL.
- If the opening is to the right, it is a Type LR.
- If the opening is on top (back), it is a Type LB.
- If there are openings on both the left and the right, it is a Type LRL.

4.2.3 Type T Conduit Bodies

Type T conduit bodies are used to provide a junction point for three intersecting conduits and are used extensively in conduit systems. A Type T conduit body is shown in *Figure 14*.

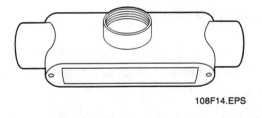

108F14.EPS

Figure 14. Type T Conduit Body

4.2.4 Type X Conduit Bodies

Type X conduit bodies are used to provide a junction point for four intersecting conduits. The removable cover provides access to the interior of the X so that wire pulling and splicing may be performed. A Type X conduit body is shown in *Figure 15*.

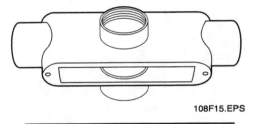

108F15.EPS

Figure 15. Type X Conduit Body

4.2.5 Threaded Weatherproof Hub

Threaded weatherproof hubs are used for conduit entering a box in a wet location. *Figure 16* shows a typical threaded weatherproof hub.

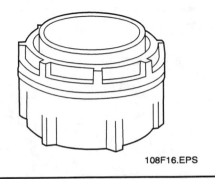

108F16.EPS

Figure 16. Threaded Weatherproof Hub

4.3.0 INSULATING BUSHINGS

An insulating bushing is either made of nonmetal or has an insulated throat. Insulating bushings are installed on the threaded end of conduit that enters a sheet metal enclosure.

4.3.1 Nongrounding Insulating Bushings

The purpose of a nongrounding insulating bushing is to protect the conductors from being damaged by the sharp edges of the threaded conduit end. *NEC Sections 345-15 and 347-12* state that where a conduit enters a box, fitting, or other enclosure, a bushing must be provided to protect the wire from abrasion unless the design of the box, fitting, or enclosure is such as to afford equivalent protection. *NEC Section 373-6(c)* references *Section 300-4(f)* that states where ungrounded conductors of No. 4 or larger enter a raceway in a cabinet or box enclosure, the conductors shall be protected by a substantial fitting providing a smoothly rounded insulating surface, unless the conductors are separated from the raceway fitting by substantial insulating material securely fastened in place. An exception is where threaded hubs or bosses that are an integral part of a cabinet, box enclosure, or raceway provide a smoothly rounded or flared entry for conductors. An insulating bushing is shown in *Figure 17*.

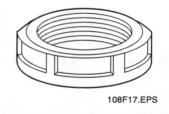

108F17.EPS

Figure 17. Insulating Bushing

4.3.2 Grounding Insulating Bushings

Grounded insulating bushings, usually called *grounding bushings*, are used to protect conductors and also have provisions for connection of an equipment grounding conductor. The ground wire, once connected to the grounding bushing, may be connected to the enclosure to which the conduit is connected. A grounding insulating bushing is shown in *Figure 18*.

BONDING SCREW

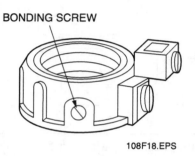

108F18.EPS

Figure 18. Grounding Insulating Bushing

4.4.0 OFFSET NIPPLES

Offset nipples are used to connect two pieces of electrical equipment in close proximity where a slight offset is required. They come in sizes ranging from ½" to 2" in diameter. See *Figure 19*.

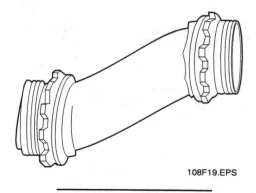

108F19.EPS

Figure 19. Offset Nipple

5.0.0   BOXES

A box is installed at each outlet, switch, or junction point for all wiring installations and branch circuits. Boxes are made from either metallic or nonmetallic material. *Figure 20* shows various boxes used in raceway systems.

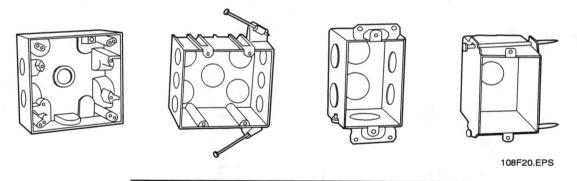

108F20.EPS

Figure 20. Various Boxes Used In Raceway Systems

5.1.0   METAL BOXES

Metal boxes are made from sheet steel. The surface is galvanized to resist corrosion and provide a continuous ground. Refer to **NEC Section 370-40** for information on thickness and grounding provisions. Metal boxes are made with removable circular sections called *pryouts* or *knockouts*. These circular sections are removed to make openings for conduit or cable connections.

5.1.1   Pryouts

In a pryout, a section is cut completely through the metal but only part of the way around, leaving solid metal tabs at two points. A slot is cut in the center of the pryout. To remove the pryout, a screwdriver is inserted into the slot and twisted to break the solid tabs (*Figure 21*).

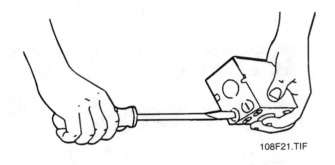

Figure 21. Pryout Removal

5.1.2 Knockouts

Knockouts are pre-punched circular sections that do not include a pryout slot. The knockout is easily removed when sharply hit by a hammer and punch, as shown in *Figure 22*.

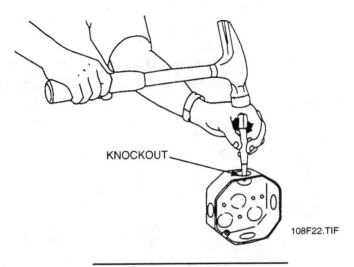

KNOCKOUT

108F22.TIF

Figure 22. Knockout Removal

Often conduit must enter boxes, cabinets, or panels that do not have pre-cut knockouts. In these cases, a knockout punch can be used to make a hole for the conduit connection. A knockout punch is shown in *Figure 23*.

5.2.0 NONMETALLIC BOXES

Nonmetallic boxes are made of PVC or Bakelite (a fiber-reinforced plastic). Nonmetallic boxes are often used in corrosive environments. *NEC Section 370-3* covers the use of nonmetallic boxes and the types of conduit, fittings, and grounding requirements for specific applications.

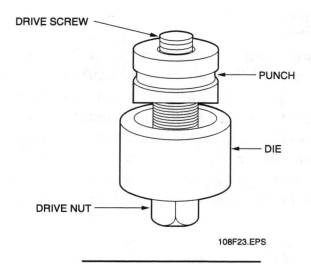

DRIVE SCREW

PUNCH

DIE

DRIVE NUT

108F23.EPS

Figure 23. Knockout Punch

6.0.0 BUSHINGS AND LOCKNUTS

Conduit is joined to boxes by connectors, adapters, threaded hubs, or locknuts.

Bushings protect the wires from the sharp edges of the conduit. Bushings are usually made of plastic or metal. Some metal bushings have a grounding screw to permit a **bonding wire** to be installed. Some different types of plastic and metal bushings are shown in *Figure 24*.

PLASTIC
INSULATING
BUSHING

METALLIC
BUSHING

INSULATED
METALLIC
BUSHING

108F24.TIF

Figure 24. Bushings

Locknuts are used on the inside and outside walls of the box to which the conduit is connected. A grounding locknut may be needed if a bonding wire is to be installed. Special sealing locknuts are also used in wet locations. Several types of locknuts are shown in *Figure 25*.

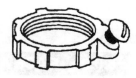

SEALING LOCKNUT

STANDARD LOCKNUT

STANDARD LOCKNUT

GROUNDING LOCKNUT

108F25.TIF

Figure 25. Locknuts

7.0.0 SEALING FITTINGS

Hazardous locations in manufacturing plants and other industrial facilities involve a wide variety of flammable gases and vapors and ignitable dusts. These hazardous substances have widely different flash points, ignition temperatures, and flammable limits requiring fittings that can be sealed. Sealing fittings are installed in conduit runs to minimize the passage of gases, vapors, or flames through the conduit and reduce the accumulation of moisture. They are required by *NEC Article 500* in hazardous locations where explosions may occur. They are also required where conduit passes from a hazardous location of one classification to another or to an unclassified location. Several types of sealing fittings are shown in *Figure 26*.

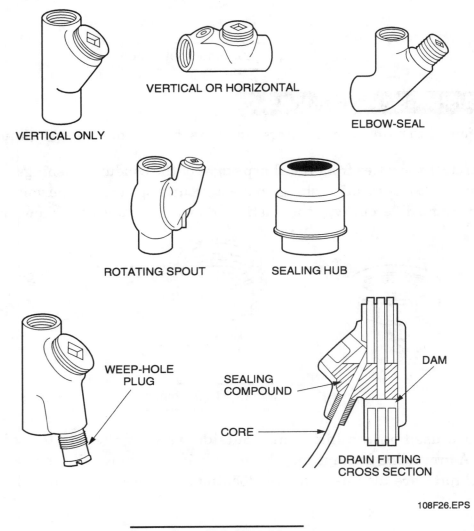

VERTICAL ONLY

VERTICAL OR HORIZONTAL

ELBOW-SEAL

ROTATING SPOUT

SEALING HUB

WEEP-HOLE PLUG

SEALING COMPOUND

CORE

DAM

DRAIN FITTING CROSS SECTION

108F26.EPS

Figure 26. Sealing Fittings

8.0.0 RACEWAY SUPPORTS

Raceway supports are available in many types and configurations. This section discusses the most common conduit supports found in electrical installations. *NEC Section 300-11* discusses the requirements for branch circuit wiring that is supported from above. Electrical

ELECTRICAL — TRAINEE TASK MODULE 26108

equipment and raceways must have their own supporting methods and may not be supported by the supporting hardware of a fire-rated roof/ceiling assembly.

8.1.0 STRAPS

Straps are used to support conduit to a surface (see *Figure 27*). The spacing of these supports must conform to the minimum support spacing requirements for each type of conduit. One- and two-hole straps are used for all types of conduit: EMT, GRC, IMC, PVC, and flex. The straps can be flexible or rigid. Two-part straps are used to secure conduit to electrical framing channels (struts). Parallel and right angle beam clamps are also used to support conduit from structural members.

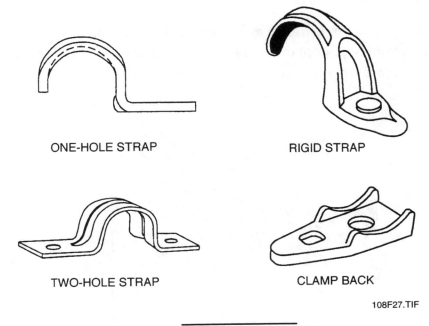

ONE-HOLE STRAP   RIGID STRAP

TWO-HOLE STRAP   CLAMP BACK

108F27.TIF

Figure 27. Straps

Clamp back straps can also be used with a backplate to maintain the ¼-inch spacing from the surface required for installations in wet locations.

8.2.0 STANDOFF SUPPORTS

The standoff support, often referred to as a *Minerallac* (the name of a manufacturer of this type of support), is used to support conduit away from the supporting structure. In the case of the one-hole and two-hole straps, the conduit must be kicked up wherever a fitting occurs. If standoff supports are used, the conduit is held away from the supporting surface, and no offsets (**kicks**) are required in the conduit at the fittings. Standoff supports may be used to support all types of conduit including GRC, IMC, EMT, PVC, and flex, as well as tubing installations. A standoff support is shown in *Figure 28*.

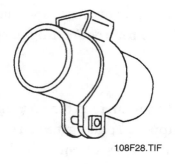

108F28.TIF

Figure 28. Standoff Support

8.3.0 ELECTRICAL FRAMING CHANNELS

Electrical framing channels or other similar framing materials are used together with Unistrut-type conduit clamps to support conduit (see *Figure 29*). They may be attached to a ceiling, wall, or other surface or be supported from a trapeze hanger.

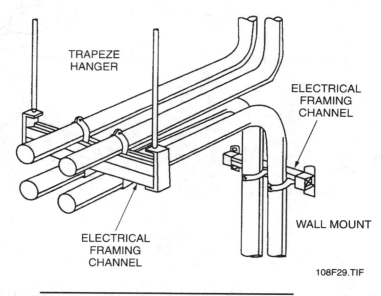

Figure 29. Electrical Framing Channels

8.4.0 BEAM CLAMPS

Beam clamps are used with suspended hangers. The raceway is attached to or laid in the hanger. The hanger is suspended by a threaded rod. One end of the threaded rod is attached to the hanger and the other end is attached to a beam clamp. The beam clamp is then attached to a beam. A beam clamp with wireway support assembly is shown in *Figure 30*.

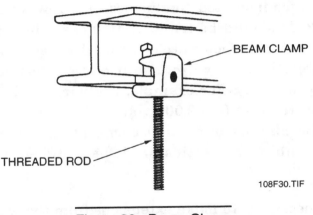

THREADED ROD

BEAM CLAMP

108F30.TIF

Figure 30. Beam Clamp

9.0.0 WIREWAYS

Wireways are sheet metal **troughs** provided with hinged or screw-on removable covers. Like other types of raceways, wireways are used for housing electric wires and cables. Wireways are available in various lengths, including 1, 2, 3, 4, 5, and 10 feet. The availability of various lengths allows runs of any exact number of feet to be made without cutting the wireway ducts. Wireways are dealt with specifically in *NEC Article 362*.

As listed in *NEC Section 362-5*, the maximum number of current-carrying conductors at any cross section of a wireway is 30. However, this number does not include conductors for signaling circuits or conductors between a motor and its starter. Also, it is important to know that the sum of the cross-sectional areas of all conductors contained at any cross section must not exceed 20% of the interior cross-sectional area of the wireway. Refer to *NEC Section 362-5* for the exceptions to this rule.

It is also noted in *NEC Section 362-7* that conductors, together with splices and taps, must not fill the wireway to more than 75% of its cross-sectional area. Each wireway is also rated for a maximum permitted conductor size for any single conductor. No conductor larger than the rated size is to be installed in any wireway.

9.1.0 AUXILIARY GUTTERS

Strictly speaking, an auxiliary gutter is a wireway that is intended to add to wiring space at switchboards, meters, and other distribution locations. Auxiliary gutters are dealt with specifically in *NEC Article 374*. Even though the component parts of wireways and auxiliary gutters are identical, you should be familiar with the differences in their use. Auxiliary gutters are used as parts of complete assemblies of apparatus such as switchboards, distribution centers, and control equipment. However, an auxiliary gutter may only contain conductors or busbars, even though it looks like a surface metal raceway which may contain devices and equipment. Unlike auxiliary gutters, wireways represent a type of wiring because they are used to carry conductors between points located considerable distances apart.

The allowable ampacities for insulated conductors in wireways are given in *NEC Tables 310-16 through 310-19*. It should be noted that these tables are used for raceways in general. These NEC tables and the notes are often used to determine if the correct materials are on hand for an installation. They are also used to determine if it is possible to add conductors in an existing wireway. For example, if a wireway is to contain not more than three insulated conductors rated at 0 to 2,000 volts at 60°C to 90°C, the table to use is *NEC Table 310-16*. Where the number of current-carrying conductors in a raceway or cable exceeds three, the allowable ampacities shall be reduced as shown in *NEC Table 310-15(b)(2)(a)*.

In many situations, it is necessary to make extensions from the wireways to wall receptacles and control devices. In these cases, *NEC Section 362-11* specifies that these extensions be made using any wiring method presented in *NEC Article 300* that includes a means for equipment grounding. Finally, as required in *NEC Section 362-12*, wireways must be marked in such a way that their manufacturer's name or trademark will be visible.

As you can see in *Figure 31*, a wide range of fittings is required for connecting wireways to one another and to fixtures such as switchboards, power panels, and conduit.

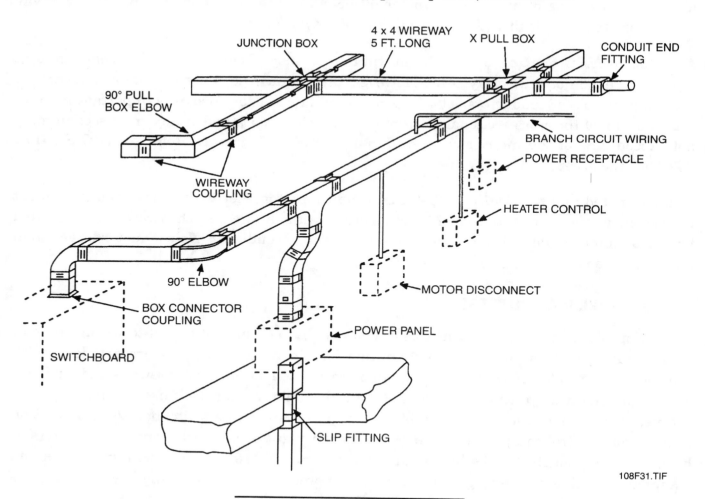

108F31.TIF

Figure 31. Wireway System Layout

9.2.0 TYPES OF WIREWAYS

Rectangular duct-type wireways come as either hinged-cover or screw-cover troughs. Typical lengths are 1, 2, 3, 4, 5, and 10 feet. Shorter lengths are also available. Raintight troughs are permitted to be used in environments where moisture is not permitted within the raceway. However, the raintight trough should not be confused with the raintight lay-in wireway which has a hinged cover. *Figure 32* shows a raintight trough with a removable side cover.

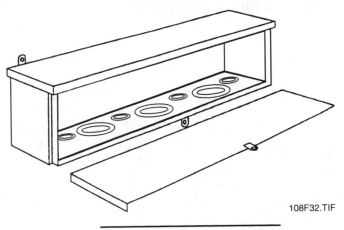

108F32.TIF

Figure 32. Raintight Trough

Wireway troughs are exposed when first installed. Whenever possible, they are mounted on the ceilings or walls, although they may sometimes be suspended from the ceiling. Note that in *Figure 33*, the trough has knockouts similar to those found on junction boxes. After the wireway system has been installed, branch circuits are brought from the distribution panels using conduit. The conduit is joined to the wireway at the most convenient knockout possible.

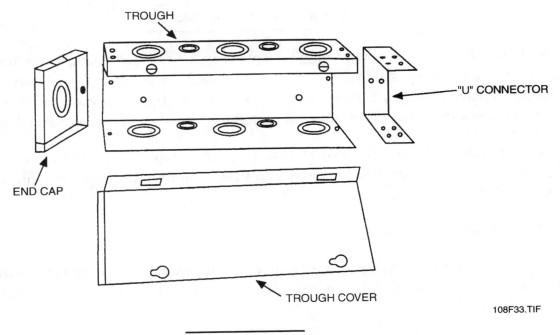

108F33.TIF

Figure 33. Trough

Wireway components such as trough crosses, 90° internal elbows, and tee connectors serve the same function as fittings on other types of raceways. The fittings are attached to the duct using slip-on connectors. All attachments are made with nuts and bolts or screws. When assembling wireways, always place the head of the bolt on the inside and the nut on the outside so that the conductors will not be resting against a sharp edge. It is usually best to assemble sections of the wireway system on the floor, and then raise the sections into position. An exploded view of a section of wireway is shown in *Figure 34*. Both the wireway fittings and the duct come with screw-on, hinged, or snap-on covers to permit conductors to be laid in or pulled through.

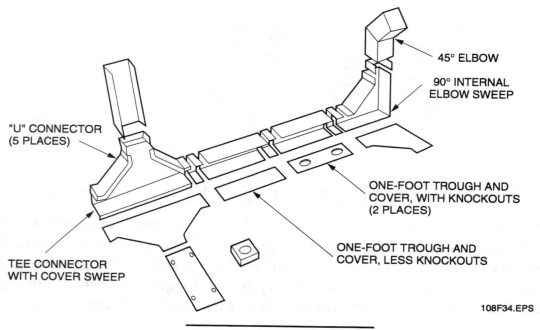

108F34.EPS

Figure 34. Wireway Sections

The NEC specifies that wireways may be used only for exposed work. Therefore, they cannot be used in underfloor installations. If they are used for outdoor work, they must be of an approved raintight construction. It is important to note that wireways must not be installed where they are subject to severe physical damage, corrosive vapors, or hazardous locations.

Wireway troughs must be installed so that they are supported at distances not exceeding 5 feet. When specially-approved supports are used, the distance between supports must not exceed 10 feet.

9.2.1 Wireway Fittings

Many different types of fittings are available for wireways, especially for use in exposed, dry locations. The following sections explain fittings commonly used in the electrical craft.

9.2.2 Connectors

Connectors are used to join wireway sections and fittings. Connectors are slipped inside the end of a wireway section and are held in place by small bolts and nuts. Alignment slots allow the connector to be moved until it is flush with the inside surface of the wireway. After the connector is in position, it can be bolted to the wireway. This helps to ensure a strong rigid connection. Connectors have a friction hinge that helps hold the wireway cover open when needed. A connector is shown in *Figure 35*.

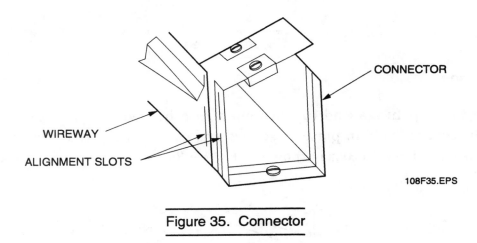

CONNECTOR

WIREWAY

ALIGNMENT SLOTS

108F35.EPS

Figure 35. Connector

9.2.3 End Plates

End plates, or closing plates, are used to seal the ends of wireways. They are inserted into the end of the wireway and fastened by screws and bolts. End plates contain knockouts so that conduit or cable may be extended from the wireway. An end plate is shown in *Figure 36*.

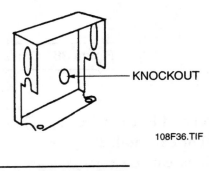

KNOCKOUT

108F36.TIF

Figure 36. End Plate

9.2.4 Tees

Tee fittings are used when a tee connection is needed in a wireway system. A tee connection is used where circuit conductors may branch in different directions. The tee fitting's covers and sides can be removed for access to splices and taps. Tee fittings are attached to other wireway sections using standard connectors. A tee is shown in *Figure 37*.

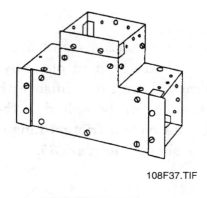

108F37.TIF

Figure 37. Tee

9.2.5 Crosses

Crosses have four openings and are attached to other wireway sections with standard connectors. The cover is held in place by screws and can be easily removed for laying in wires or for making connections. A cross is shown in *Figure 38*.

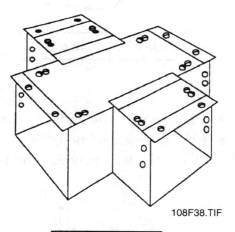

108F38.TIF

Figure 38. Cross

9.2.6 Elbows

Elbows are used to make a bend in the wireway. They are available in angles of 22½°, 45°, or 90°, and are either internal or external. They are attached to wireway sections with standard connectors. Covers and sides can be removed for wire installation. The inside corners of elbows are rounded to prevent damage to conductor insulation. An inside elbow is shown in *Figure 39*.

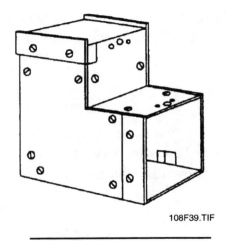

108F39.TIF

Figure 39. 90° Inside Elbow

9.2.7 Telescopic Fittings

Telescopic or slip fittings may be used between lengths of wireway. Slip fittings are attached to standard lengths by setscrews and usually adjust from ½ inch to 11½ inches. Slip fittings have a removable cover for installing wires and are similar in appearance to a nipple.

9.3.0 WIREWAY SUPPORTS

Wireways should be securely supported where run horizontally at each end and at intervals of no more than 5 feet or for individual lengths greater than 5 feet at each end or joint, unless listed for other support intervals. In no case shall the support distance be greater than 10 feet, in accordance with *NEC Section 362-8*. If possible, wireways can be mounted directly to a surface. Otherwise, wireways are supported by hangers or brackets.

9.3.1 Suspended Hangers

In many cases, the wireway is supported from a ceiling, beam, or other structural member. In such installations, a suspended hanger (*Figure 40*) may be used to support the wireway.

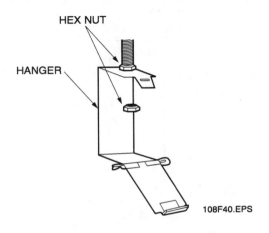

HEX NUT

HANGER

108F40.EPS

Figure 40. Suspended Hanger

The wireway is attached to or laid in the hanger. The hanger is suspended by a threaded rod. One end of the rod is attached to the hanger with hex nuts. The other end of the rod is attached to a beam clamp or anchor.

9.3.2 Gusset Brackets

Another type of support used to mount wireways is a gusset bracket. This is an L-type bracket that is mounted to a wall. The wireway rests on the bracket and is attached by screws or bolts. A gusset bracket is shown in *Figure 41*.

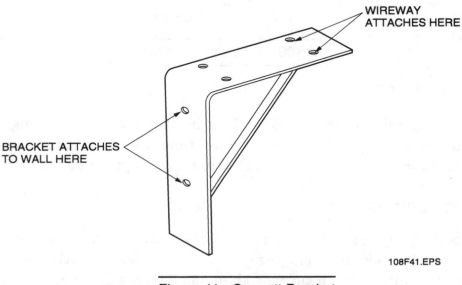

108F41.EPS

Figure 41. Gussett Bracket

9.3.3 Standard Hangers

Standard hangers are made in two pieces. The two pieces are combined in different ways for different installation requirements. The wireway is attached to the hanger by bolts and nuts. A standard hanger is shown in *Figure 42*.

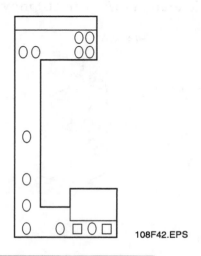

108F42.EPS

Figure 42. Standard Hanger

9.3.4 Wireway Hangers

When a larger wireway must be suspended, a wireway hanger may be used. A wireway hanger is made by suspending a piece of strut from a ceiling, beam, or other structural member. The strut is suspended by threaded rods attached to beam clamps or other ceiling anchors, as shown in *Figure 43*.

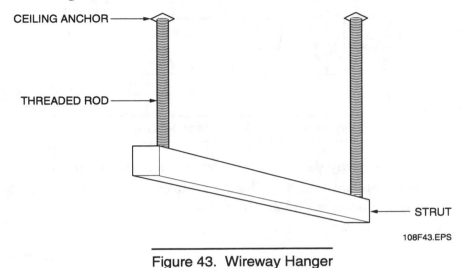

CEILING ANCHOR

THREADED ROD

STRUT

108F43.EPS

Figure 43. Wireway Hanger

9.4.0 OTHER TYPES OF RACEWAYS

In this section, other types of raceways will be discussed. Depending on the particular purpose for which they are intended, raceways include enclosures such as surface metal and nonmetallic raceways, and underfloor raceways.

9.4.1 Surface Metal And Nonmetallic Raceways

Surface metal raceways consist of a wide variety of special raceways designed primarily to carry power and communications wiring to locations on the surface of ceilings or walls of building interiors.

Installation specifications of both surface metal raceways and surface nonmetallic raceways are listed in detail in **NEC Article 352**. All these raceways must be installed in dry, interior locations. The number of conductors, their amperage, and the allowable cross-sectional area of the conductors, as well as regulations for combination raceways, are specified in **NEC Tables 310-16 through 310-19 and NEC Article 352**.

One use of surface metal raceways is to protect conductors that run to non-accessible outlets.

Surface metal and nonmetallic raceways have been divided into subgroups based on the specific purpose for which they are intended. There are three small surface raceways that are primarily used for extending power circuits from one point to another. In addition, there are six larger surface raceways that have a much wider range of applications. Typical cross sections of the first three smaller raceways are shown in *Figure 44*.

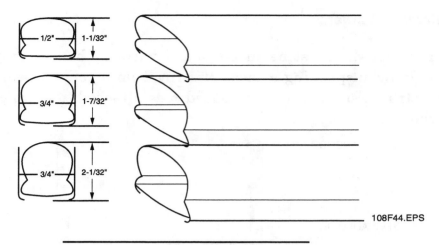

Figure 44. Smaller Surface Raceways

Additional surface metal raceway designs are referred to as *pancake raceways*, because their flat cross sections resemble pancakes. Their primary use is to extend power, lighting, telephone, or signal wire to locations away from the walls of a room without embedding them under the floor. A pancake raceway is shown in *Figure 45*.

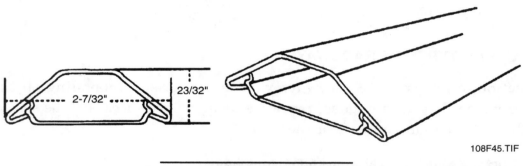

Figure 45. Pancake Raceway

There are also surface metal raceways available that house two or three different conductor raceways. These are referred to as *twinduct* or *tripleduct*. These raceways permit different circuits, such as power and signal, to be placed within the same raceway.

The number and types of conductors permitted to be installed and the capacity of a particular surface raceway must be calculated and matched with NEC requirements, as discussed previously. *NEC Tables 310-16 through 310-19* are used for surface raceways in the same manner in which they are used for wireways. For surface raceway installations with more than three conductors in each raceway, particular reference must be made to *NEC Table 310-15(b)(2)(a)*.

9.4.2 Plugmold Multi-Outlet Systems

Manufacturers offer a wide variety of plugmold multi-outlet surface raceways. Their function is to hold receptacles and other devices within the raceway. When the surface raceways are used in this manner, the assembly is referred to as a *multi-outlet assembly*.

Plugmold systems are either wired in the field or come pre-wired from the factory. *Figure 46* shows typical plugmold cross sections, wiring configurations, and some of the available fittings for the system.

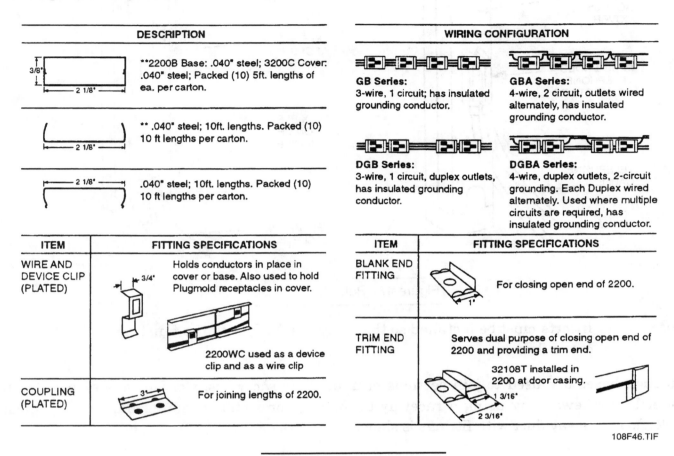

Figure 46. Plugmold System

9.4.3 Pole Systems

There are many situations in which power and other electric circuits have to be carried from overhead wiring systems to devices that are not located near existing wall outlets or control circuits. This type of wiring is typically used in open office spaces where cubicles are provided by temporary dividers. Poles are used to accomplish this. Some common manufacturers' names for these poles include *Tele-Power poles, Quick-E poles,* and *Walkerpoles.* The poles usually come in lengths suitable for 10-, 12-, or 15-foot ceilings. *Figure 47* shows two views of the pole base in a typical system.

9.4.4 Underfloor Systems

Underfloor raceway systems were developed to provide a practical means of bringing conductors for lighting, power, and signaling to cabinets and consoles. Underfloor raceways are available in 10-foot lengths and widths of 4 and 8 inches. The sections are made with inserts spaced every 24 inches. The inserts can be removed for outlet installation. These are explained in *NEC Article 354*.

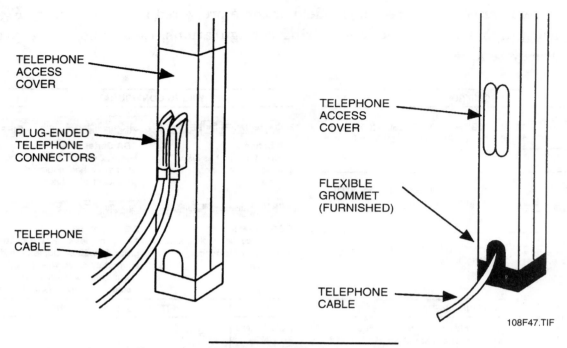

Figure 47. Pole Base Details

Note: Inserts must be installed so that they are flush with the finished grade of the floor.

Junction boxes are used to join sections of underfloor raceways. Conduit is also used with underfloor raceways by using a raceway-to-conduit connector (conduit adapter). A typical underfloor raceway duct with fittings is shown in *Figure 48*.

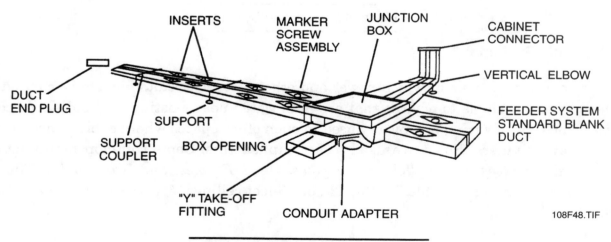

Figure 48. Underfloor Raceway Duct

This wiring method makes it possible to place a desk or table in any location where it will always be over, or very near to, a duct line. The wiring method for lighting and power between cabinets and the raceway junction boxes may be conduit, underfloor raceway, wall elbows, and cabinet connectors. ***NEC Article 354*** covers the installation of underfloor raceways.

9.4.5 Cellular Metal Floor Raceways

A cellular metal floor raceway is a type of floor construction designed for use in steel-frame buildings. In these buildings, the members supporting the floor between the beams consist of sheet steel rolled into shapes. These shapes are combined to form cells, or closed passageways, which extend across the building. The cells are of various shapes and sizes, depending upon the structural strength required. The cells of this type of floor construction form the raceways, as shown in *Figure 49*.

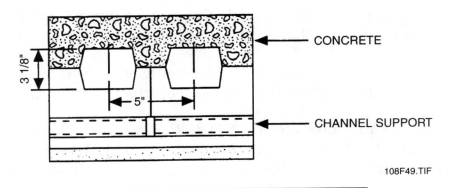

Figure 49. Cross Section Of A Cellular Floor

Connections to the cells are made using headers which extend across the cells. A header connects only to those cells which are to be used as raceways for conductors. A junction box or access fitting is necessary at each joint where a header connects to a cell. Two or three separate headers, connecting to different sets of cells, may be used for different systems. For example, light and power, signaling systems, and public telephones would each have a separate header. A special elbow fitting is used to extend the headers up to the distribution equipment on a wall or column. **NEC Article 356** covers the installation of cellular metal floor raceways.

9.4.6 Cellular Concrete Floor Raceways

The term *precast cellular concrete floor* refers to a type of floor used in steel-frame, concrete-frame, and wall-bearing construction. In this type of system, the floor members are precast with hollow voids which form smooth, round cells. The cells form raceways which can be adapted, using fittings, for use as underfloor raceways. A precast cellular concrete floor is fire-resistant and requires no further fireproofing. The precast reinforced concrete floor members form the structural floor and are supported by beams or bearing walls. Connections to the cells are made with headers which are secured to the precast concrete floor. **NEC Article 358** covers the installation of cellular concrete floor raceways.

10.0.0 CABLE TRAYS

Cable trays function as a support for conductors and tubing (see *NEC Article 318*). A cable tray has the advantage of easy access to conductors, and thus lends itself to installations where the addition or removal of conductors is a common practice. Cable trays are fabricated from aluminum, steel, and fiberglass. Cable trays are available in two basic forms: ladder and trough. Ladder tray, as the name implies, consists of two parallel channels connected by rungs. Trough consists of two parallel channels (side rails) having a corrugated, ventilated bottom, or a corrugated, solid bottom. (There is also a special center rail cable tray available for use in light-duty applications such as telephone and sound wiring. We will discuss this type of cable tray in more detail in Level 2.)

Cable trays are commonly available in 12- and 24-foot lengths. They are usually available in widths of 6, 9, 12, 18, 24, 30, and 36 inches, and load depths of 4, 6, and 8 inches.

Cable trays may be used in most electrical installations. Cable trays may be used in air handling ceiling space, but only to support the wiring methods permitted in such spaces by *NEC Section 300-22(c)*. Also, cable trays may be used in Class 1, Division 2 locations according to *NEC Section 501-4*. Cable trays may also be used above a suspended ceiling that is not used as an air handling space. Some manufacturers offer an aluminum cable tray that is coated with PVC for installation in caustic environments. A typical cable tray system with fittings is shown in *Figure 50*.

Wire and cable installation in cable trays is defined by the NEC. Read *NEC Article 318* to become familiar with the requirements and restrictions made by the NEC for safe installation of wire and cable in a cable tray.

Metallic cable trays that support electrical conductors must be grounded as required by *NEC Article 250*. Where steel and aluminum cable tray systems are used as an equipment grounding conductor, all of the provisions of *NEC Section 318-7(b)* must be complied with.

WARNING!   Do not stand on, climb in, or walk on a cable tray.

10.1.0 CABLE TRAY FITTINGS

Cable tray fittings are part of the cable tray system and provide a means of changing the direction or dimension of the different trays. Some of the uses of horizontal and vertical tees, horizontal and vertical bends, horizontal crosses, reducers, barrier strips, covers, and box connectors are shown in *Figure 50*.

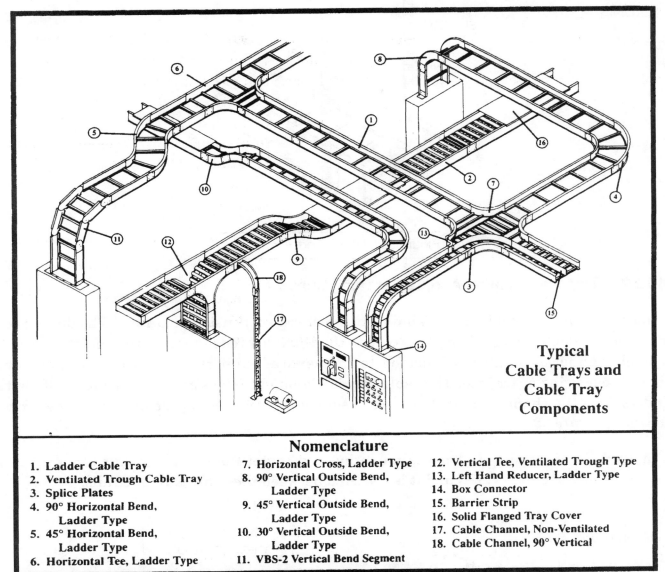

Nomenclature

1. Ladder Cable Tray
2. Ventilated Trough Cable Tray
3. Splice Plates
4. 90° Horizontal Bend, Ladder Type
5. 45° Horizontal Bend, Ladder Type
6. Horizontal Tee, Ladder Type
7. Horizontal Cross, Ladder Type
8. 90° Vertical Outside Bend, Ladder Type
9. 45° Vertical Outside Bend, Ladder Type
10. 30° Vertical Outside Bend, Ladder Type
11. VBS-2 Vertical Bend Segment
12. Vertical Tee, Ventilated Trough Type
13. Left Hand Reducer, Ladder Type
14. Box Connector
15. Barrier Strip
16. Solid Flanged Tray Cover
17. Cable Channel, Non-Ventilated
18. Cable Channel, 90° Vertical

108F50.TIF

Figure 50. Cable Tray System

10.2.0 CABLE TRAY SUPPORTS

Cable trays are usually supported in one of five ways: direct rod suspension, trapeze mounting, center hung, wall mounting, and pipe rack mounting.

10.2.1 Direct Rod Suspension

The direct rod suspension method of supporting cable tray uses threaded rods and hanger clamps. One end of the threaded rod is connected to the ceiling or other overhead structure. The other end is connected to hanger clamps that are attached to the cable tray side rails. A direct rod suspension assembly is shown in *Figure 51*.

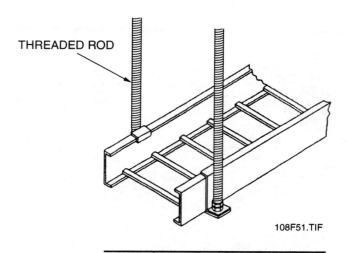

THREADED ROD

108F51.TIF

Figure 51. Direct Rod Suspension

10.2.2   Trapeze Mounting And Center Hung Support

Trapeze mounting of cable tray is similar to direct rod suspension mounting. The difference is in the method of attaching the cable tray to the threaded rods. A structural member, usually a steel channel or strut, is connected to the vertical supports to provide an appearance similar to a swing or trapeze. The cable tray is mounted to the structural member. Often, the underside of the channel or strut is used to support conduit. A trapeze mounting assembly is shown in *Figure 52*.

A method that is similar to trapeze mounting is a center hung tray support (*Figure 52*). In this case, only one rod is used and it is centered between the cable tray side rails.

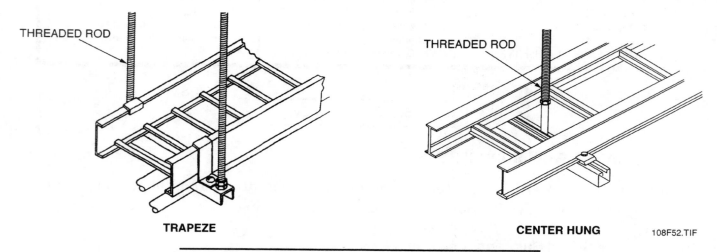

THREADED ROD

THREADED ROD

TRAPEZE

CENTER HUNG

108F52.TIF

Figure 52. Trapeze Mounting And Center Hung Support

10.2.3 Wall Mounting

Wall mounting is accomplished by supporting the cable tray with structural members attached to the wall. This method of support is often used in tunnels and other underground or sheltered installations where large numbers of conductors interconnect equipment that is separated by long distances. A wall mounting assembly is shown in *Figure 53*.

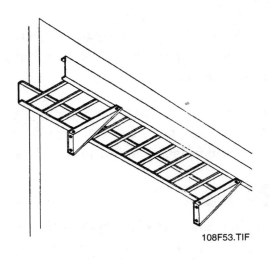

108F53.TIF

Figure 53. Wall Mounting

10.2.4 Pipe Rack Mounting

Pipe racks are structural frames used to support piping that interconnects equipment in outdoor industrial facilities. Usually, some space on the rack is reserved for conduit and cable tray. Pipe rack mounting of cable tray is often used when power distribution and electrical wiring is routed over a large area.

11.0.0 STORING RACEWAYS

Proper and safe methods of storing conduit, wireways, raceways, and cable trays may sound like a simple task, but improper storage techniques can result in wasted time and damage to the raceways, as well as personal injury. There are correct ways to store raceways that will help avoid costly damage, save time in identifying stored raceways, and reduce the chance of personal injury.

Pipe racks are commonly used for storing conduit. The racks provide support to prevent bending, sagging, distorting, scratching, or marring of conduit surfaces. Most racks have compartments where different types and sizes of conduit can be separated for ease of identification and selection. The storage compartments in racks are usually elevated to help avoid damage that might occur at floor level. Conduit that is stored at floor level is easily damaged by people and other materials or equipment in the area.

The ends of stored conduit should be sealed to help prevent contamination and damage. Conduit ends can be capped, taped, or plugged.

Always inspect raceway before storing it to make sure that it is clean and not damaged. It is discouraging to get raceway for a job and find that it is dirty or damaged. Also, make sure that the raceway is stored securely so that when someone comes to get it for a job, it will not fall in any way that could cause injury.

To prevent contamination and corrosion of stored raceway, it should be covered with a tarpaulin or other suitable covering. It should also be separated from non-compatible materials such as hazardous chemicals.

Wireways, surface metal raceways, and cable trays should always be stored off the ground on boards in an area where people will not step on it and equipment will not run over it. Stepping or running over raceway bends the metal and makes it unusable.

12.0.0   HANDLING RACEWAYS

Raceway is made to strict specifications. It can easily be damaged by careless handling. From the time raceway is delivered to a job site until the installation is complete, use proper and safe handling techniques. These are a few basic guidelines for handling raceway that will help avoid damaging or contaminating it:

- Never drag raceway off of a delivery truck or off other lengths of raceway.
- Never drag raceway on the ground or floor. Dragging raceway can cause damage to the ends.
- Keep the thread protection caps on when handling or transporting conduit raceway.
- Keep raceway away from any material that might contaminate it during handling.
- Flag the ends of long lengths of raceway when transporting it to the job site.
- Never drop or throw raceway when handling it.
- Never hit raceway against other objects when transporting it.
- Always use two people when carrying long pieces of raceway. Make sure that you both stay on the same side and that the load is balanced. Each person should be about ¼ of the length of the raceway from the end. Lift and put down the raceway at the same time.

13.0.0   DUCTING

In the common vocabulary of the electrical trade, a duct is a single enclosed raceway, or runway, through which conductors or cables can be led. Basically, ducting is a system of ducts. However, underground duct systems include manholes, transformer vaults, and risers.

There are several reasons for running power lines underground rather than overhead. In some situations, an overhead high-voltage line would be dangerous, or the space may not be adequate. For aesthetic reasons, architectural plans may require buried lines throughout a subdivision or a planned community. Tunnels may already exist, or be planned, for carrying steam or water lines. In any of these situations, underground installations are appropriate. Underground cables may be buried directly in the ground or run through conduit.

In underground construction, a duct system provides a safe passageway for power lines, communication cables, or both. In buildings, underfloor raceways and cellular floor raceways are built to provide ducting so that electricity will be available throughout a large area. As an electrician, you need to know the approved methods of constructing underground ducting. You also need to know how to avoid potential electrical hazards, both in original construction and maintenance. It is essential to understand the requirements and limitations imposed on running wires through underfloor and cellular floor raceways and ducts.

14.0.0   UNDERGROUND SYSTEMS

There are five different ways to install cable underground:

- Duct line
- Conductors located in tunnels
- Conductors buried directly in the earth
- PVC
- GRC

The method used will depend on the materials which are available, the number of conductors to be pulled, and the type of wiring to be done (service drop, branch circuit, commercial, etc.).

14.1.0   DUCT LINE

A duct line consists of at least one subway placed in a trench and covered with earth. Conduit, in some cases, can be classified as a *subway*. Subways may come in a single duct line or multiple duct lines (2, 3, 4, and 6 subways per section). The depth in which the duct will be placed is determined using **NEC Table 300-5**. The conduit subways are encased in concrete or other materials. This provides good mechanical strength and allows for power losses from the cable to be dissipated into the earth.

In underground cable installations, a duct is a buried conduit through which a cable passes. Manholes are set at intervals in an underground duct run. Manholes provide access through throats (sometimes called *chimneys*). At ground level, or street surface level, a manhole cover closes off the manhole area tightly. An individual cable length running underground normally terminates at a manhole, where it is spliced to another length of cable. A duct line may consist of a single conduit or several, each carrying a cable length from one manhole to the next.

Manholes provide room for installing lengths of cable in conduit lines. They are also used for maintenance work and for performing tests. Workers enter a manhole from above. In a two-way manhole, cables enter and leave in only two directions. There are also three-way and four-way manholes. Often manholes are located at the intersection of two streets so that they can be used for cables leaving in four directions. Manholes are usually constructed of brick or concrete. Their design must provide room for drainage and for workers to move around inside them. A similar opening known as a *handhole* is sometimes provided for splicing on lateral two-way duct lines.

Transformer vaults house power transformers, voltage regulators, network protectors, meters, and circuit breakers. A cable may end at a transformer vault. Other cables end at a customer's substation or terminate as risers which connect with overhead lines.

14.2.0   DUCT MATERIALS

Underground duct lines can be made of fiber, vitrified tile, iron conduit, plastic, or poured concrete. The inside diameter of the ducting for a specific job is determined by the size of the cable that will be drawn into the duct. Sizes from two to six inches (inside diameter) are available for most types of ducting.

CAUTION: Be careful when working with unfamiliar duct materials. In older installations, asbestos/cement duct may have been used. You must be certified to remove or disturb asbestos.

14.3.0   PLASTIC CONDUIT

Plastic conduit may be made of PVC (polyvinyl chloride), PE (polyethylene), or styrene. Since this type of conduit is available in lengths up to 20 feet, fewer couplings are needed than with other types of ducting. Plastic conduit is popular because it is easy to install, requires less labor than other types of conduit, and is low in cost.

14.4.0   MONOLITHIC CONCRETE DUCT

Monolithic concrete duct is poured at the job site. Multiple duct lines can be formed using rubber tubing cores on spacers. The cores may be removed after the concrete has set. A die containing steel tubes, known as a *boat*, can also be used to form ducts. It is pulled slowly through the trench on a track as concrete is poured from the top. Poured concrete ducting made by either method is relatively expensive, but offers the advantage of creating a very clean duct interior with no residue that can decay. The rubber core method is especially useful for curving or turning part of a duct system.

14.5.0   CABLE-IN-DUCT

One of the most popular duct types is the cable-in-duct. This type of duct comes from the manufacturer with cables already installed. The duct comes in a reel and can be laid in the trench with ease. The installed cables can be withdrawn in the future, if necessary. This type of duct, due to the form in which it comes, reduces the need for fittings and couplings. It is most frequently used for street lighting systems.

15.0.0   MAKING A CONDUIT-TO-BOX CONNECTION

A proper conduit-to-box connection is shown in *Figure 54*.

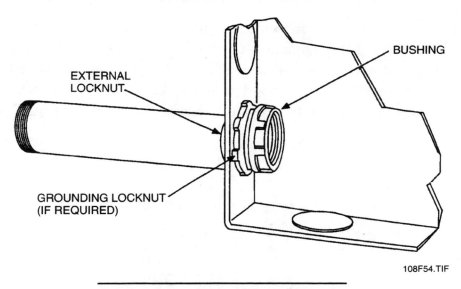

108F54.TIF

Figure 54.   Conduit-To-Box Connection

In order to make a good connection, use the following procedure:

Step 1   Thread the external locknut onto the conduit. Run the locknut to the bottom of the threads.

Step 2   Insert the conduit into the box opening.

Step 3   If an inside locknut or grounding locknut is required, screw it onto the conduit inside the box opening.

Step 4   Screw the bushing onto the threads projecting into the box opening. Make sure the bushing is tightened as much as possible.

Step 5   Tighten the external locknut to secure the conduit to the box.

It is important that the bushings and locknuts fit tightly against the box. For this reason, the conduit must enter straight into the box (*Figure 55*). This may require that a box offset or kick be made in the conduit.

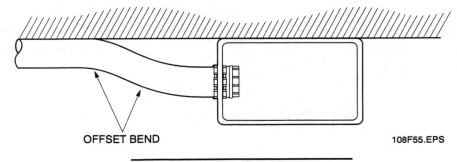

OFFSET BEND

108F55.EPS

Figure 55.   Correct Entrance Angle

16.1.0   MASONRY AND CONCRETE FLUSH-MOUNT CONSTRUCTION

In a reinforced concrete construction environment, the conduit and boxes must be embedded in the concrete to achieve a flush surface. Ordinary boxes may be used, but special concrete boxes are preferred and are available in depths up to six inches. These boxes have special ears by which they are nailed to the wooden forms for the concrete. When installing them, stuff the boxes tightly with paper to prevent concrete from seeping in. *Figure 56* shows an installed box.

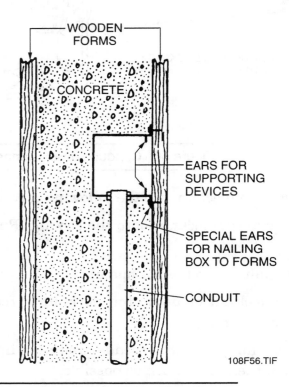

Figure 56. Concrete Flush-Mount Installation

Flush construction can also be done on existing concrete walls, but this requires chiseling a channel and box opening, anchoring the box and conduit, and then resealing the wall.

To achieve flush construction with masonry walls, the most acceptable method is for the electrician to work closely with the mason laying the blocks. When the construction blocks reach the convenience outlet elevation, boxes are made up as shown in *Figure 57*. The figure shows a raised tile ring or box device cover.

Note: The electrician must work with the mason to ensure the box is
             properly grouted and sealed.

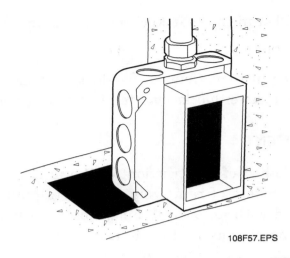

108F57.EPS

Figure 57. Concrete Outlet Box

Figure 58 shows the use of a 4-S extension ring installed to bring the box to the masonry surface. *Figure 59* shows a masonry box that needs no extension or deep plaster ring to bring it to the surface.

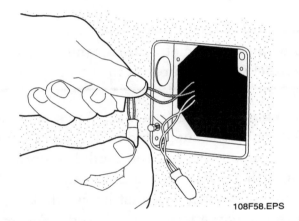

108F58.EPS

Figure 58. 4-S Extension Ring Used To Bring The Box To the Masonry Surface

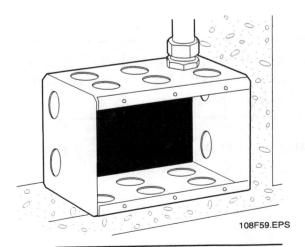

108F59.EPS

Figure 59. Three-Gang Switch Box

Note: EMT should be installed in the rear knockout of this masonry switch box.

Sections of conduit are then coupled in short (four- or five-foot) lengths. This is done because it is impractical for the mason to maneuver blocks over ten-foot sections of conduit.

16.2.0   METAL STUD ENVIRONMENT

Metal stud walls are a popular method of construction for the interior walls of commercial buildings. Metal stud framing consists of relatively thin metal channel studs, usually constructed of galvanized steel and with an overall dimension the same as standard 2 x 4 wooden studs. Wiring in this type of construction is relatively easy when compared to masonry.

EMT conduit is the most common type of raceway specified for metal stud wiring. Metal studs usually have some number of pre-punched holes that can be used to route the conduit. If a prepunched hole is not located where it needs to be, holes can be easily punched in the metal stud with a hole cutter or knockout punch.

WARNING!   Cutting or punching metal studs can create sharp edges. Avoid contact which can result in cuts.

Boxes can be secured to the metal stud using self-tapping screws or one of the many types of box supports available. EMT conduit is supported by the metal studs using conduit straps or other approved methods. It is important that the conduit be properly supported to facilitate pulling the conductors through the tubing. Boxes are mounted on the metal studs so that the box will be flush with the finished walls. You must know what the finished wall thickness is going to be to properly secure the boxes to the metal studs. For example, if the finished wall will be ⅝-inch drywall, then the box must be fastened so that it protrudes ⅝ of an inch from the metal stud.

WARNING!   When using a screw gun or cordless drill to mount boxes to studs, keep the hand holding the box away from the gun/drill to avoid injury.

Figure 60 shows several examples of clips known as *caddy-fastening devices* that are used in metal stud environments.

108F60.TIF

Figure 60. Caddy-Fastening Devices

16.3.0 WOOD FRAME ENVIRONMENT

At one time, the use of rigid conduit in partitions and ceilings was a laborious and time consuming operation. Thinwall conduit makes an easier and far quicker job, largely because of the types of fittings that are specially adapted to it.

Figure 61 shows two methods of running thinwall conduit in these locations: boring timbers and notching them. When boring, holes must be drilled large enough for the tubing to be inserted between the studs. The tubing is cut rather short, calling for multiple couplings. EMT can be bowed quite a bit while threading through holes in studs. Boring is the preferred method.

WARNING!   Always wear safety goggles when boring wood.

NEC Section 300-4 addresses the requirements to prevent physical damage to conductors and cabling in wood members. By keeping the edge of the drilled hole 1¼ inches from the closest edge of the stud, nails are not likely to penetrate the stud far enough to damage the cables. The building codes provide maximum requirements for bored or notched holes in studs.

The exception in the NEC permits IMC, RMC, RNMC, and EMT to be installed through bored holes or laid in notches less than 1¼ inches from the nearest edge without a steelplate or bushing.

Because of its weakening effect upon the structure, notching should be resorted to only where absolutely necessary. Notches should be as narrow as possible and in no case deeper than ⅙ the stock of a bearing timber. A bearing timber supports floor joists or other weight. An additional requirement is for the notch to be covered with a steel reinforcement bracket. This bracket aids in retaining the original strength of the timber.

Note: Always check with the architect before notching.

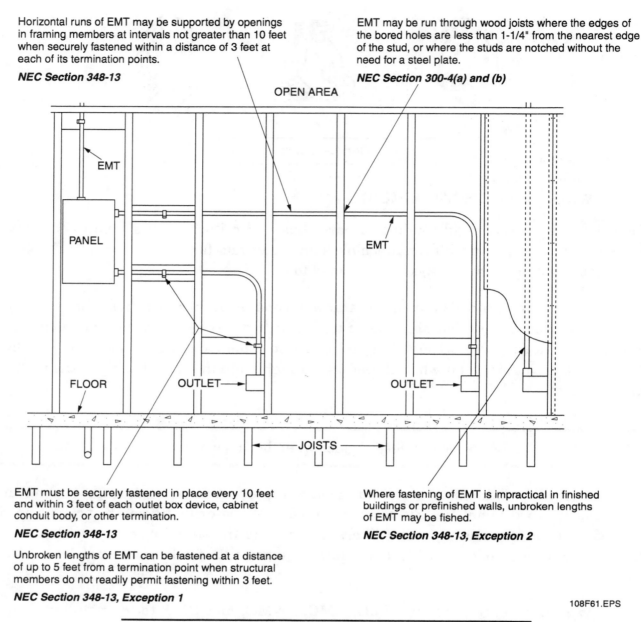

Horizontal runs of EMT may be supported by openings in framing members at intervals not greater than 10 feet when securely fastened within a distance of 3 feet at each of its termination points.

NEC Section 348-13

EMT may be run through wood joists where the edges of the bored holes are less than 1-1/4" from the nearest edge of the stud, or where the studs are notched without the need for a steel plate.

NEC Section 300-4(a) and (b)

OPEN AREA

EMT

PANEL

EMT

FLOOR

OUTLET →

OUTLET →

JOISTS

EMT must be securely fastened in place every 10 feet and within 3 feet of each outlet box device, cabinet conduit body, or other termination.

NEC Section 348-13

Unbroken lengths of EMT can be fastened at a distance of up to 5 feet from a termination point when structural members do not readily permit fastening within 3 feet.

NEC Section 348-13, Exception 1

Where fastening of EMT is impractical in finished buildings or prefinished walls, unbroken lengths of EMT may be fished.

NEC Section 348-13, Exception 2

108F61.EPS

Figure 61. Installing Wire Or Conduit In A Wood Frame Building

16.4.0   STEEL ENVIRONMENT

Electrical installations in buildings where steel beams are the structural framework are most often found in industrial buildings and warehouses. This type of construction is typically of the pre-engineered building where beams and other supports are pre-cut and pre-drilled so that erection of the building is fast and simple.

The interior of the building will in most cases be unfinished, and the wiring will be supported by the metal beams and purlings. Beams and purlings should not be drilled through; consequently, the conduit is supported from the metal beams by anchoring devices designed especially for that purpose. The supports attach to the beams or supports and have clamps to secure the conduit to the structure. All conduit runs should be plumb since they are exposed. Bends should be correct and have a neat and orderly appearance.

Since steel construction usually takes place in buildings where load handling and moving large and heavy items is common, rigid metal conduit is often required. If a large number of conduits are run along the same path, strut-type systems are used. These systems are sometimes referred to as *Unistrut* systems (Unistrut is a manufacturer of these systems). Another manufacturer of strut systems is *B-Line* systems. Both are very similar. These systems use a channel-type member that can support conduits from the ceiling by using threaded rod supports for the channel, as shown in *Figure 62*. Strut channel can also be secured to masonry walls to support vertical runs of conduit, wireways, and various types of boxes.

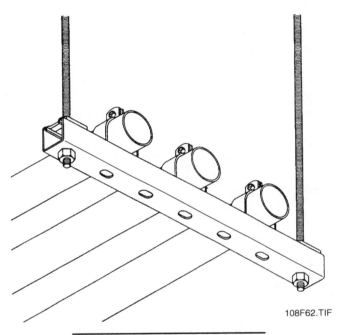

108F62.TIF

Figure 62. Steel Strut System

SUMMARY

This module discussed the various types of raceways, boxes, and fittings, including their uses and procedures for installation. The primary purpose of raceways is to house electric wire used for power distribution, communication, or electronic signal transmission. Raceways provide protection to the wiring and even a means of identifying one type of wire from another when run adjacent to each other. This process requires proper planning to allow for current needs, future expansion, and a neat and orderly appearance.

References

For advanced study of topics covered in this task module, the following books are suggested:

Benfield Conduit Bending Manual, Latest Edition, McGraw-Hill Publishing Company, New York, NY.

National Electrical Code Handbook, Latest Edition, National Fire Protection Association, Quincy, MA.

REVIEW QUESTIONS

1. _____ is the lightest duty and most widely used non-flexible metal conduit available for enclosing and protecting electrical wiring.
   a. Electrical metallic tubing
   b. Rigid metal conduit
   c. Aluminum conduit
   d. Plastic-coated GRC

2. Because the wall thickness of _____ is much less than that of rigid conduit, it is often referred to as *thinwall conduit*.
   a. intermediate metal conduit
   b. electrical metallic tubing
   c. rigid non-metallic conduit
   d. galvanized rigid steel conduit

3. The type of conduit that provides the best physical protection for the wire inside is _____.
   a. flexible metal conduit
   b. rigid metal conduit
   c. electrical metallic tubing
   d. intermediate metal conduit

4. Bending IMC is often easier than bending the same size and quantity of rigid metal conduit because IMC has a slightly _____.
   a. larger outside diameter
   b. larger internal diameter
   c. smaller outside diameter
   d. smaller internal diameter

5. Flexible metal conduit _____.
   a. is made from a single strip of steel or aluminum
   b. can be used in wet locations if lead-covered conductors are installed
   c. can be used in underground locations
   d. is available in sizes up to six inches in diameter

6. Liquid-tight flexible metal conduit can be used in _____ locations.
   a. dry
   b. wet
   c. underground
   d. outdoor

7. A Type LB conduit body has the cover on _____.
   a. the left
   b. the right
   c. the back
   d. both sides

8. When conduit is joined to metal boxes, _____ protect the wires from the sharp edges of the conduit.
   a. washers
   b. locknuts
   c. bushings
   d. couplings

9. _____ are rigid rectangular raceways used for housing electric wires and cables.
   a. Troughs
   b. Gutters
   c. Pull boxes
   d. Conduits

10. Wireways must not be installed where they are subject to _____.
    a. physical damage
    b. corrosion and sunlight
    c. hazardous locations
    d. physical damage, corrosive vapors, and hazardous locations

11. Pancake raceways are a type of _____.
    a. rigid tubing
    b. flexible metal conduit
    c. surface metal raceway
    d. surface nonmetallic raceway

12. The statement that best describes the use of surface metal raceway systems is _____.
    a. a raceway system designed for concealed wiring within walls and partitions
    b. a raceway system suited for use in wet or damp locations
    c. a raceway system mainly used for making the transition from a masonry surface to a wood surface
    d. a raceway system designed for exposed installations on the surface of walls and ceilings

13. Surface metal raceways are designed primarily to protect conductors that run to devices _____ in a room.
    a. located near the walls and outlets
    b. located on the surface of the walls and ceilings
    c. not located near the walls and outlets
    d. not located near the walls and ceiling

14. When storing raceway, it is best to _____.
    a. lay it on the floor so that there is no chance that it will fall and injure someone
    b. store all raceway together regardless of the type of raceway material
    c. stand it on end so that it does not collect moisture
    d. cover it with a tarpaulin or other suitable covering to prevent contamination and corrosion

15. When transporting metal conduit, it is best to _____.
    a. remove your gloves so that you do not contaminate the conduit
    b. remove the thread protection caps so that they do not get lost
    c. place the metal conduit inside a larger raceway to keep it from bending
    d. keep the thread protection caps on to prevent damage to the threads

ANSWERS TO REVIEW QUESTIONS

Answer		**Section**
1.	a	3.2.1
2.	b	3.2.1
3.	b	3.2.2
4.	b	3.2.7
5.	a	3.2.10
6.	b	3.2.9
7.	c	4.2.2
8.	c	6.0.0
9.	a	Terms/9.0.0
10.	d	9.2.0
11.	c	9.4.1
12.	d	9.4.1
13.	b	9.4.1
14.	d	11.0.0
15.	d	12.0.0

NCCER CRAFT TRAINING USER UPDATES

The NCCER makes every effort to keep these manuals up-to-date and free of technical errors. We appreciate your help in this process. If you have an idea for improving this manual, or if you find an error, a typographical mistake, or an inaccuracy in the NCCER's Craft Training Manuals, please write us, using this form or a photocopy. Be sure to include the exact module number, page number, a description of the problem, and the correction, if possible. Your input will be brought to the attention of the Technical Review Committee. Thank you for your assistance.

Instructors – If you found that additional materials were necessary in order to teach this module effectively, please let us know so that we may include them in the Equipment/Materials list in the Instructor's Guide.

Write: Curriculum Development and Revision Department
National Center for Construction Education and Research
P.O. Box 141104
Gainesville, FL 32614-1104

Fax: 352-334-0932

Craft _____ Module Name _____

Copyright Date _____ Module Number _____ Page Number(s) _____

Description of Problem _____

_____

_____

_____

(Optional) Correction of Problem _____

_____

_____

(Optional) Your Name and Address _____

_____

_____

_____

notes

Conductors

Module 26109

Electrical Trainee Task Module 26109

NATIONAL
CENTER FOR
CONSTRUCTION
EDUCATION AND
RESEARCH

CONDUCTORS

OBJECTIVES

Upon completion of this module, the trainee will be able to:

1. Explain the various sizes and gauges of wire in accordance with American Wire Gauge standards.
2. Identify insulation and jacket types according to conditions and applications.
3. Describe voltage ratings of conductors and cables.
4. Read and identify markings on conductors and cables.
5. Use the tables in the NEC to determine the ampacity of a conductor.
6. State the purpose of stranded wire.
7. State the purpose of compressed conductors.
8. Describe the different materials from which conductors are made.
9. Describe the different types of conductor insulation.
10. Describe the color coding of insulation.
11. Describe instrumentation control wiring.
12. Describe the equipment required for pulling wire through conduit.
13. Describe the procedure for pulling wire through conduit.
14. Install conductors in conduit.
15. Pull conductors in a conduit system.

Prerequisites

Successful completion of the following Task Modules is required before beginning study of this Task Module: Core Curricula; Electrical Level 1, Modules 26101 through 26108.

Required Trainee Materials

1. Trainee Task Module
2. Copy of the latest edition of the *National Electrical Code*
3. Appropriate Personal Protective Equipment

Note: The designations "National Electrical Code," "NE Code," and "NEC," where used in this document, refer to the *National Electrical Code®*, which is a registered trademark of the National Fire Protection Association, Quincy, MA. All National Electrical Code (NEC) references in this module refer to the 1999 edition of the NEC.

Course Map

This course map shows all of the task modules in the first level of the Electrical curricula. The suggested training order begins at the bottom and proceeds up. Skill levels increase as a trainee advances on the course map. The training order may be adjusted by the local Training Program Sponsor.

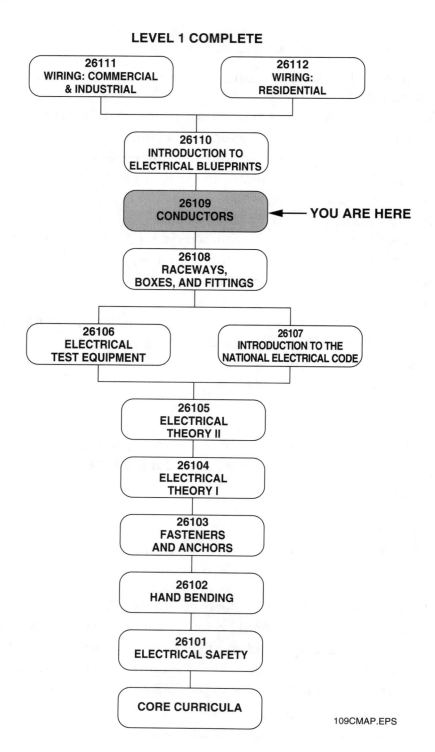

LEVEL 1 COMPLETE

26111
WIRING: COMMERCIAL & INDUSTRIAL

26112
WIRING: RESIDENTIAL

26110
INTRODUCTION TO ELECTRICAL BLUEPRINTS

26109
CONDUCTORS ← YOU ARE HERE

26108
RACEWAYS, BOXES, AND FITTINGS

26106
ELECTRICAL TEST EQUIPMENT

26107
INTRODUCTION TO THE NATIONAL ELECTRICAL CODE

26105
ELECTRICAL THEORY II

26104
ELECTRICAL THEORY I

26103
FASTENERS AND ANCHORS

26102
HAND BENDING

26101
ELECTRICAL SAFETY

CORE CURRICULA

109CMAP.EPS

TABLE OF CONTENTS

TABLE OF CONTENTS (Continued)

Terms Introduced In This Module

Ampacity: The current in amperes a conductor can carry continuously under the conditions of use without exceeding its temperature rating.

Capstan: The turning drum of the cable puller on which the rope is wrapped and pulled.

Fish tape: A hand device used to pull a wire through a conduit run.

Mouse: A cylinder of foam rubber that fits inside the conduit and is then propelled by compressed air or vacuumed through the conduit run, pulling a line or tape.

Wire grip: A device used to link pulling rope to cable during a pull.

1.0.0   INTRODUCTION

As an electrician, you will be required to select the proper wire and/or cable for a job. You will also be required to pull this wire or cable through conduit runs in order to terminate it. This module will examine the different types of conductors and conductor insulation. It will also examine how these conductors are rated and classified by the NEC and the different methods used for pulling these conductors through conduit runs.

2.0.0   CONDUCTORS AND INSULATION

The term *conductor* is used in two ways. It is used to describe the current-carrying portion of a wire or cable, and it is used to describe the wire or cable composed of a current-carrying portion and an outer covering (insulation). In this module, the term *conductor*, if not specified otherwise, will be used to describe the wire assembly which includes the insulation and the current-carrying portion of the wire. Conductors are uniquely identified by size and insulation material. Size refers to the physical size of the current-carrying portion of the conductor.

NEC Table 310-13 presents application and construction data on the wide range of 600-volt insulated, individual conductors recognized by the NEC, with the appropriate letter designation used to identify each type of insulated conductor. Important data that should be noted in this table is as follows:

*   The designation for one thousand circular mils is *kcmil*, which has been substituted for the long-time designation *MCM* in this table and throughout the NEC.
*   Type MI (mineral insulated) cable may have either a copper or an alloy steel sheath.
*   Type RHW-2 is a conductor insulation made of moisture- and heat-resistant rubber with a 90°C (194°F) rating for use in dry and wet locations.
*   Type XHHW-2 is a moisture- and heat-resistant crosslinked synthetic polymer with a 90°C (194°F) rating for use in dry and wet locations.

- The suffix *LS* designates a conductor insulation as low-smoke producing and flame retardant. For example, Type THHN/LS is a THHN conductor with a limited smoke-producing characteristic.
- Type THHW is a moisture- and heat-resistant insulation rated at 75°C (167°F) for wet locations and 90°C (194°F) for dry locations. This is similar to THWN and THHN without the outer nylon covering but with thicker insulation.
- All insulations using asbestos have been deleted from *NEC Table 310-13* because they are no longer made.

2.1.0 AMPACITY

Ampacity is the current in amperes a conductor can carry continuously under the conditions of use without exceeding its temperature rating. The ampacity of conductors for given conditions of use are listed in *NEC Tables 310-16 through 310-19*.

2.1.1 NEC Ampacity Tables

NEC Table 310-16 covers conductors rated up to 2,000 volts where not more than three conductors are installed in a raceway or cable or are directly buried in the earth, based on an ambient temperature of 30°C (86°F).

NEC Table 310-17 covers both copper conductors and aluminum or copper-clad aluminum conductors up to 2,000 volts where conductors are used as single conductors in free air, based on an ambient temperature of 30°C (86°F).

NEC Tables 310-18 and 19 apply to conductors rated at 150°C to 250°C (302°F to 482°F), used either in raceway or cable or as single conductors in free air, based on an ambient temperature of 40°C (104°F).

Example:

Determine the ampacity of a No. 12 Cu (copper) THW conductor.

Solution:

25 amps (from *NEC Table 310-16*).

2.2.0 UNDERGROUND

Any conductor used in a wet location (see definition under *Wet Location* in *NEC Article 100*) must be designated as suitable for wet locations. Any conduit run underground is assumed to be subject to water infiltration and is, therefore, a wet location, requiring the use of only the listed conductor types.

2.2.1 Direct Burial

Direct burial conductors should be trench-laid without crossovers; slightly snaked to allow for possible earth settlement, movement, or heaving due to frost action; and have cushions and covers of sand or screened fill to protect conductors against sharp objects in trenches or backfill.

2.3.0 WIRE SIZE

Wire sizes are expressed in gauge numbers. The standard system of wire sizes in the United States is the American Wire Gauge (AWG) system.

2.3.1 AWG System

The AWG system uses numbers to identify the different sizes of wire and cable (*Figure 1*). The larger the number, the smaller the cross-sectional area of the wire. The larger the cross-sectional area of the current-carrying portion of a conductor, the higher the amount of current the wire can conduct. The AWG numbers range from 50 to 1; then 0, 00, 000, and 0000 [one aught (1/0), two aught (2/0), three aught (3/0), and four aught (4/0)]. Any wire larger than 0000 is identified by its area in circular mils. Wire sizes smaller than No. 18 AWG are usually solid, but may be stranded in some cases. Wire sizes of No. 6 AWG or larger are stranded.

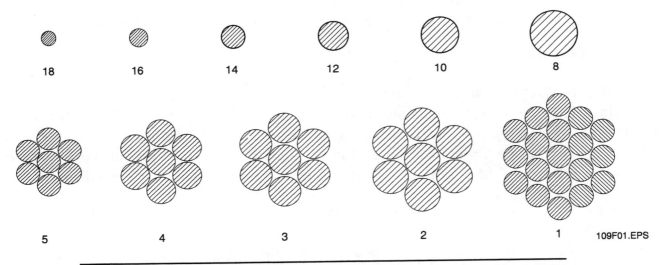

109F01.EPS

Figure 1. Comparison Of Wire Sizes (Enlarged) From No. 18 To No. 1 AWG

For wire sizes larger than No. 16 AWG, the wire size is marked on the insulation (*Figure 2*).

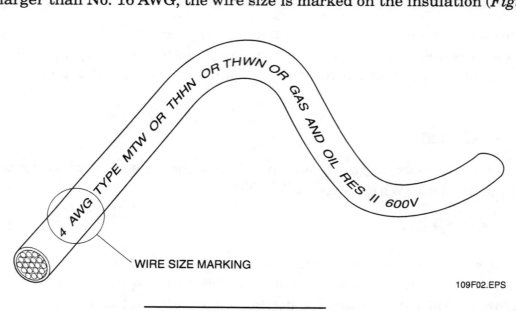

WIRE SIZE MARKING

109F02.EPS

Figure 2. Wire Size Marking

NEC Chapter 9, Table 8 has descriptive information on wire sizes. Again, note that all wires smaller than No. 6 are available as solid or stranded. Wire sizes of No. 6 or larger are shown only as stranded. Solid wire larger than No. 6 is manufactured; however, the NEC only permits the use of solid wire in a raceway for sizes smaller than No. 8 (*NEC Section 310-3*).

2.3.2 Stranding

According to *NEC Chapter 9, Table 8*, wire sizes No. 18 to No. 2 have seven strands; wire sizes No. 1 to No. 4/0 have 19 strands; and wire sizes between 250 kcmil and 500 kcmil have 37 strands. The purpose of stranding is to increase the flexibility of the wire. Terminating solid wire sizes larger than No. 8 in pull boxes, disconnect switches, and panels would not only be very difficult, but might also result in damage to equipment and wire insulation.

Pulling solid wire in conduit around bends could pose a major problem and cause damage to equipment for wire sizes larger than No. 8. The reason for choosing 7, 19, and 37 strands for stranded conductors is that it is necessary to provide a flexible, almost round conductor. In order for a conductor to be flexible, individual strands must not be too large. *Figure 3* shows how these conductors are configured.

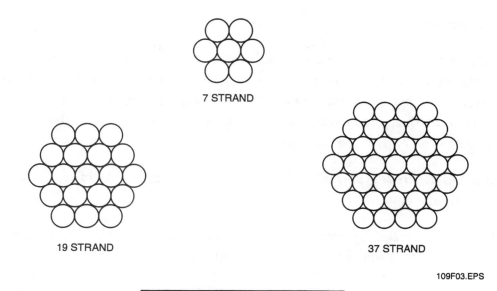

Figure 3. Strand Configurations

2.3.3 Compressed Conductors

A relatively new entry in the scheme of conductors is compressed aluminum conductors *(Figure 4)*. Compressed aluminum conductors are those that have been compressed to reduce the air space between the strands.

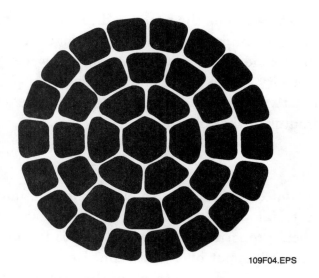

Figure 4. Compressed Conductor

The purpose of a compressed conductor is to reduce the overall diameter of the cable so that it may be installed in a conduit that is smaller than that required for standard conductors of the same wire size. Compressed conductors are especially useful when increasing the ampacity of an existing service.

2.3.4   Circular Mils

A circular mil is a circle which has a diameter of 1 mil. A mil is 0.001 inches. When a wire size is 250 kcmil, the cross-sectional area of the current- carrying portion of the wire is the same as 250,000 circles having a diameter of 0.001 inches. This may seem to be a rather clumsy way of sizing wire at first; however, the alternative would be to size the wire as a function of its cross-sectional area expressed in square inches.

According to *NEC Chapter 9, Table 8*, the cross-sectional area of a 250 kcmil conductor is 0.260 square inches.

If a conductor is to be sized by cross-sectional area, it is much easier to express the wire size in circular mils (or thousands of circular mils) than in square inches.

2.4.0   CONDUCTOR MATERIAL

The most common conductor material is copper. Copper is used because of its excellent conductivity (low resistance), ease of use, and value. The value of a material as an ingredient for wire is determined by several factors, including conductivity, cost, availability, and workability.

2.4.1   Conductivity

Conductivity is a word that describes the ease (or difficulty) of travel presented to an electric current by a conductor. If a conductor has a low resistance, it has a high conductivity. Platinum is one of the best conductors since it has very low resistance and high conductivity. Copper has high conductivity and a much lower price than platinum. Aluminum, another material with good conductivity, is also a good choice for conductor material. The conductivity of aluminum is approximately ⅔ that of copper.

2.4.2   Cost

Cost is always an issue that contributes to the selection of a material to be used for a given application. Often, a material which has low cost may be selected as a conductor material even though it has physical properties that are inferior to the more expensive material. Such is the case in the selection of copper over platinum. Here, the cost of platinum is very high, and very little thinking is required to determine that copper is a better choice. The choice between copper and aluminum is often more difficult to make.

2.4.3   Availability

The availability of some material is often a concern when selecting components for a job. As applied to wire, the mining industry often controls the availability of raw materials which could produce shortages of some material. The availability of a substance such as copper or aluminum affects the price of the finished product (copper or aluminum wire).

2.4.4 Workability

It is a good idea to select a material that requires less expense for tools and is easier to work with. Aluminum conductors are lighter than copper conductors of the same size. They are also much more flexible than copper conductors and, in general, are easier to work with. However, terminating aluminum conductors often requires special tools and treatment of termination surfaces with an anti-oxidation material. Splicing and terminating aluminum conductors often requires a higher degree of training on the part of the electrician than do similar efforts with copper wire. This is partly due to the fact that aluminum expands and contracts with heat more than copper.

2.5.0 CONDUCTOR INSULATION

The first attempt to insulate wire was made in the early 1800s during the development of the telegraph. This insulation was designed to provide physical protection rather than electrical protection. Electrical insulation was not an important issue because the telegraph operated at low-voltage DC. This early form of insulation was a substance composed of tarred hemp or cotton fiber and shellac and was used primarily for weatherproofing long distance distribution lines to mines, industrial sites, and railroads.

Some early electrical distribution systems utilized the knob-and-tube technique of installing wire. The wire was often bare and was pulled between and wrapped around ceramic knobs that were affixed to the building structure. When it was necessary to pull wire through structural members, it was pulled through ceramic tubes. The structural member (usually wood) was drilled, the tube was pressed into the hole, and the wire was pulled through the hole in the tube. As dangerous as this may appear, older homes still exist that have knob-and-tube wiring that was installed in the early 1900s and is still operational. Knob-and-tube wiring was revised to use insulated conductors and was in use up to 1957 in some areas.

The grounded or neutral conductor in overhead services may be bare. Furthermore, the concentric grounded conductor in Type SE cable may be bare when used as a service-entrance cable. However, all current-carrying conductors (including the grounded conductor) must be insulated when used on the inside of buildings, or after the first overcurrent protection device.

2.5.1 Thermoplastic

Thermoplastic is a popular and effective insulation material used on conductors for the present-day market. The following thermoplastics are widely used as insulation materials:

- *Polyvinyl chloride (PVC)* – The base material used for the manufacture of TW and THW insulation.
- *Polyethylene (PE)* – An excellent weatherproofing material used primarily for insulation of control and communications wiring. It is not used for high-voltage conductors (those exceeding 5,000 volts).

- *Cross-linked polyethylene (XLP)* – An improved PE with superior heat- and moisture-resistant qualities. Used for THHN, THWN, and XHHW wiring as well as most high-voltage cables.
- *Nylon* – Primarily used as jacketing material. THHN building wire has an outer coating of nylon.
- *Teflon* – A high-temperature insulation. Widely used for telephone wiring in a plenum (where other insulated conductors require conduit routing).

2.5.2 Letter Coding

Conductor insulation as applied to building wire is coded by letters. The letters generally, but not always, indicate the type of insulation or its environmental rating. The types of conductor insulation described in this module will be those indicated at the top of *NEC Table 310-16*. The various insulation designations are:

LETTER	DESCRIPTION
A	Asbestos
B	Braid
E	Ethylene or Entrance
F	Fluorinated or Feeder
H	Heat-Rated or Flame-Retardant
I	Impregnated
N	Nylon
P	Propylene
R	Rubber
S	Silicon or Synthetic
T	Thermoplastic
U	Underground
V	Varnished Cambric
W	Weather-Rated
X	Cross-Linked Polyethylene
Z	Modified Ethylene Tetrafluorethylene
TW	Weather-Rated Thermoplastic (60°C/140°F)
UF	Underground Feeder
FEPW	Weather-Rated Fluorinated Ethylene Propylene
RH	Heat-Rated Rubber (75°C/167°F)
RHW	Weather-Rated, Heat-Rated Rubber (75°C/167°F)
THW	Weather-Rated, Heat-Rated Thermoplastic (75°C/167°F)
THWN	Weather-Rated, Heat-Rated Thermoplastic with Nylon Cover
XHHW	Heat-Rated, Flame-Retardant, Weather-Rated, Cross-Linked Polyethylene
USE	Underground
ZW	Weather-Rated Modified Ethylene Tetrafluorethylene

TA	Thermoplastic and Asbestos
TBS	Thermoplastic Braided Silicon
SA	Silicon Asbestos
AVB	Asbestos Varnished and Braided
SIS	Synthetic Heat-Resistant
FEP	Fluorinated Ethylene Propylene
RHH	Flame-Retardant Heat-Rated Rubber

2.5.3 Color Coding

A color code is used to help identify wires by the color of the insulation. Understanding the color code makes it easier to install and properly connect the wires. When two or more wires are used together, as in a cable, the color code is as follows:

- *Two-conductor cable* – One white wire, one black wire, and a grounding wire (usually bare)
- *Three-conductor cable* – One white, one black, one red, and a grounding wire
- *Four-conductor cable* – Same as three-conductor cable plus fourth wire (blue)
- *Five-conductor cable* – Same as four-conductor cable plus fifth wire (yellow)

The grounding conductor may be either green or green with a yellow stripe. Color codes are shown in *Figure 5*.

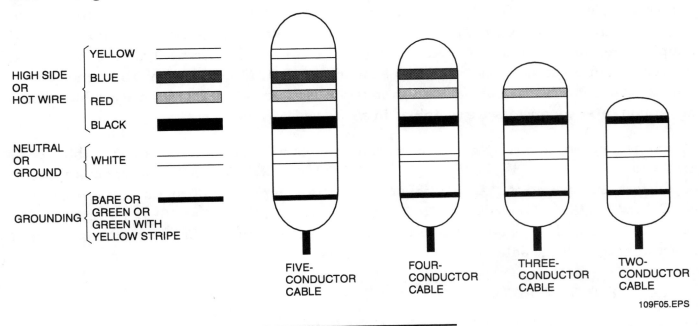

Figure 5. Insulator Color Codes

The NEC does not require color coding of ungrounded conductors in many cases. The ungrounded conductors may be any color with the exception of white, grey, or green; however, it is a good practice to color code conductors as described here. In fact, many construction specifications require color coding. Furthermore, on a four-wire, delta-connected secondary where the midpoint of one phase is grounded to supply lighting and similar loads, the phase conductor having the higher voltage to ground must be identified by an outer finish that is orange in color, by tagging, or by other effective means. Such identification must be placed at each point where a connection is made if the grounded conductor is also present. In most cases, orange tape is used at all termination points when such a condition exists.

2.5.4 Wire Ratings

Conductor selection is based largely on the temperature rating of the wire. This requirement is extremely important and is the basis of safe operation for insulated conductors. As shown in *NEC Table 310-13*, conductors have various ratings, (60°C, 75°C, 90°C, etc.). Since *NEC Tables 310-16 through 310-19* are based on an assumed ambient temperature of 30°C (86°F), conductor ampacities are based on the ambient temperature plus the heat (I^2R) produced by the conductor while carrying current. Therefore, the type of insulation used on the conductor is the first consideration in determining the maximum permitted conductor ampacity.

For example, a No. 3/0 THW copper conductor for use in a raceway has an ampacity of 200 according to *NEC Table 310-16*. In a 30°C ambient temperature, the conductor is subjected to this temperature when it carries no current. Since a THW-insulated conductor is rated at 75°C, this leaves 45°C (75 – 30) for increased temperature due to current flow. If the ambient temperature exceeds 30°C, the conductor maximum load-current rating must be reduced proportionally (see *Correction Factors* at the bottom of *NEC Table 310-16*) so that the total temperature (ambient plus conductor temperature rise due to current flow) will not exceed the temperature rating of the conductor insulation (60°C, 75°C, etc.). For the same reason, conductor maximum load current ratings must be reduced below the ampacity values where more than three conductors are contained in a raceway or cable.

Using the ampacity tables – An important step in circuit design is the selection of the type of conductor to be used (TW, THW, THWN, RHH, THHN, XHHW, etc.). The various types of conductors are covered in *NEC Article 310*, and the ampacities of conductors are given in *NEC Tables 310-16 through 310-19* for the varying conditions of use (e.g., in a raceway, in open air, at normal or higher-than-normal ambient temperatures). Conductors must be used in accordance with the data in these tables and notes.

2.6.0 FIXTURE WIRES

Fixture wire is used for the interior wiring of fixtures and for wiring fixtures to a power source. Guidelines concerning fixture wire are given in *NEC Article 402*. The list of approved types of fixture wire is given in *NEC Table 402-3*. *Figure 6* shows one example of fixture wire. The wires are composed of insulated conductors with or without an outer jacket. The conductors range in size from No. 18 to No. 10 AWG.

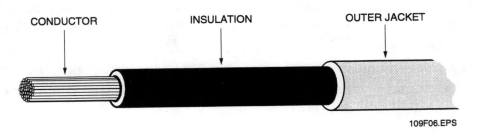

CONDUCTOR   INSULATION   OUTER JACKET

109F06.EPS

Figure 6. Fixture Wire

The decision of which fixture wire to use depends primarily upon the operating temperature that is expected within the fixture. Therefore, it is the character of the insulation that will determine the wire selected. For instance, fixture wires insulated with perfluoroalkoxy (PAF) or extruded polytetrafluorethylene (PTF) would be selected if the operating temperature of the fixture is expected to reach a maximum of 482°F. This is the highest operating temperature allowed for any fixture wire.

As indicated by **NEC Section 402-3**, fixture wires are suitable for service at 600 volts unless otherwise specified in **NEC Table 402-3**. The allowed ampacities of fixture wire are given in **NEC Table 402-5**.

Although the primary use for fixture wire is the internal wiring of fixtures, several of the wires listed in **NEC Table 402-3** may be used for wiring remote-control, signaling, or power-limited circuits in accordance with **NEC Section 725-15**. Fixture wires may never be used as substitutes for branch circuit conductors.

2.6.1 Heating Cables

Heating cables are another important group of special wires that are commonly used in the building industry. The purpose of these cables is to produce heat when energized. There are three general categories of heating cables:

- Interior space-heating cables
- De-icing and snow-melting cables
- Pipeline and vessel heating cables

These heating cables consist of a long insulated wire, cabled heating wire, or resistance wire which produces heat in the wire when it is connected to a power source. The ends of the resistance wire are connected to non-heating lead wires which extend to a thermostat or junction box.

Guidelines pertaining to heating cable used for interior space heating are given in **NEC Article 424**. The NEC considers the term *heating equipment* to include any factory-manufactured heating devices such as unit heaters, boilers, and local heating systems, as well as heating cables. Heating cables are specifically covered in **NEC Article 424, Part E**.

Heating cables meant for installation in ceilings are available from the factory in unit lengths from 75 to 1,800 feet, capacities ranging from 200 to 5,000 watts, and voltage ratings of 120, 208, and 240 volts. The cables are usually rated at 2¾ watts per linear foot. The nonheating lead wires are at least seven feet in length and are color coded to indicate the voltage rating. The insulation on the heating wire is designed to be resistant to high temperatures, water absorption, aging, and chemical action.

Heating cables meant for installation in floors have capacities that vary with the cable dimensions and heating wire resistance. They are typically rated at voltages from 120 to 600 volts. There are two popular floor heating cables. In the first type, the resistance wire is covered with a sheath of polyvinyl chloride (PVC). In the second, the resistance wire is encased in a mineral insulation and then covered with a copper sheath. This construction is identical to that of Type MI (mineral-insulated, metal-sheathed cable). In contrast to ceiling heating cables, which are available in specific unit lengths, these heating cables are sold in random lengths which must be cut and properly terminated at the job site according to the job's heating requirements.

NEC Article 424, Part E specifies guidelines for installing floor and ceiling heating cables; it should be read carefully. Generally, ceiling heating cables are stapled to the ceiling with at least 1½ inches between adjacent runs. According to the NEC, the cables must not come within eight inches of recessed lighting fixtures or within six inches of other metallic materials in the ceiling. Floor heating cables are usually embedded in concrete, with at least a one-inch space between adjacent runs of cable. Spacing must also be maintained between the cable and other metallic objects in the floor, unless a grounded metal-clad cable is used.

Guidelines pertaining to de-icing and snow-melting cables are given in *NEC Article 426*. These cables are similar in construction to interior floor heating cables in that they consist of resistance wires insulated with a sheath of PVC or with mineral insulation and a copper sheath. Most de-icing and snow-melting cables are installed by embedding them in concrete or asphalt. As with floor heating cables, these cables are rated at various voltages and must be cut to the desired length.

Guidelines pertaining to pipeline and vessel heating cables are given in *NEC Article 427*. These cables are generally available in two types of construction. The cable can either be insulated resistance wire with nonheating lead wires attached to each end, or it can be in the form of a flat heating tape. The tape consists of a single piece of heating wire doubled over so that the halves of the wire run parallel to each other. The wire is encased in a plastic enclosure, thus producing an assembly which resembles a tape. The nonheating lead wires are attached to the open end of the tape.

The obvious purpose of pipeline and vessel heating cables is to prevent the fluid in the pipeline or vessel from falling below a specified minimum temperature. Flat heating tapes are often used in buildings under construction or not completely enclosed. They can be used on water pipes to keep the water from freezing in cold weather.

2.7.0 CABLES

Cables are two or more insulated wires and a grounding wire covered by an outer jacket or sheath. Cable is usually classified by the type of covering it has, either nonmetallic (i.e., plastic) or metallic, also called *armored cable*.

Cable may also be classified according to where it can be used (see **NEC Table 400-4**). Because water is such a good conductor of electricity, moisture on conductors can cause power loss or short circuits. For this reason, cables are classified for either dry, damp, or wet locations. Cables can also be classified regarding exposure to sunlight and rough use.

2.7.1 Cable Markings

All cables are marked to show important properties and uses. Cable markings show the wire size, number of conductors, cable type, and voltage rating. In addition, a marking may be included to signify approved service or applications. This information is printed on nonmetallic cable (*Figure 7*). On metallic cable, marking information is usually included on a tag.

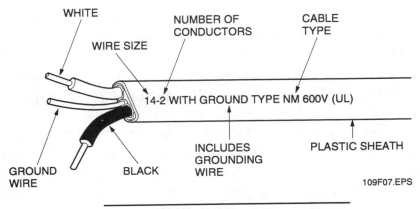

Figure 7. Nonmetallic Cable Markings

2.7.2 Nonmetallic-Sheathed Cable

Nonmetallic-sheathed cable (Type NM and Type NMC) is widely used for branch circuits and feeders in residential and commercial systems. See *Figure 8*. Both types are commonly called *Romex*, even though the cable manufacturer only calls Type NM cable Romex. Guidelines for the use of nonmetallic-sheathed cable are given in **NEC Article 336**. This cable consists of two or three insulated conductors and one bare conductor enclosed in a nonmetallic sheath. The conductors may be wrapped individually with paper, and the spaces between the conductors may be filled with jute or paper to protect the conductors and help the cable keep its shape. The sheath covering both Type NM cable and Type NMC cable is flame-retardant and moisture-resistant. The sheath covering Type NMC cable has the additional characteristics of being fungus- and corrosion-resistant.

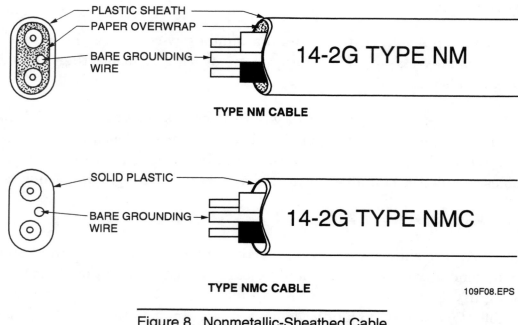

TYPE NM CABLE

Labels on upper diagram: PLASTIC SHEATH, PAPER OVERWRAP, BARE GROUNDING WIRE, 14-2G TYPE NM

TYPE NMC CABLE

Labels on lower diagram: SOLID PLASTIC, BARE GROUNDING WIRE, 14-2G TYPE NMC

109F08.EPS

Figure 8. Nonmetallic-Sheathed Cable

NEC Article 336 lists the allowed and prohibited uses for Type NM cable and Type NMC cable. Both are allowed to be installed in either exposed or concealed work. The primary difference in their applied uses is that Type NM cable is suitable for dry locations only, whereas Type NMC is permitted for dry, moist, damp, or corrosive locations. Since you will probably work with these two cables often, read *NEC Article 336* carefully.

Guidelines for the use of Type SE (service-entrance) cable are given in *NEC Article 338*. The NEC contains no specifications for the construction of this cable; it is left to UL to determine what types of cable should be approved for this purpose. Currently, service-entrance cable is labeled in sizes No. 12 AWG and larger for copper, and No. 10 AWG and larger for aluminum or copper-clad aluminum, with Types RH, RHW, RHH, or XHHW conductors. If the type designation for the conductor is marked on the outside surface of the cable, the temperature rating of the cable corresponds to the rating of the individual conductor. When this marking does not appear, the temperature rating of the cable is 75°C (167°F). Type SE cable is for above-ground installation.

When used as a service-entrance cable, Type SE must be installed as specified in *NEC Article 230*. Service-entrance cable may also be used as feeder and branch circuit cable. Guidelines for the use of service-entrance cable are given in *NEC Section 338-3*.

2.7.3   Type UF Cable

Guidelines for the use of Type UF (underground feeder and branch circuit) cable are given in *NEC Article 339*. Type UF cable is very similar in appearance, construction, and use to Type NMC cable. The main difference between these two cables is that Type UF cable is suitable for direct burial, whereas Type NMC cable is not.

2.7.4   Type NMS Cable

Refer to **NEC Sections 336-4(c), 336-5, and 336-30** for the applications of Type NMS cable. Type NMS cable is a form of nonmetallic-sheathed cable that contains a factory assembly of power, communications, and signaling conductors enclosed within a moisture-resistant, flame-retardant sheath.

2.7.5   Type MV Cable

Type MV (medium-voltage) cable is covered in **NEC Article 326**. It consists of one or more insulated conductors encased in an outer jacket. This cable is suitable for use with voltages ranging from 2,001 to 35,000 volts. It may be installed in wet and dry locations and may be buried directly in the earth.

2.7.6   High-Voltage Shielded Cable

Shielding of high-voltage cables protects the conductor assembly against surface discharge or burning due to corona discharge in ionized air, which can be destructive to the insulation and jacketing.

Electrostatic shielding of cables makes use of both nonmetallic and metallic materials (*Figures 9* and *10*).

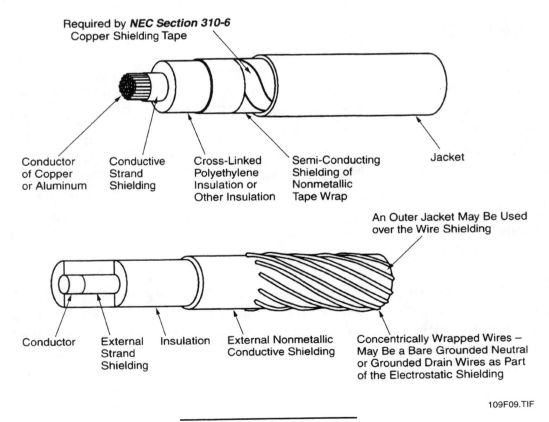

109F09.TIF

Figure 9. Metallic Shielding

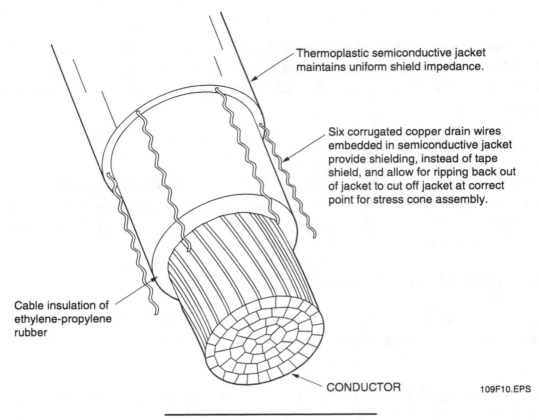

Thermoplastic semiconductive jacket maintains uniform shield impedance.

Six corrugated copper drain wires embedded in semiconductive jacket provide shielding, instead of tape shield, and allow for ripping back out of jacket to cut off jacket at correct point for stress cone assembly.

Cable insulation of ethylene-propylene rubber

CONDUCTOR

109F10.EPS

Figure 10. Nonmetallic Shielding

2.7.7 Channel Wire Assemblies

Channel wire assemblies (Type FC) comprise an entire wiring system which includes the cable, cable supports, splicers, circuit taps, fixture hangers, and fittings (*Figure 11*). Guidelines for the use of this system are given in ***NEC Article 363***. Type FC cable is a flat cable assembly with three or four parallel No. 10 special stranded copper conductors. The assembly is installed in an approved U-channel surface metal raceway with one side open. Tap devices can be inserted anywhere along the run. Connections from the tap devices to the flat cable assembly are made by pin-type contacts when the tap devices are fastened in place. The pin-type contacts penetrate the insulation of the cable assembly and contact the multi-stranded conductors in a matched phase sequence. These taps can then be wired to lighting fixtures or power outlets (*Figure 12*).

As indicated in ***NEC Section 363-3***, this wiring system is suitable for branch circuits which only supply small appliances and lights. This system is suitable for exposed wiring only and may not be concealed within the building structure. It is ideal for quick branch circuit wiring at field installations.

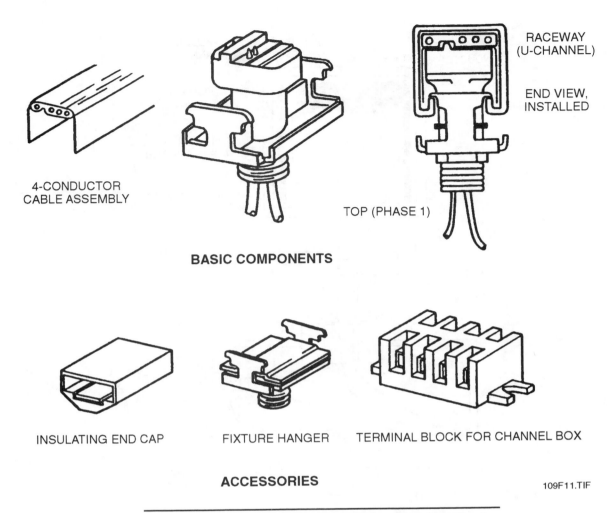

4-CONDUCTOR
CABLE ASSEMBLY

BASIC COMPONENTS

RACEWAY
(U-CHANNEL)

END VIEW,
INSTALLED

TOP (PHASE 1)

INSULATING END CAP   FIXTURE HANGER   TERMINAL BLOCK FOR CHANNEL BOX

ACCESSORIES

109F11.TIF

Figure 11. Channel Wire Components And Accessories

2.7.8   Flat Conductor Cable

Type FCC (flat conductor) cable comprises an entire branch wiring system similar in many respects to Type FC flat conductor assemblies. Guidelines for the use of this system are given in **NEC Article 328**. Type FCC cable consists of three to five flat conductors placed edge-to-edge, separated, and enclosed in a moisture-resistant and flame-retardant insulating assembly. Accessories include cable connectors, terminators, power source adapters, and receptacles.

This wiring system has been designed to supply floor outlets in office areas and other commercial and institutional interiors. It is meant to be run under carpets so that no floor drilling is required. This system is also suitable for wall mounting. As indicated in **NEC Article 328**, telephone and other communications circuits may share the same enclosure as Type FCC flat cable. The main advantage of the system is its ease of installation. It is the ideal wiring system for use when remodeling or expanding existing office facilities.

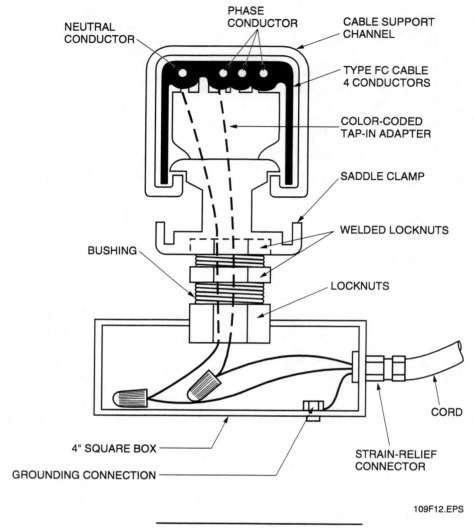

NEUTRAL CONDUCTOR

PHASE CONDUCTOR

CABLE SUPPORT CHANNEL

TYPE FC CABLE 4 CONDUCTORS

COLOR-CODED TAP-IN ADAPTER

SADDLE CLAMP

WELDED LOCKNUTS

BUSHING

LOCKNUTS

4" SQUARE BOX

GROUNDING CONNECTION

CORD

STRAIN-RELIEF CONNECTOR

109F12.EPS

Figure 12. Type FC Connection

2.7.9 Type TC Cable

Guidelines for the use of Type TC (power and control tray) cable are given in **NEC Article 340**. Type TC cable consists of two or more insulated conductors twisted together, with or without associated bare or fully-insulated grounding conductors, and covered with a nonmetallic jacket. The cables are rated at 600 volts. The cable is listed in conductor sizes No. 18 AWG to 2,000 kcmil copper or No. 12 AWG to 2,000 kcmil aluminum or copper-clad aluminum (*Figure 13*).

As the *T* in the letter designator indicates, this cable is tray cable. It can be used in cable trays and raceways. It may also be buried directly if the sheathing material is suitable for this use. Type TC cable is also good for use in sunlight when indicated by the cable markings.

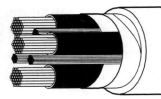

TRAY CABLE (UL)

109F13.EPS

Figure 13. Type TC Cable

2.7.10 Type USE Cable

Type USE cable is for underground installation including burial directly in the earth. Type USE cable in sizes No. 4/0 AWG and smaller with all conductors insulated is suitable for all of the underground uses for which Type UF cable is permitted by the NEC.

Type USE cable may consist of either single conductors or a multi-conductor assembly provided with a moisture-resistant covering, but it is not required to have a flame-retardant covering. This type of cable may have a bare copper conductor cabled with the assembly. Furthermore, Type USE single, parallel, or cabled conductor assemblies recognized for underground use may have a bare copper concentric conductor applied. These constructions do not require an outer overall covering. Guidelines for the use of Type USE cable are specified in *NEC Article 338*. See *Figure 14*.

When used as a service-entrance cable, Type USE cable must be installed as specified in *NEC Article 230*. Take the time to read *NEC Article 230* to ensure proper installation. Type USE service-entrance cable may also be used as feeder and branch circuit cable. Guidelines for this use of service-entrance cable are given in *NEC Section 338-3*.

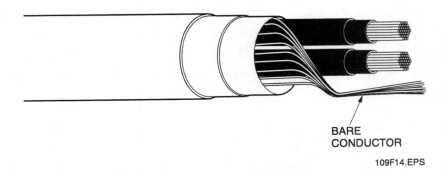

BARE
CONDUCTOR

109F14.EPS

Figure 14. Type USE Cable

2.8.0 INSTRUMENTATION CONTROL WIRING

Instrumentation control wiring links the field-sensing, controlling, printout, and operating devices that form an electronic instrumentation control system. The style and size of instrumentation control wiring must be matched to a specific job.

Instrumentation control wiring usually has two or more insulated conductor wires. These wires may also have a shield and a ground wire. An outer layer called the *jacket* protects the wiring (*Figure 15*). Instrumentation conductor wires come in pairs. The number of pairs in a multi-conductor cable depends on the size of the wire used. A multi-pair cable may have as many as 60 pairs of conductor wires.

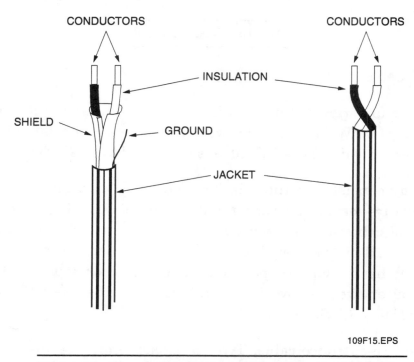

109F15.EPS

Figure 15. Two-Conductor Instrumentation Control Wiring

2.8.1   Shields

Shields are provided on instrumentation control wiring to protect the electrical signals traveling through the conductors from electrical interference or noise. Shields are usually constructed of aluminum foil bonded to a plastic film (*Figure 16*). If the wiring is not properly shielded, electrical noise may cause erratic or erroneous control signals, false indications, and improper operation of control devices.

2.8.2   Grounding

A ground wire is a bare copper wire used to provide continuous contact with a specified grounding terminal. A ground wire allows connection of all the instruments within a loop to a common grounding system. In some electronic systems, the grounding wire is called a *drain wire*. Always refer to the loop diagram to determine whether or not the ground wire is to be terminated.

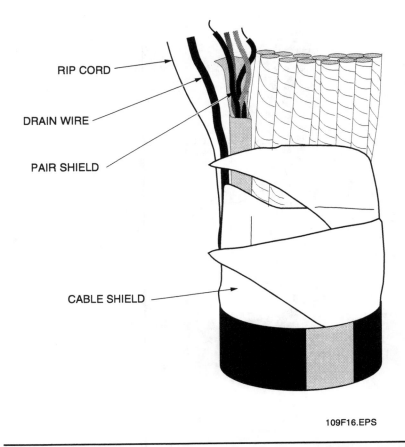

RIP CORD

DRAIN WIRE

PAIR SHIELD

CABLE SHIELD

109F16.EPS

Figure 16. Multi-Conductor Instrumentation Control Wiring With Shields

Usually, instrumentation is not grounded at both ends of the wire. This is to prevent unwanted ground loops in the system. If the ground is not to be connected at the end of the wire you are installing, do not remove the ground wire. Fold it back and tape it to the cable. This is called *floating the ground*. This is done in case the ground at the other end ever develops a problem. In that case, that ground can be removed and the wire taped to the cable can be untaped and installed.

2.8.3 Jackets

A plastic jacket covers and protects the components within the wire. Polyethylene (PE) and polyvinyl chloride (PVC) jackets are the most commonly used (*Figure 17*). Some jackets have a nylon rip cord that allows the jacket to be peeled back without the use of a knife or cable cutter. This eliminates nicking of the conductor insulation when preparing for termination.

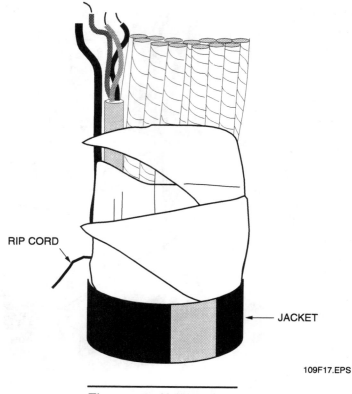

RIP CORD

JACKET

109F17.EPS

Figure 17. Wire Jacket

3.0.0   INSTALLING CONDUCTORS IN CONDUIT SYSTEMS

Conductors are installed in all types of conduit by pulling them through the conduit. This is done by using **fish tape**, pull lines, and pulling equipment.

3.1.0   FISH TAPE

Fish tape can be made of flexible steel or nylon and is available in coils of 25 to 200 feet. It should be kept on a reel to avoid twisting. Fish tape has a hook or loop on one end to attach to the conductors to be pulled (*Figure 18*). Broken or damaged fish tape should not be used. To prevent electrical shock, fish tape should not be used near or in live circuits.

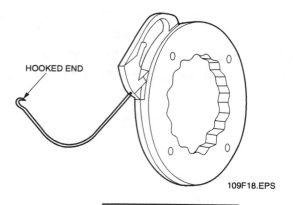

HOOKED END

109F18.EPS

Figure 18. Fish Tape

Fish tape is fed through the conduit from its reel. The tape usually enters at one outlet or junction box and is fed through to another outlet or junction box (*Figure 19*).

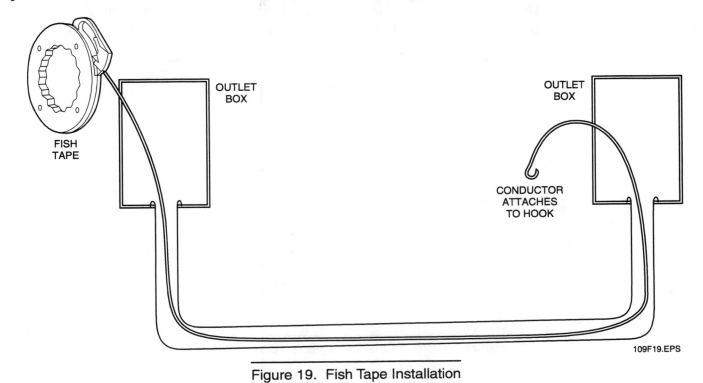

Figure 19. Fish Tape Installation

3.1.1 Power Conduit Fishing Systems

String lines can be installed by using different types of power systems. The power system is similar to an industrial vacuum cleaner and pulls a string or rope attached to a piston-like plug (sometimes called a **mouse**) through the conduit. Once the string emerges at the opposite end, either the conductor or a pull rope is then attached and pulled through the conduit, either manually or with power tools. See *Figure 20*.

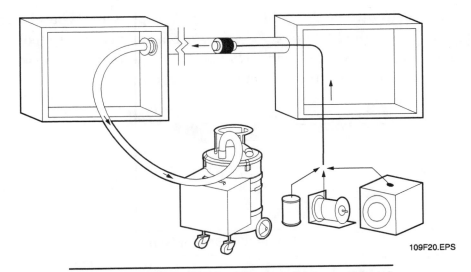

Figure 20. Power Fishing System (Vacuum Mode)

The hose connection on these vacuum systems can also be reversed to push the mouse through the conduit as shown in *Figure 21*. In other words, the system can either suck or blow the mouse through the conduit, depending on which method is best in a given situation. In either case, a fish tape is then attached to the string for retrieving through the conduit.

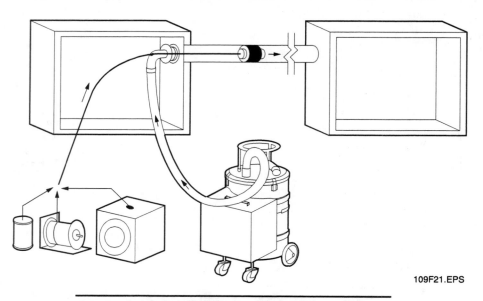

109F21.EPS

Figure 21. Power Fishing System (Blower Mode)

3.1.2 Connecting Wire To A Stringline

Once the string is installed in the conduit run, a fish tape is connected to it and pulled back through the conduit. Conductors are then attached to the hooked end of the fish tape or else connected to a basket grip. In most cases, all required conductors are pulled at one time.

3.2.0 WIRE GRIPS

Wire grips are used to attach the cable to the pull tape. One type of wire grip used is a basket grip (sometimes called *Chinese Fingers*). A basket grip is a steel mesh basket that slips over the end of a large wire or cable (*Figure 22*). The fish tape hooks onto the end and the pull on the fish tape tightens the basket over the conductor.

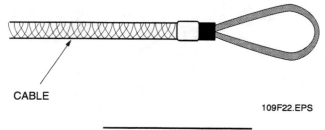

CABLE

109F22.EPS

Figure 22. Basket Grip

3.3.0 PULL LINES

After the tape has been inserted into the conduit run, you must determine if the pull on the wire will be easy or difficult. If the pull is going to be difficult because of bends in the conduit or the size of the conductors, or if several conductors are to be pulled together, a pull line should be used.

A pull line is usually made of nylon or some other synthetic fiber. It is made with a factory-spliced eye for easy connection to fish tape or conductors.

WARNING!   When using pull lines, exercise extreme caution and *never* stand in a direct line with the pulling rope. If the rope breaks, the line will whip back with great force. This can result in serious injury or death.

3.4.0 SAFETY PRECAUTIONS

The following are several important safety precautions that will help to reduce the chance of being injured while pulling cable.

- To avoid electrical shock, never use fish tape near or in live circuits. If wires must be pulled into boxes which contain live circuits, use rubber blankets over the exposed live circuitry.
- Read and understand both the operating and safety instructions for the pull system before pulling cable.
- When moving reels of cable, avoid back strain by using your legs to lift (rather than your back) and asking for help with heavy loads. Also, when manually pulling wire, spread your legs to maintain your balance and do not stretch.
- Select a rope that has a pulling load rating greater than the estimated forces required for the pull.
- Use only low-stretch rope such as multiplex and double-braided polyester for cable pulling. High-stretch ropes store energy much like a stretched rubber band. If there is a failure of the rope, pulling grip, conductors, or any other component in the pulling system, this potential energy will suddenly be unleashed. The whipping action of a rope can cause considerable damage, serious injury, or death.
- Inspect the rope thoroughly before use. Make sure there are no cuts or frays in the rope. Remember, the rope is only as strong as its weakest point.
- When designing the pull, keep the rope confined in conduit wherever possible. Should the rope break or any other part of the pulling system fail, releasing the stored energy in the rope, the confinement in the conduit will work against the whipping action of the rope by playing out much of this energy within the conduit.
- Do not stand in a direct line with the pulling rope.
- Wrap up the pulling rope after use to prevent others from tripping over it.

3.5.0 PULLING EQUIPMENT

Many types of pulling equipment are available to help pull conductors through conduit. Pulling equipment can be operated both manually and electrically. A manually-operated puller is used mainly for smaller pulling jobs where hand pulling is not possible or practical (*Figure 23*). It is also used in many locations where hand pulling would put an unnecessary strain on the conductors because of the angle of the pull involved.

109F23.EPS

Figure 23. Manual Wire Puller

Electrically-driven power pullers are used where long runs, several bends, or large conductors are involved (*Figure 24*).

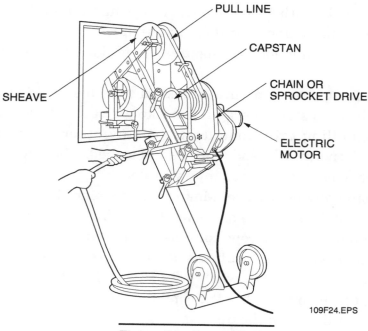

PULL LINE

CAPSTAN

CHAIN OR
SPROCKET DRIVE

ELECTRIC
MOTOR

SHEAVE

109F24.EPS

Figure 24. Power Puller

The main parts of a power puller are the electric motor, the chain or sprocket drive, the **capstan**, the sheave, and the pull line.

The pull line is routed over the sheave to ensure a straight pull. The pull line is wrapped around the capstan two or three times to provide a good grip on the capstan. The capstan is driven by the electric motor and does the actual pulling. The pull line is unwound by hand at the same speed at which the capstan is pulling. This eliminates the need for a large spool on the puller to wind the pull line.

Attachments to power pullers, such as special application sheaves and extensions, are available for most pulling jobs (*Figure 25*). Follow the manufacturer's instructions for setup and operation of the puller.

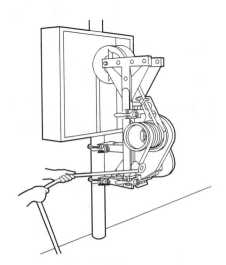
UP PULL THROUGH EXPOSED CONDUIT

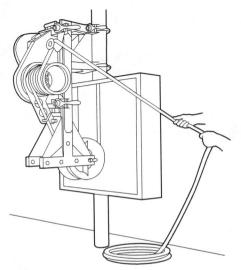
DOWN PULL THROUGH EXPOSED CONDUIT

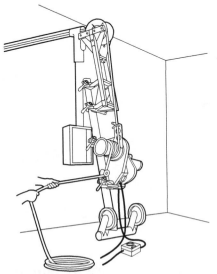
HORIZONTAL PULL WITH PIPE ADAPTER

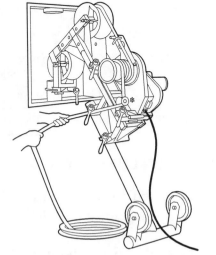
FLUSH-MOUNTED PULL WITH PIPE ADAPTER

109F25.EPS

Figure 25. Power Puller Uses

CAUTION: Before using power pullers, a qualified person must verify the amount of pull or tension that can be withstood by the conductors being pulled.

3.6.0   FEEDING CONDUCTORS INTO CONDUIT

After the fish tape or pull line is attached to the conductors, they must be pulled back through the conduit. As the fish tape is pulled, the attached conductors must be properly fed into the conduit.

Usually, more than one conductor is fed into the conduit during a wire pull. It is important to keep the conductors straight and parallel, and free from kinks, bends, and crossovers. Conductors that are allowed to cross each other will form a bulge and make pulling difficult. This could also damage the conductors.

Spools and rolls of conductors must be set up so that they unwind easily, without kinks and bends.

When several conductors must be fed into the conduit at the same time, a reel cart is used (*Figure 26*). The reel cart will allow the spools to turn freely and help prevent the wires from tangling.

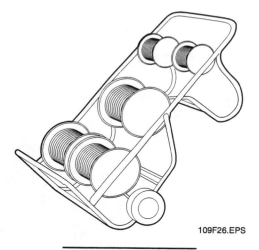

109F26.EPS

Figure 26.   Reel Cart

3.7.0   CONDUCTOR LUBRICATION

When conductors are fed into long runs of conduit or conduit with several bends, the wires are lubricated with a compound designed for wire lubrication.

Several types of formulated compounds designed for wire lubrication are available in either dry powder, paste, or gel form. These compounds must be noncorrosive to the insulation material of the conductor and to the conduit itself. The compounds are applied by hand to the conductors as they are fed into the conduit. Avoid excess lubrication.

3.8.0 CONDUCTOR TERMINATION

The amount of free conductor at each junction or outlet box must meet certain NEC specifications. For example, there must be sufficient free conductor so that bends or terminations inside of the box, cabinet, or enclosure may be made to a radius as specified in the NEC. The NEC specifies a minimum of six inches for connections made to wiring devices or for splices. Where conductors pass through junction or pull boxes, enough slack should be provided for splices at a later date.

When a box is used as a pull box, the conductors are not necessarily spliced. They may merely enter the pull box via one conduit run and exit via another conduit run. The purpose of a pull box, as the name suggests, is to facilitate pulling conductors on long runs. A junction box, however, is used not only to facilitate pulling conductors through the raceway system, but it also provides an enclosure for splices in the conductors.

SUMMARY

Knowing the different types of conductors, how they are rated, and what they are used for will help you when selecting conductors for a specific job. Pulling wires and cables through conduit systems is an important part of your job as an electrician. The more you learn about the concepts involved in pulling cable, the safer and more efficient you will be.

References

For advanced study of topics covered in this task module, the following book is suggested:

National Electrical Code Handbook, Latest Edition, National Fire Protection Association, Quincy, MA.

REVIEW QUESTIONS

1. Where can the ampacity ratings of conductors be found in the NEC?
   a. *NEC Chapter 1*
   b. *NEC Articles 348 through 352*
   c. *NEC Tables 310-16 through 310-19*
   d. *NEC Chapter 9*

2. How many wires are contained in the three configurations of stranded cable?
   a. 7, 19, and 37
   b. 5, 7, and 9
   c. 3, 5, and 7
   d. 6, 18, and 36

3. What is the purpose of stranding?
   a. To stiffen the wire
   b. To increase the conductivity of the wire
   c. To decrease the conductivity of the wire
   d. To increase the flexibility of the wire

4. What is the most common conductor material?
   a. Silver
   b. Aluminum
   c. Copper
   d. Platinum

5. Which of the following does *not* help to determine how good a material will be for the construction of wire?
   a. Availability
   b. Cost
   c. Conductivity
   d. Molecular weight

6. What letter must be included in the marking of a conductor if it is to be used in a wet, outdoor application?
   a. D
   b. O
   c. I
   d. W

7. What service conductor may be green or green with a yellow stripe?
   a. The grounding conductor of a multi-conductor cable
   b. The neutral conductor of a multi-conductor cable
   c. The hot conductor of a multi-conductor cable
   d. Both a and b

8. What are the colors of insulation on the conductors for a three-conductor cable?
   a. One white, one red, and one black
   b. Two white and one red
   c. One white, one red, one black, and a grounding conductor
   d. One green, one white, one blue, and a grounding conductor

9. Type NMC cable is suitable for _____ locations.
   a. dry
   b. damp
   c. corrosive
   d. all of the above

10. Type USE cable can be used for _____.
    a. above-ground installation only
    b. underground installation within a special PVC pipeline
    c. underground installation including direct burial
    d. none of the above

notes

ANSWERS TO REVIEW QUESTIONS

Answer		Section
1.	c	2.1.0
2.	a	2.3.2
3.	d	2.3.2
4.	c	2.4.0
5.	d	2.4.0
6.	d	2.5.2
7.	a	2.5.3
8.	c	2.5.3
9.	d	2.7.2
10.	c	2.7.10

NCCER CRAFT TRAINING USER UPDATES

The NCCER makes every effort to keep these manuals up-to-date and free of technical errors. We appreciate your help in this process. If you have an idea for improving this manual, or if you find an error, a typographical mistake, or an inaccuracy in the NCCER's Craft Training Manuals, please write us, using this form or a photocopy. Be sure to include the exact module number, page number, a description of the problem, and the correction, if possible. Your input will be brought to the attention of the Technical Review Committee. Thank you for your assistance.

Instructors – If you found that additional materials were necessary in order to teach this module effectively, please let us know so that we may include them in the Equipment/Materials list in the Instructor's Guide.

Write: Curriculum Development and Revision Department
National Center for Construction Education and Research
P.O. Box 141104
Gainesville, FL 32614-1104
Fax: 352-334-0932

Craft _____ Module Name _____

Copyright Date _____ Module Number _____ Page Number(s) _____

Description of Problem _____

_____

_____

_____

(Optional) Correction of Problem _____

_____

_____

(Optional) Your Name and Address _____

_____

_____

notes

Introduction to Electrical Blueprints

Module 26110

Electrical Trainee Task Module 26110

INTRODUCTION TO ELECTRICAL BLUEPRINTS

OBJECTIVES

Upon completion of this module, the trainee will be able to:

1. Explain the basic layout of a blueprint.
2. Describe the information included in the title block of a blueprint.
3. Identify the types of lines used on blueprints.
4. Identify common symbols used on blueprints.
5. Understand the use of architect's and engineer's scales.
6. Interpret electrical drawings, including site plans, floor plans, and detail drawings.
7. Read equipment schedules found on electrical blueprints.
8. Describe the type of information included in electrical specifications.

Prerequisites

Successful completion of the following Task Modules is required before beginning study of this Task Module: Core Curricula; Electrical Level 1, Modules 26101 through 26109.

Required Trainee Materials

1. Trainee Task Module
2. Copy of the latest edition of the *National Electrical Code*
3. Appropriate Personal Protective Equipment

Note: The designations "National Electrical Code," "NE Code," and "NEC," where used in this document, refer to the *National Electrical Code®*, which is a registered trademark of the National Fire Protection Association, Quincy, MA. *All National Electrical Code (NEC) references in this module refer to the 1999 edition of the NEC.*

Course Map

This course map shows all of the task modules in the first level of the Electrical curricula. The suggested training order begins at the bottom and proceeds up. Skill levels increase as a trainee advances on the course map. The training order may be adjusted by the local Training Program Sponsor.

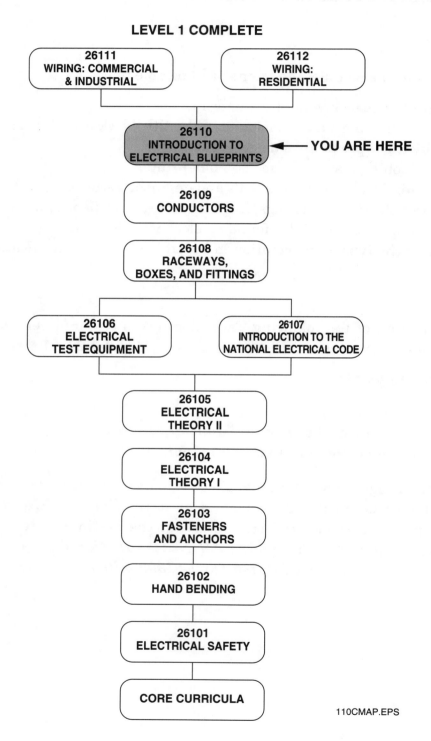

LEVEL 1 COMPLETE

26111
WIRING: COMMERCIAL & INDUSTRIAL

26112
WIRING: RESIDENTIAL

26110
INTRODUCTION TO ELECTRICAL BLUEPRINTS ← **YOU ARE HERE**

26109
CONDUCTORS

26108
RACEWAYS, BOXES, AND FITTINGS

26106
ELECTRICAL TEST EQUIPMENT

26107
INTRODUCTION TO THE NATIONAL ELECTRICAL CODE

26105
ELECTRICAL THEORY II

26104
ELECTRICAL THEORY I

26103
FASTENERS AND ANCHORS

26102
HAND BENDING

26101
ELECTRICAL SAFETY

CORE CURRICULA

110CMAP.EPS

TABLE OF CONTENTS

Trade Terms Introduced In This Module

Architectural drawings: Working drawings consisting of plans, elevations, details, and other information necessary for the construction of a building. Architectural drawings usually include:

- A site (plot) plan indicating the location of the building on the property
- Floor plans showing the walls and partitions for each floor or level
- Elevations of all exterior faces of the building
- Several vertical cross sections to indicate clearly the various floor levels and details of the footings, foundations, walls, floors, ceilings, and roof construction
- Large-scale detail drawings showing such construction details as may be required

Block diagram: A single-line diagram used to show electrical equipment and related connections. See *power-riser diagram*.

Blueprint: An exact copy or reproduction of an original drawing.

Detail drawing: An enlarged, detailed view taken from an area of a drawing and shown in a separate view.

Dimensions: Sizes or measurements printed on a drawing.

Electrical drawing: A means of conveying a large amount of exact, detailed information in an abbreviated language. Consists of lines, symbols, dimensions, and notations to accurately convey an engineer's designs to electricians who install the electrical system on a job.

Elevation drawing: An architectural drawing showing height, but not depth; usually the front, rear, and sides of a building or object.

Floor plan: A drawing of a building as if a horizontal cut were made through a building at about window level, and the top portion removed. The floor plan is what would appear if the remaining structure were viewed from above.

One-line diagram: A drawing that shows, by means of lines and symbols, the path of an electrical circuit or system of circuits along with the various circuit components. Also called a *single-line diagram*.

Plan view: A drawing made as though the viewer were looking straight down (from above) on an object.

Power-riser diagram: A single-line block diagram used to indicate the electric service equipment, service conductors and feeders, and subpanels. Notes are used on power-riser diagrams to identify the equipment; indicate the size of conduit; show the number, size, and type of conductors; and list related materials. A panelboard schedule is usually included with power-riser diagrams to indicate the exact components (panel type and size), along with fuses, circuit breakers, etc., contained in each panelboard.

Scale: On a drawing, the size relationship between an object's actual size and the size it is drawn. Scale also refers to the measuring tool used to determine this relationship.

Schedule: A systematic method of presenting equipment lists on a drawing in tabular form.

Schematic drawing: A cutaway drawing that shows the inside of an object or building.

Shop drawing: A drawing that is usually developed by manufacturers, fabricators, or contractors to show specific dimensions and other pertinent information concerning a particular piece of equipment and its installation methods.

Site plan: A drawing showing the location of a building or buildings on the building site. Such drawings frequently show topographical lines, electrical and communication lines, water and sewer lines, sidewalks, driveways, and similar information.

Written specifications: A written description of what is required by the owner, architect, and engineer in the way of materials and workmanship. Together with working drawings, the specifications form the basis of the contract requirements for construction.

1.0.0   INTRODUCTION TO BLUEPRINT READING

In all large construction projects and in many of the smaller ones, an architect is commissioned to prepare complete working drawings and specifications for the project. These drawings usually include:

- A **site plan** indicating the location of the building on the property.
- **Floor plans** showing the walls and partitions for each floor or level.
- Elevations of all exterior faces of the building.
- Several vertical cross sections to indicate clearly the various floor levels and details of the footings, foundation, walls, floors, ceilings, and roof construction.
- Large-scale detail drawings showing such construction details as may be required.

For projects of any consequence, the architect usually hires consulting engineers to prepare structural, electrical, and mechanical drawings, with the latter encompassing pipe-fitting, instrumentation, plumbing, heating, ventilating, and air conditioning drawings.

1.1.0 SITE PLAN

This type of plan of the building site looks as if the site is viewed from an airplane and shows the property boundaries, the existing contour lines, the new contour lines (after grading), the location of the building on the property, new and existing roadways, all utility lines, and other pertinent details. The drawing **scale** is also shown. Descriptive notes may also be found on the site (plot) plan listing names of adjacent property owners, the land surveyor, and the date of the survey. A legend or symbol list is also included so that anyone who must work with the site plan can readily read the information. See *Figure 1*.

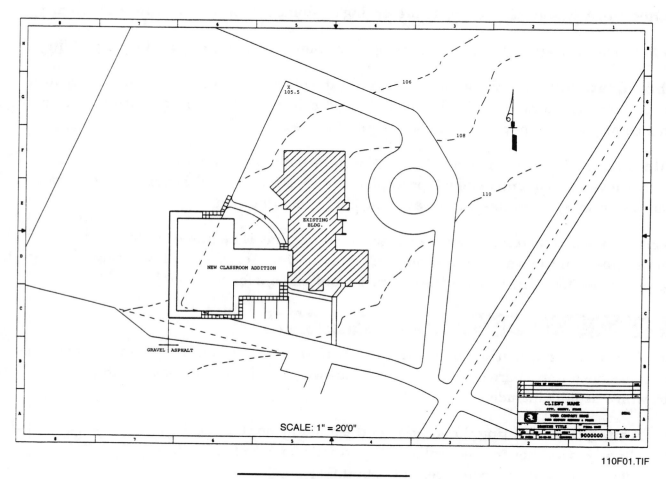

110F01.TIF

Figure 1. Typical Site Plan

1.2.0 FLOOR PLANS

The **plan view** of any object is a drawing showing the outline and all details as seen when looking directly down on the object. It shows only two **dimensions**, length and width. The floor plan of a building is drawn as if a horizontal cut were made through the building—at about window height—and then the top portion removed to reveal the bottom part. See *Figure 2*.

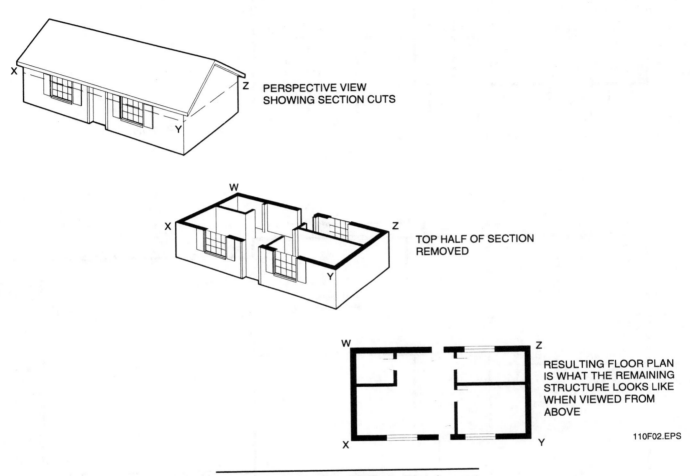

PERSPECTIVE VIEW
SHOWING SECTION CUTS

TOP HALF OF SECTION
REMOVED

RESULTING FLOOR PLAN
IS WHAT THE REMAINING
STRUCTURE LOOKS LIKE
WHEN VIEWED FROM
ABOVE

110F02.EPS

Figure 2. Principles Of Floor Plan Layout

If a plan view of a home's basement is needed, the part of the house above the middle of the basement windows is imagined to be cut away. By looking down on the uncovered portion, every detail and partition can be seen. Likewise, imagine the part above the middle of the first floor windows being cut away. A drawing which looks straight down at the remaining part would be called the *first floor plan* or *lower level*. A cut through the second floor windows would be called the *second floor plan* or *upper level*. See *Figure 3*.

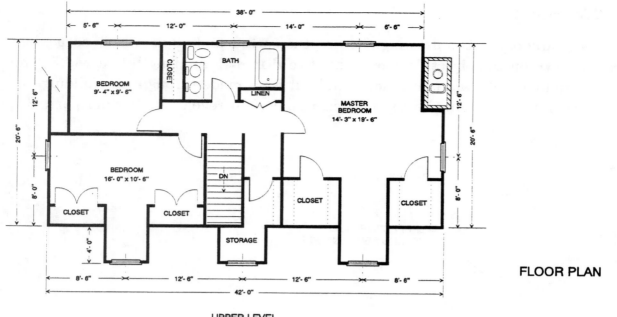

FLOOR PLAN

UPPER LEVEL

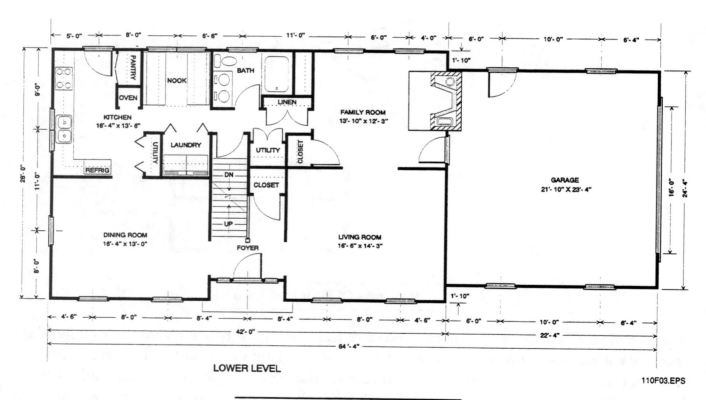

LOWER LEVEL

110F03.EPS

Figure 3. Floor Plans Of A Building

1.3.0 ELEVATIONS

The *elevation* is an outline of an object that shows heights and may show the length or width of a particular side, but not depth. *Figures 4* and *5* show **elevation drawings** for a building.

Note: These elevation drawings show the heights of windows, doors, and porches, the pitch of roofs, etc., because all of these measurements cannot be shown conveniently on floor plans.

FRONT ELEVATION

REAR ELEVATION   110F04.EPS

Figure 4. Front And Rear Elevations

LEFT ELEVATION

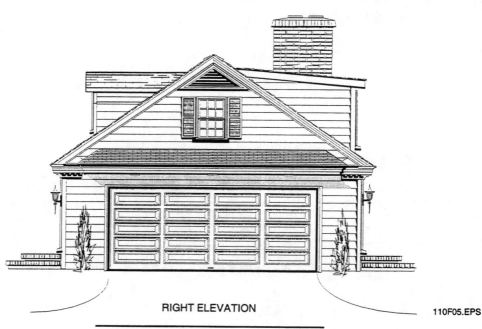

RIGHT ELEVATION

110F05.EPS

Figure 5. Left And Right Elevations

1.4.0 SECTIONS

A section or sectional view (*Figure 6*) is a cutaway view that allows the viewer to see the inside of a structure. The point on the plan or elevation showing where the imaginary cut has been made is indicated by the section line, which is usually a dashed line. The section line shows the location of the section on the plan or elevation. It is necessary to know which of the cutaway parts is represented in the sectional drawing. To show this, arrow points are placed at the ends of the section lines.

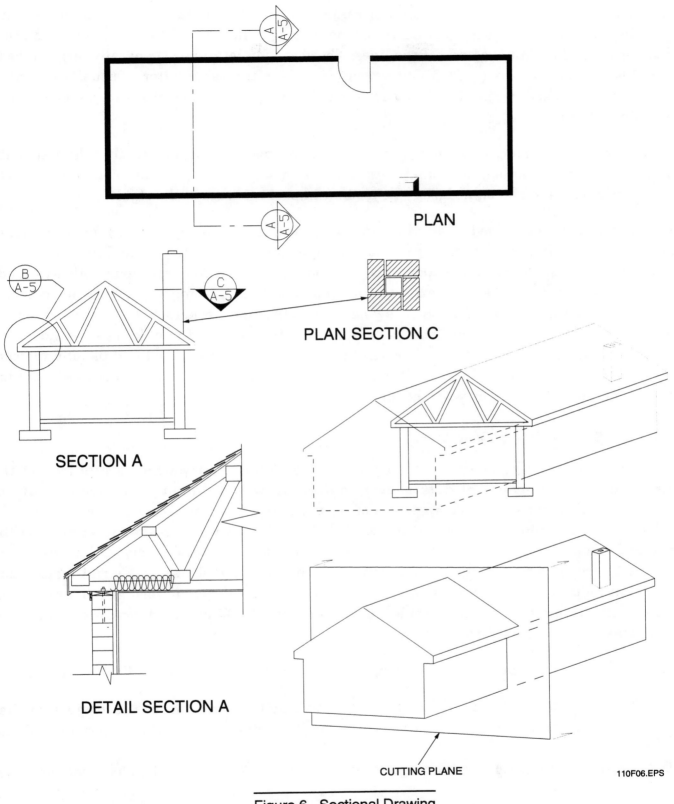

PLAN

PLAN SECTION C

SECTION A

DETAIL SECTION A

CUTTING PLANE

110F06.EPS

Figure 6. Sectional Drawing

In **architectural drawings**, it is often necessary to show more than one section on the same drawing. The different section lines must be distinguished by letters, numbers, or other designations placed at the ends of the lines. These section letters are generally large so as to stand out on the drawings. To further avoid confusion, the same letter is usually placed at each end of the section line. The section is named according to these letters (e.g., Section A-A, Section B-B, and so forth).

A longitudinal section is taken lengthwise while a cross section is usually taken straight across the width of an object. Sometimes, however, a section is not taken along one straight line. It is often taken along a zigzag line to show important parts of the object.

A sectional view, as applied to architectural drawings, is a drawing showing the building, or portion of a building, as though it were cut through on some imaginary line. This line may be either vertical (straight up and down) or horizontal. Wall sections are nearly always made vertically so that the cut edge is exposed from top to bottom. In some ways, the wall section is one of the most important of all the drawings to construction workers, because it answers the questions as to how a structure should be built. The floor plans of a building show how each floor is arranged, but the wall sections tell how each part is constructed and usually indicate the material to be used. The electrician needs to know this information when determining wiring methods that comply with the NEC.

1.5.0 ELECTRICAL DRAWINGS

Electrical drawings show in a clear, concise manner exactly what is required of the electricians. The amount of data shown on such drawings should be sufficient, but not overdone. This means that a complete set of electrical drawings could consist of only one 8½" x 11" sheet, or it could consist of several dozen 24" x 36" (or larger) sheets, depending on the size and complexity of a given project. A **shop drawing**, for example, may contain details of only one piece of equipment, while a set of working drawings for an industrial installation may contain dozens of drawing sheets detailing the electrical system for lighting and power, along with equipment, motor controls, wiring diagrams, **schematic diagrams**, equipment **schedules**, and a host of other pertinent data.

In general, the electrical working drawings for a given project serve three distinct functions:

- They provide electrical contractors with an exact description of the project so that materials and labor may be estimated to project a total cost of the project for bidding purposes.
- They provide workers on the project with instructions as to how the electrical system is to be installed.
- They provide a map of the electrical system once the job is completed to aid in maintenance and troubleshooting for years to come.

Electrical drawings from consulting engineering firms will vary in quality from sketchy, incomplete drawings to neat, precise drawings that are easy to understand. Few, however, will cover every detail of the electrical system. Therefore, a good knowledge of installation practices must go hand-in-hand with interpreting electrical working drawings.

Sometimes electrical contractors will have electrical drafters prepare special supplemental drawings for use by the contractors' employees. On certain projects, these supplemental drawings can save supervision time in the field once the project has begun.

2.0.0   BLUEPRINT LAYOUT

Although a strong effort has been made to standardize drawing practices in the building construction industry, **blueprints** prepared by different architectural or engineering firms will rarely be identical. Similarities, however, will exist between most sets of blueprints, and with a little experience, you should have no trouble interpreting any set of drawings that might be encountered.

Most drawings used for building construction projects will be drawn on sheets ranging from 11" x 17" to 24" x 36" in size. Each drawing sheet will have border lines framing the overall drawing and one or more title blocks, as shown in *Figure 7*. The type and size of title blocks varies with each firm preparing the drawings. In addition, some drawing sheets will also contain a revision block near the title block, and perhaps an approval block. This information is normally found on each drawing sheet, regardless of the type of project or the information contained on the sheet.

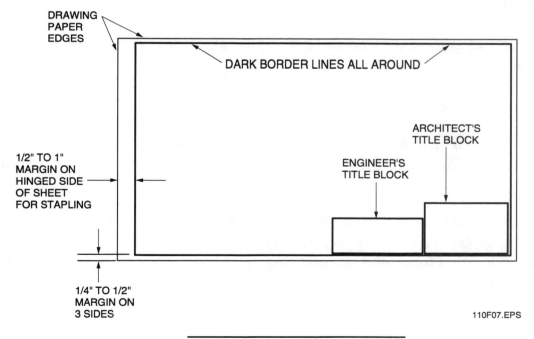

110F07.EPS

Figure 7. Typical Blueprint Layout

2.1.0 TITLE BLOCK

The architect's title block for a blueprint is usually boxed in the lower right-hand corner of the drawing sheet; the size of the block varies with the size of the drawing and with the information required. See *Figure 8*.

Figure 8. Typical Architect's Title Block

In general, the title block of an electrical drawing should contain the following information:

- Name of the project
- Address of the project
- Name of the owner or client
- Name of the architectural firm
- Date of completion
- Scale(s)
- Initials of the drafter, checker, and designer, with dates under each
- Job number
- Sheet number
- General description of the drawing

Every architectural firm has its own standard for drawing titles, and they are often preprinted directly on the tracing paper or else printed on a sticker which is placed on the drawing.

Often, the consulting engineering firm will also be listed, which means that an additional title block will be applied to the drawing, usually next to the architect's title block. *Figure 9* shows completed architectural and engineering title blocks as they appear on an actual drawing.

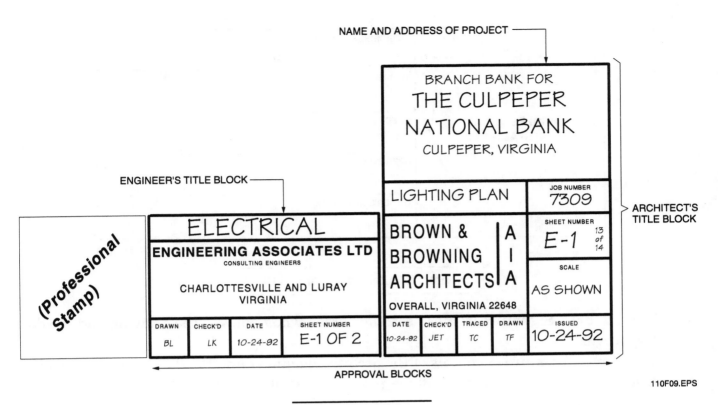

Figure 9. Title Blocks

2.2.0 APPROVAL BLOCK

The approval block, in most cases, will appear on the drawing sheet as shown in *Figure 10*. The various types of approval blocks (drawn, checked, etc.) will be initialed by the appropriate personnel. This type of approval block is usually part of the title block and appears on each drawing sheet.

COMM. NO.	DATE	DRAWN	CHECKED	REVISED
7215	9/6/92	GK	GLC	

110F10.EPS

Figure 10. Typical Approval Block

On some projects, authorized signatures are required before certain systems may be installed, or even before the project begins. An approval block such as the one shown in *Figure 11* indicates that all required personnel have checked the drawings for accuracy, and that the set meets with everyone's approval. Such an approval block usually appears on the front sheet of the blueprint set and may include:

- *Professional stamp* – Registered seal of approval by the architect or consulting engineer.
- *Design supervisor* – Signature of the person who is overseeing the design.
- *Drawn (by)* – Signature or initials of the person who drafted the drawing and the date it was completed.

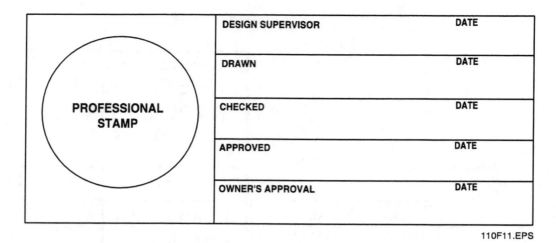

110F11.EPS

Figure 11. Alternate Approval Block

- *Checked (by)* – Signature or initials of the person who reviewed the drawing and the date of approval.
- *Approved* – Signature or initials of the architect/engineer and the date of the approval.
- *Owner's approval* – Signature of the project owner or the owner's representative along with the date signed.

2.3.0   REVISION BLOCK

Sometimes electrical drawings will have to be partially redrawn or modified during the construction of a project. It is extremely important that such modifications are noted and dated on the drawings to ensure that the workers have an up-to-date set of drawings to work from. In some situations, sufficient space is left near the title block for dates and descriptions of revisions, as shown in *Figure 12*. In other cases, a revision block is provided (again, near the title block), as shown in *Figure 13*.

Note: Architects, engineers, designers, and drafters has their own methods of showing revisions, so expect to find deviations from those shown.

CAUTION: When a set of electrical working drawings has been revised, always make certain that the most up-to-date set is used for all future layout work. Either destroy the old, obsolete set of drawings or else clearly mark on the affected sheets, *Obsolete Drawing—Do Not Use.* Also, when working with a set of working drawings and **written specifications** for the first time, thoroughly check each page to see if any revisions or modifications have been made to the originals. Doing so can save much time and expense to all concerned with the project.

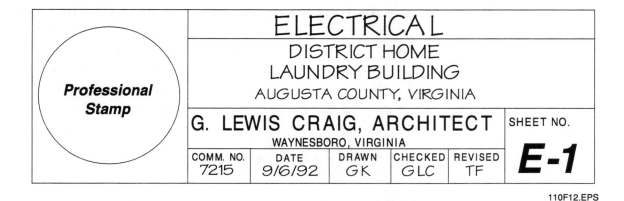

Figure 12. One Method Of Showing Revisions On Working Drawings

REVISIONS				
REV	DESCRIPTION	DR	APPD	DATE
1	FIXTURE NO. 3 IN. LIGHTING-FIXTURE SCHDL	GK	GLC	10/12/92

ELECTRICAL

DISTRICT HOME
LAUNDRY BUILDING

AUGUSTA COUNTY, VIRGINIA

G. LEWIS CRAIG, ARCHITECT SHEET NO.
WAYNESBORO, VIRGINIA

COMM. NO.	DATE	DRAWN	CHECKED	REVISED	**E-1**
7215	9/6/92	GK	GLC	TF	

Professional Stamp

110F13.EPS

Figure 13. Alternative Method Of Showing Revisions On Working Drawings

3.0.0   DRAFTING LINES

You will encounter many types of drafting lines. To specify the meaning of each type of line, contrasting lines can be made by varying the width of the lines or breaking the lines in a uniform way.

Figure 14 shows common lines used on architectural drawings. However, these lines can vary. Architects and engineers have strived for a common standard for the past century, but unfortunately, their goal has yet to be reached. Therefore, you will find variations in lines and symbols from drawing to drawing, so always consult the legend or symbol list when referring to any drawing. Also, carefully inspect each drawing to ensure that line types are used consistently.

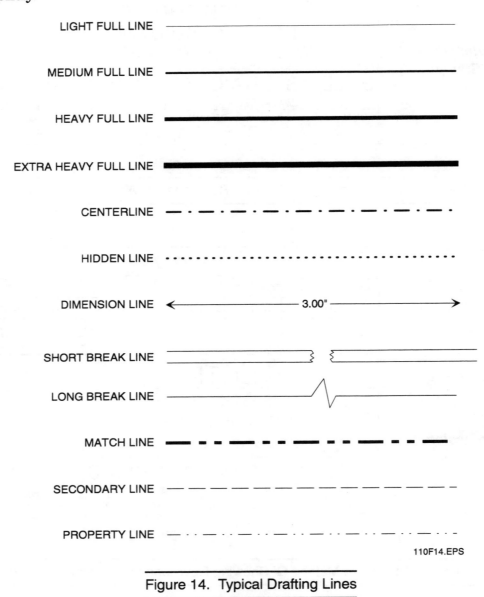

110F14.EPS

Figure 14. Typical Drafting Lines

The drafting lines shown in *Figure 14* are used as follows:

- *Light full line* – This line is used for section lines, building background (outlines), and similar uses where the object to be drawn is secondary to the system being shown (e.g., HVAC or electrical).
- *Medium full line* – This type of line is frequently used for hand lettering on drawings. It is further used for some drawing symbols, circuit lines, etc.

- *Heavy full line* – This line is used for borders around title blocks, schedules, and for hand lettering drawing titles. Some types of symbols are frequently drawn with a heavy full line.
- *Extra heavy full line* – This line is used for border lines on architectural/engineering drawings.
- *Centerline* – A centerline is a broken line made up of alternately-spaced long and short dashes. It indicates the centers of objects such as holes, pillars, or fixtures. Sometimes, the centerline indicates the dimensions of a finished floor.
- *Hidden line* – A hidden line consists of a series of short dashes that are closely and evenly spaced. It shows the edges of objects that are not visible in a particular view. The object outlined by hidden lines in one drawing is often fully pictured in another drawing.
- *Dimension line* – These are thin lines used to show the extent and direction of dimensions. The dimension is usually placed in a break inside of the dimension lines. Normal practice is to place the dimension lines outside the object's outline. However, it may sometimes be necessary to draw the dimensions inside the outline.
- *Short break line* – This line is usually drawn freehand and is used for short breaks.
- *Long break line* – This line, which is drawn partly with a straightedge and partly with freehand zigzags, is used for long breaks.
- *Match line* – This line is used to show the position of the cutting plane. Therefore, it is also called the *cutting plane line*. A match or cutting plane line is a heavy line with long dashes alternating with two short dashes. It is used on drawings of large structures to show where one drawing stops and the next drawing starts.
- *Secondary line* – This line is frequently used to outline pieces of equipment or to indicate reference points of a drawing that are secondary to the drawing's purpose.
- *Property line* – This is a light line made up of one long and two short dashes that are alternately spaced. It indicates land boundaries on the site plan.

Other uses of the lines just mentioned include the following:

- *Extension lines* – Extension lines are lightweight lines that start about ⅟₁₆ inch away from the edge of an object and extend out. A common use of extension lines is to create a boundary for dimension lines. Dimension lines meet extension lines with arrowheads, slashes, or dots. Extension lines that point from a note or other reference to a particular feature on a drawing are called *leaders*. They usually end in either an arrowhead or a dot and may include an explanatory note at the end.
- *Section lines* – These are often referred to as *cross-hatch lines*. Drawn at a 45° angle, these lines show where an object has been cut away to reveal the inside.
- *Phantom lines* – Phantom lines are solid, light lines that show where an object will be installed. A future door opening or a future piece of equipment can be shown with phantom lines.

3.1.0 ELECTRICAL DRAFTING LINES

Besides the architectural lines shown in *Figure 14*, consulting electrical engineers, designers, and drafters use additional lines to represent circuits and their related components. Again, these lines may vary from drawing to drawing, so check the symbol list or legend for the exact meaning of lines on the drawing with which you are working. *Figure 15* shows lines used on some electrical drawings.

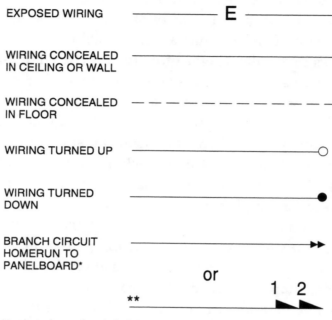

* Number of arrowheads indicate number of circuits. A number at each arrowhead may be used to identify circuit numbers.

** Half arrowheads are sometimes used for homeruns to avoid confusing them with drawing callouts.

110F15.EPS

Figure 15. Electrical Drafting Lines

4.0.0 ELECTRICAL SYMBOLS

The electrician must be able to correctly read and understand electrical working drawings. This includes a thorough knowledge of electrical symbols and their applications.

An electrical symbol is a figure or mark that stands for a component used in the electrical system. *Figure 16* shows a list of electrical symbols that are currently recommended by the American National Standards Institute (ANSI). It is evident from this list of symbols that many have the same basic form, but, because of some slight difference, their meaning changes. For example, the receptacle symbols in *Figure 17* each have the same basic form (a circle) but the addition of a line or an abbreviation gives each an individual meaning. A good procedure to follow in learning symbols is to first learn the basic form and then apply the variations for obtaining different meanings.

SWITCH OUTLETS

Single-Pole Switch	S
Double-Pole Switch	S$_2$
Three-Way Switch	S$_3$
Four-Way Switch	S$_4$
Key-Operated Switch	S$_K$
Switch w/Pilot	S$_P$
Low-Voltage Switch	S$_L$
Switch & Single Receptacle	⊖$_S$
Switch & Duplex Receptacle	⊖$_S$
Door Switch	S$_D$
Momentary Contact Switch	S$_{MC}$

RECEPTACLE OUTLETS

Single Receptacle	⊖
Duplex Receptacle	⊖
Triplex Receptacle	⊕
Split-Wired Duplex Recep.	⬤
Single Special Purpose Recep.	⊖△
Duplex Special Purpose Recep.	⊖△
Range Receptacle	⊖$_R$
Special Purpose Connection or Provision for Connection. Subscript letters indicate Function (DW - Dishwasher; CD - Clothes Dryer, etc.)	⬤$_{DW}$
Clock Receptacle w/Hanger	Ⓒ
Fan Receptacle w/Hanger	Ⓕ
Single Floor Receptacle	⊡

Note: A numeral or letter within the symbol or as a subscript keyed to the list of symbols indicates type of receptacle or usage.

LIGHTING OUTLETS

	Ceiling	Wall
Surface Fixture	○	⊸○
Surface Fixt. w/Pull Switch	○ PS	⊸○ PS
Recessed Fixture	Ⓡ	⊸Ⓡ
Surface or Pendant Fluorescent Fixture	▭○	
Recessed Fluor. Fixture	▭○R	
Surface or Pendant Continuous Row Fluor. Fixtures	▭○	
Recessed Continuous Row Fluorescent Fixtures	▭○ R	
Surface Exit Light	Ⓧ	⊸Ⓧ
Recessed Exit Light	⊗R	⊸⊗R
Blanked Outlet	Ⓑ	⊸Ⓑ
Junction Box	Ⓙ	⊸Ⓙ

CIRCUITING

Wiring Concealed in Ceiling or Wall	————
Wiring Concealed in Floor	– – – –
Wiring Exposed	··········

Branch Circuit Homerun to Panelboard. Number of arrows indicates number of circuits in run. Note: Any circuit without further identification is 2-wire. A greater number of wires is indicated by cross lines as shown below. Wire size is sometimes shown with numerals placed above or below cross lines.

———/// 3-Wire

———//// 4-Wire

Figure 16. ANSI Electrical Symbols

110F16.EPS

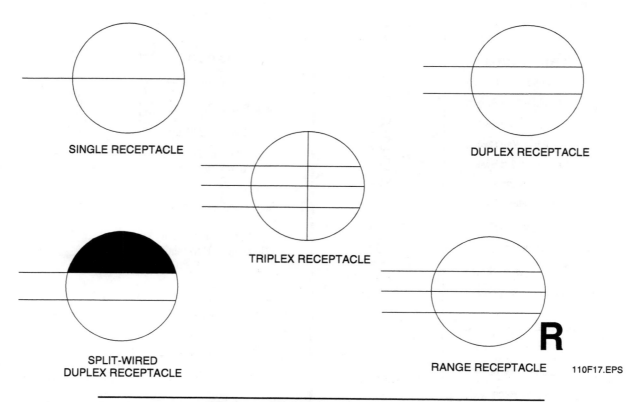

SINGLE RECEPTACLE

DUPLEX RECEPTACLE

TRIPLEX RECEPTACLE

SPLIT-WIRED
DUPLEX RECEPTACLE

RANGE RECEPTACLE

110F17.EPS

Figure 17. Various Receptacle Symbols Used On Electrical Drawings

It would be much simpler if all architects, engineers, electrical designers, and drafters used the same symbols; however, this is not the case. Although standardization is getting closer to a reality, existing symbols are still modified, and new symbols are created for almost every new project.

The electrical symbols described in the following paragraphs represent those found on actual electrical working drawings throughout the United States and Canada. Many are similar to those recommended by ANSI and the Consulting Engineers Council/US; others are not. Understanding how these symbols were devised will help you to interpret unknown electrical symbols in the future.

Some of the symbols used on electrical drawings are abbreviations, such as WP for weatherproof and AFF for above finished floor. Others are simplified pictographs, such as (A) in *Figure 18* for a double floodlight fixture or (B) for an infrared electric heater with two quartz lamps.

In some cases, the symbols are combinations of abbreviations and pictographs, such as (C) in *Figure 18* for a fusible safety switch, (D) for a nonfusible safety switch, and (E) for a double-throw safety switch. In each example, a pictograph of a switch enclosure has been combined with an abbreviation: F (fusible), DT (double-throw), and NF (nonfusible), respectively.

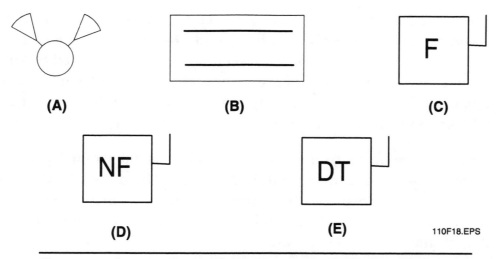

Figure 18. General Types Of Symbols Used On Electrical Drawings

Lighting outlet symbols have been devised that represent incandescent, fluorescent, and high-intensity discharge lighting; a circle usually represents an incandescent fixture, and a rectangle is used to represent a fluorescent fixture. These symbols are designed to indicate the physical shape of a particular fixture, and while the circles representing incandescent lamps are frequently enlarged somewhat, symbols for fluorescent fixtures are usually drawn as close to scale as possible. The type of mounting used for all lighting fixtures is usually indicated in a lighting fixture schedule, which is shown on the drawings or in the written specifications.

The type of lighting fixture is identified by a numeral placed inside a triangle or other symbol, and placed near the fixture to be identified. A complete description of the fixtures identified by the symbols must be given in the lighting fixture schedule and should include the manufacturer, catalog number, number and type of lamps, voltage, finish, mounting, and any other information needed for proper installation of the fixture.

Switches used to control lighting fixtures are also indicated by symbols (usually the letter S followed by numerals or letters to define the exact type of switch). For example, S_3 indicates a three-way switch; S_4 identifies a four-way switch; and S_P indicates a single-pole switch with a pilot light.

Main distribution centers, panelboards, transformers, safety switches, and other similar electrical components are indicated by electrical symbols on floor plans and by a combination of symbols and semipictorial drawings in riser diagrams.

A detailed description of the service equipment is usually given in the panelboard schedule or in the written specifications. However, on small projects, the service equipment is sometimes indicated only by notes on the drawings.

Circuit and feeder wiring symbols are getting closer to being standardized. Most circuits concealed in the ceiling or wall are indicated by a solid line; a broken line is used for circuits concealed in the floor or ceiling below; and exposed raceways are indicated by short dashes or else the letter E placed in the same plane with the circuit line at various intervals. The number of conductors in a conduit or raceway system may be indicated in the panelboard schedule under the appropriate column, or the information may be shown on the floor plan.

Symbols for communication and signal systems, as well as symbols for light and power, are drawn to an appropriate scale and accurately located with respect to the building. This reduces the number of references made to the architectural drawings. Where extreme accuracy is required in locating outlets and equipment, exact dimensions are given on larger-scale drawings and shown on the plans.

Each different category in an electrical system is usually represented by a basic distinguishing symbol. To further identify items of equipment or outlets in the category, a numeral or other identifying mark is placed within the open basic symbol. In addition, all such individual symbols used on the drawings should be included in the symbol list or legend. The electrical symbols shown in *Figure 19* were modified by a consulting engineering firm for use on a small industrial electrical installation. The symbols shown in *Figure 20* are those recommended by the Consulting Engineers Council/US. You should become familiar with these symbols.

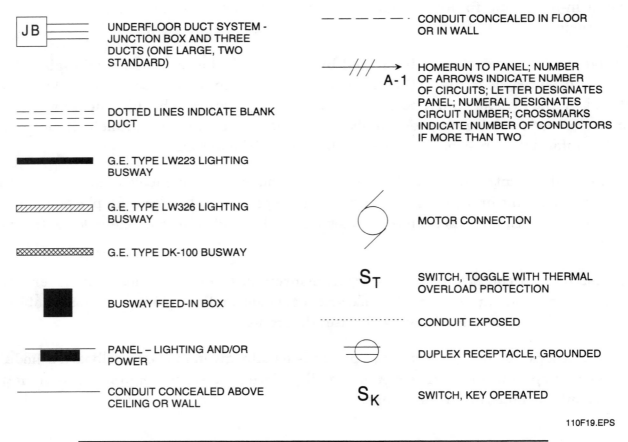

110F19.EPS

Figure 19. Electrical Symbols Used By One Consulting Engineering Firm

ELECTRICAL — TRAINEE TASK MODULE 26110

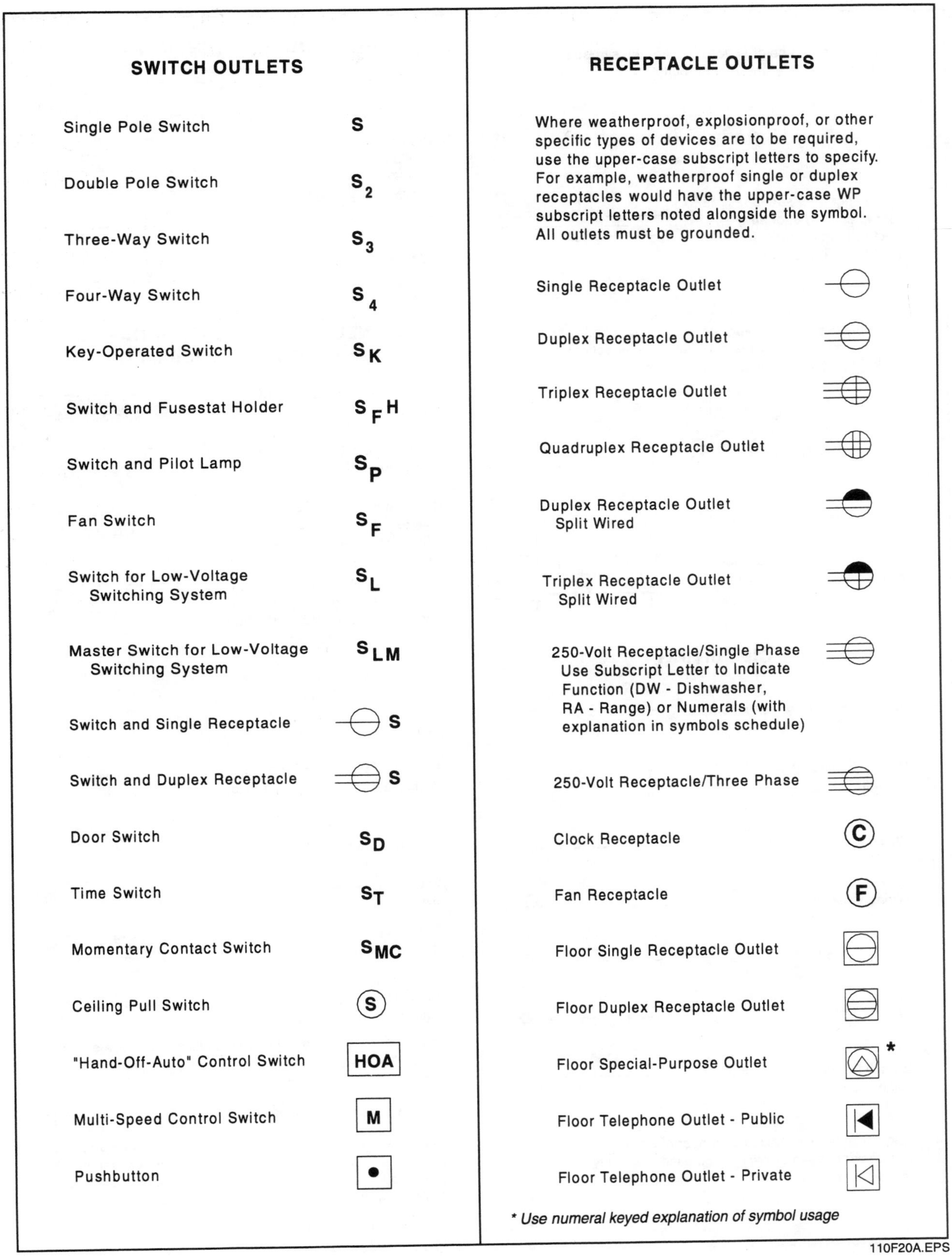

SWITCH OUTLETS

Single Pole Switch	S
Double Pole Switch	S_2
Three-Way Switch	S_3
Four-Way Switch	S_4
Key-Operated Switch	S_K
Switch and Fusestat Holder	$S_F H$
Switch and Pilot Lamp	S_P
Fan Switch	S_F
Switch for Low-Voltage Switching System	S_L
Master Switch for Low-Voltage Switching System	S_{LM}
Switch and Single Receptacle	⊖ S
Switch and Duplex Receptacle	⊜ S
Door Switch	S_D
Time Switch	S_T
Momentary Contact Switch	S_{MC}
Ceiling Pull Switch	ⓈS
"Hand-Off-Auto" Control Switch	HOA
Multi-Speed Control Switch	M
Pushbutton	●

RECEPTACLE OUTLETS

Where weatherproof, explosionproof, or other specific types of devices are to be required, use the upper-case subscript letters to specify. For example, weatherproof single or duplex receptacles would have the upper-case WP subscript letters noted alongside the symbol. All outlets must be grounded.

Single Receptacle Outlet	
Duplex Receptacle Outlet	
Triplex Receptacle Outlet	
Quadruplex Receptacle Outlet	
Duplex Receptacle Outlet Split Wired	
Triplex Receptacle Outlet Split Wired	
250-Volt Receptacle/Single Phase Use Subscript Letter to Indicate Function (DW - Dishwasher, RA - Range) or Numerals (with explanation in symbols schedule)	
250-Volt Receptacle/Three Phase	
Clock Receptacle	Ⓒ
Fan Receptacle	Ⓕ
Floor Single Receptacle Outlet	
Floor Duplex Receptacle Outlet	
Floor Special-Purpose Outlet	*
Floor Telephone Outlet - Public	
Floor Telephone Outlet - Private	

Use numeral keyed explanation of symbol usage

110F20A.EPS

Figure 20. Recommended Electrical Symbols (1 Of 7)

Example of the use of several floor outlet symbols to identify a 2, 3, or more gang outlet:

Underfloor duct and junction box for triple, double, or single duct system as indicated by the number of parallel lines

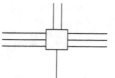

Example of the use of various symbols to identify the location of different types of outlets or connections for underfloor duct or cellular floor systems:

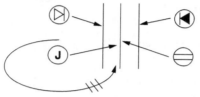

Cellular Floor Heater Duct

CIRCUITING

Wiring Exposed (not in conduit)	——— E ———
Wiring Concealed in Ceiling or Wall	———————
Wiring Concealed in Floor	— — — — — —
Wiring Existing*	··················
Wiring Turned Up	————○
Wiring Turned Down	————●
Branch Circuit Homerun to Panelboard	2 1 ⟶⟶

Number of arrows indicates number of circuits. (A number at each arrow may be used to identify the circuit number.)**

BUS DUCTS AND WIREWAYS

Trolley Duct***	T	T
Busway (Service, Feeder or Plug-in)***	B	B
Cable Trough Ladder or Channel***	C	C
Wireway***	W	W

PANELBOARDS, SWITCHBOARDS AND RELATED EQUIPMENT

Flush Mounted Panelboard and Cabinet***

Surface Mounted Panelboard and Cabinet***

Switchboard, Power Control Center, Unit Substation (Should be drawn to scale)***

Flush Mounted Terminal Cabinet (In small scale drawings the TC may be indicated alongside the symbol)***

Surface Mounted Terminal Cabinet (In small scale drawings the TC may be indicated alongside the symbol)***

Pull Box (Identify in relation to Wiring System Section and Size)

Motor or Other Power Controller May be a starter or contactor***

Externally Operated Disconnection Switch***

Combination Controller and Disconnection Means***

*Note: Use heavy-weight line to identify service and feeders. Indicate empty conduit by notation CO.

**Note: Any circuit without further identification indicates two-wire circuit. For a greater number of wires, indicate with cross lines, e.g.:

3 wires

4 wires, etc.

Neutral wire may be shown longer. Unless indicated otherwise, the wire size of the circuit is the minimum size required by the specification. Identify different functions of wiring system (e.g., signaling system) by notation or other means.

***Identify by Notation or Schedule

110F20B.EPS

Figure 20. Recommended Electrical Symbols (2 Of 7)

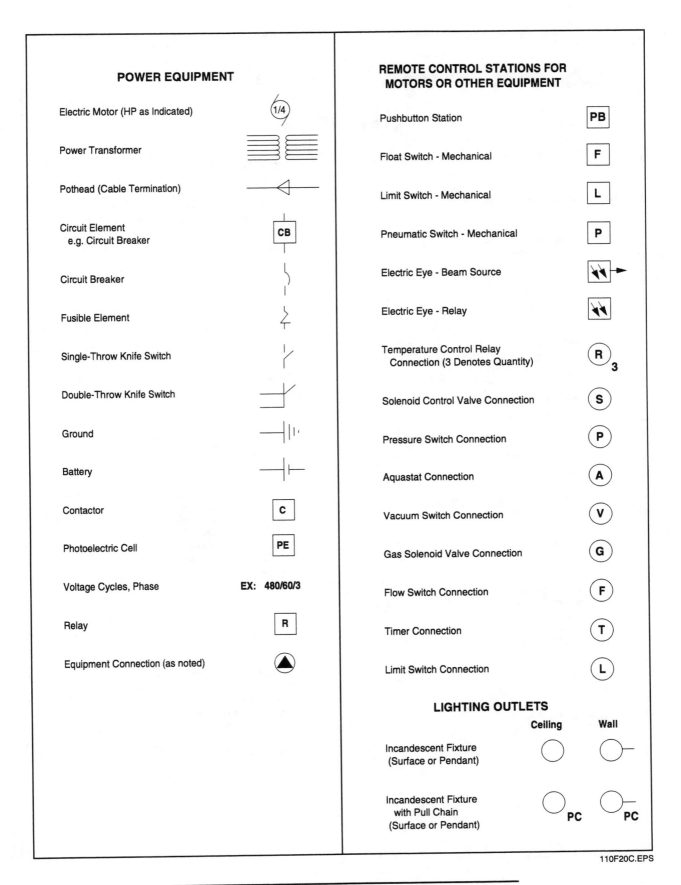

Figure 20. Recommended Electrical Symbols (3 Of 7)

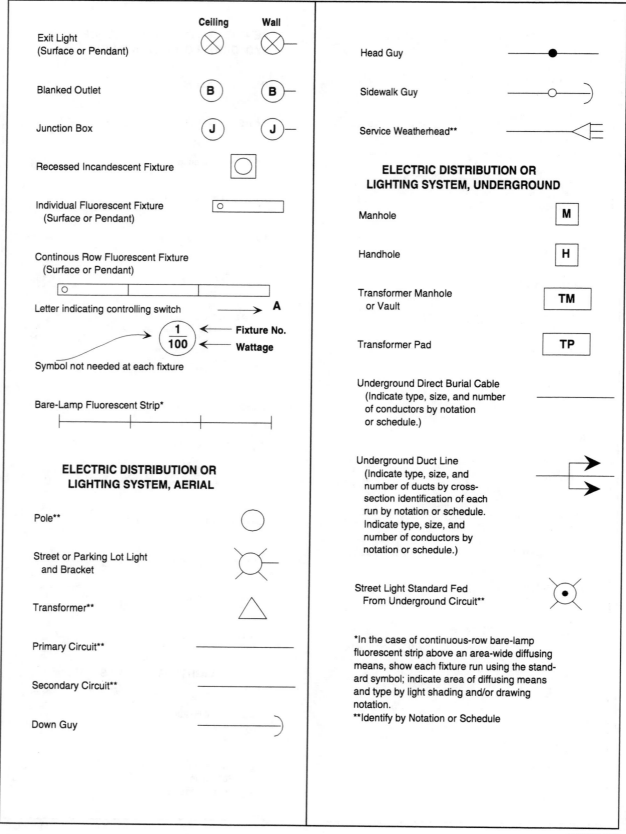

Figure 20. Recommended Electrical Symbols (4 Of 7)

110F20D.EPS

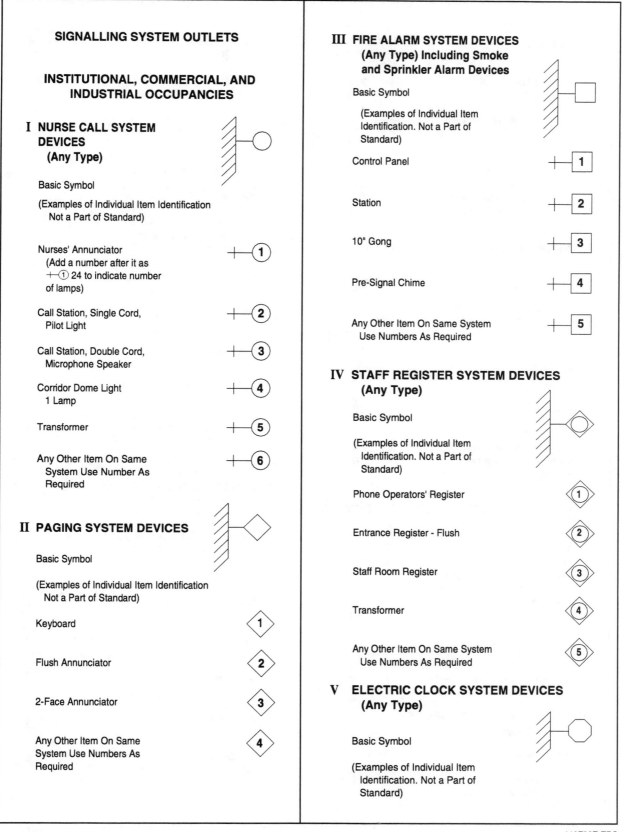

SIGNALLING SYSTEM OUTLETS

INSTITUTIONAL, COMMERCIAL, AND
INDUSTRIAL OCCUPANCIES

I NURSE CALL SYSTEM DEVICES (Any Type)

Basic Symbol

(Examples of Individual Item Identification Not a Part of Standard)

Nurses' Annunciator
(Add a number after it as
—① 24 to indicate number of lamps)

Call Station, Single Cord, Pilot Light

Call Station, Double Cord, Microphone Speaker

Corridor Dome Light 1 Lamp

Transformer

Any Other Item On Same System Use Number As Required

II PAGING SYSTEM DEVICES

Basic Symbol

(Examples of Individual Item Identification Not a Part of Standard)

Keyboard

Flush Annunciator

2-Face Annunciator

Any Other Item On Same System Use Numbers As Required

III FIRE ALARM SYSTEM DEVICES (Any Type) Including Smoke and Sprinkler Alarm Devices

Basic Symbol

(Examples of Individual Item Identification. Not a Part of Standard)

Control Panel

Station

10" Gong

Pre-Signal Chime

Any Other Item On Same System Use Numbers As Required

IV STAFF REGISTER SYSTEM DEVICES (Any Type)

Basic Symbol

(Examples of Individual Item Identification. Not a Part of Standard)

Phone Operators' Register

Entrance Register - Flush

Staff Room Register

Transformer

Any Other Item On Same System Use Numbers As Required

V ELECTRIC CLOCK SYSTEM DEVICES (Any Type)

Basic Symbol

(Examples of Individual Item Identification. Not a Part of Standard)

110F20E.EPS

Figure 20. Recommended Electrical Symbols (5 Of 7)

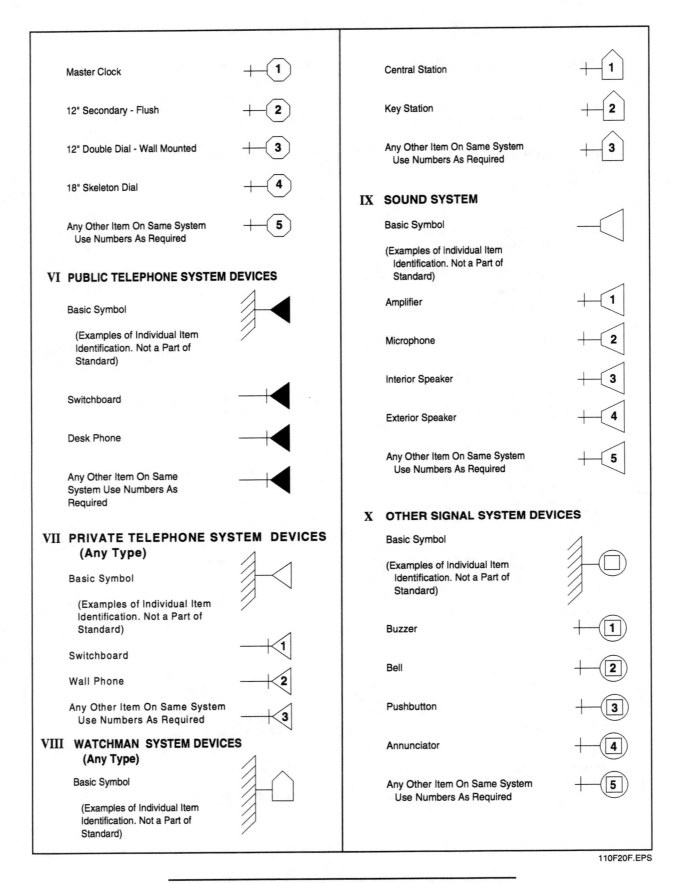

Master Clock 1

12" Secondary - Flush 2

12" Double Dial - Wall Mounted 3

18" Skeleton Dial 4

Any Other Item On Same System Use Numbers As Required 5

VI PUBLIC TELEPHONE SYSTEM DEVICES

Basic Symbol

(Examples of Individual Item Identification. Not a Part of Standard)

Switchboard

Desk Phone

Any Other Item On Same System Use Numbers As Required

VII PRIVATE TELEPHONE SYSTEM DEVICES (Any Type)

Basic Symbol

(Examples of Individual Item Identification. Not a Part of Standard)

Switchboard 1

Wall Phone 2

Any Other Item On Same System Use Numbers As Required 3

VIII WATCHMAN SYSTEM DEVICES (Any Type)

Basic Symbol

(Examples of Individual Item Identification. Not a Part of Standard)

Central Station 1

Key Station 2

Any Other Item On Same System Use Numbers As Required 3

IX SOUND SYSTEM

Basic Symbol

(Examples of Individual Item Identification. Not a Part of Standard)

Amplifier 1

Microphone 2

Interior Speaker 3

Exterior Speaker 4

Any Other Item On Same System Use Numbers As Required 5

X OTHER SIGNAL SYSTEM DEVICES

Basic Symbol

(Examples of Individual Item Identification. Not a Part of Standard)

Buzzer 1

Bell 2

Pushbutton 3

Annunciator 4

Any Other Item On Same System Use Numbers As Required 5

110F20F.EPS

Figure 20. Recommended Electrical Symbols (6 Of 7)

RESIDENTIAL OCCUPANCIES

Signalling system symbols for use in identifying standardized residential-type signal system items on residential drawings where a descriptive symbol list is not included on the drawing. When other signal system items are to be identified, use the above basic symbols for such items together with a descriptive symbol list.

Pushbutton

Buzzer

Bell

Combination Bell - Buzzer

Chime

Annunciator

Electric Door Opener

Maid's Signal Plug

Interconnection Box

Bell-Ringing Transformer

Outside Telephone

Interconnecting Telephone

Television Outlet

110F20G.EPS

Figure 20. Recommended Electrical Symbols (7 Of 7)

5.0.0   SCALE DRAWINGS

In most electrical drawings, the components are so large that it would be impossible to draw them actual size. Consequently, drawings are made to some reduced *scale*; that is, all the distances are drawn smaller than the actual dimensions of the object itself, with all dimensions being reduced in the same proportion. For example, if a floor plan of a building is to be drawn to a scale of ¼" = 1'–0", each ¼" on the drawing would equal 1 foot on the building itself; if the scale is ⅛" = 1'–0", each ⅛" on the drawing equals 1 foot on the building, and so forth.

When architectural and engineering drawings are produced, the selected scale is very important. Where dimensions must be held to extreme accuracy, the scale drawings should be made as large as practical with dimension lines added. Where dimensions require only reasonable accuracy, the object may be drawn to a smaller scale (with dimension lines possibly omitted).

In dimensioning drawings, the dimensions written on the drawing are the actual dimensions of the building, not the distances that are measured on the drawing. To further illustrate this point, look at the floor plan in *Figure 21*; it is drawn to a scale of ½" = 1' – 0". One of the walls is drawn to an actual length of 3½" on the drawing paper, but since the scale is ½" = 1' – 0" and since 3½" contains 7 halves of an inch (7 x ½ = 3½"), the dimension shown on the drawing will therefore be 7' – 0" on the actual building.

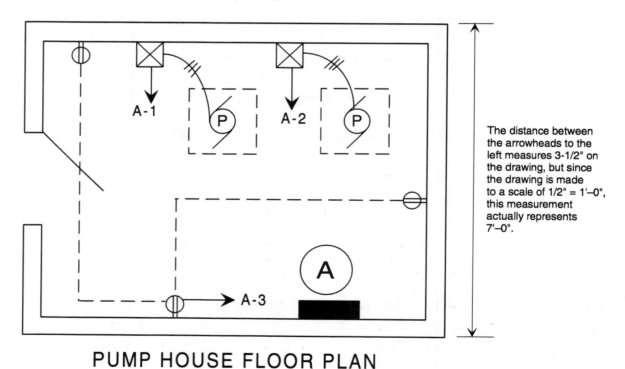

The distance between the arrowheads to the left measures 3-1/2" on the drawing, but since the drawing is made to a scale of 1/2" = 1'–0", this measurement actually represents 7'–0".

PUMP HOUSE FLOOR PLAN

1/2" = 1'–0"

110F21.EPS

Figure 21. Typical Floor Plan Showing Drawing Scale

As shown in the previous example, the most common method of reducing all the dimensions (in feet and inches) in the same proportion is to choose a certain distance and let that distance represent one foot. This distance can then be divided into twelve parts, each of which represents an inch. If half inches are required, these twelfths are further subdivided into halves, etc. Now the scale represents the common foot rule with its subdivisions into inches and fractions, except that the scaled foot is smaller than the distance known as a foot and, likewise, its subdivisions are proportionately smaller.

When a measurement is made on the drawing, it is made with the reduced foot rule or scale; when a measurement is made on the building, it is made with the standard foot rule. The most common reduced foot rules or scales used in electrical drawings are the architect's scale and the engineer's scale. Drawings may sometimes be encountered that use a metric scale, but using this scale is similar to using the architect's or engineer's scales.

5.1.0 ARCHITECT'S SCALE

Figure 22 shows two configurations of architect's scales. The one on the top is designed so that 1" = 1' – 0", and the one on the bottom has graduations spaced to represent ⅛" = 1' – 0".

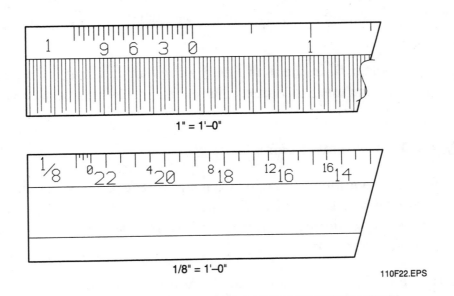

1" = 1'–0"

1/8" = 1'–0"

110F22.EPS

Figure 22. Two Different Configurations Of Architect's Scales

Note that on the one-inch scale in *Figure 23*, the longer marks to the right of the zero (with a numeral beneath) represent feet. Therefore, the distance between the zero and the numeral 1 equals one foot. The shorter mark between the zero and 1 represents ½ of a foot, or six inches.

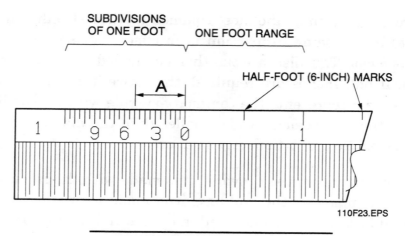

Figure 23. One-Inch Architect's Scale

Referring again to *Figure 23*, look at the marks to the left of the zero. The numbered marks are spaced three scaled inches apart and have the numerals 0, 3, 6, and 9 for use as reference points. The other lines of the same length also represent scaled inches, but are not marked with numerals. In use, you can count the number of long marks to the left of the zero to find the number of inches, but after some practice, you will be able to tell the exact measurement at a glance. For example, the measurement A represents five inches because it is the fifth inch mark to the left of the zero; it is also one inch mark short of the six-inch line on the scale.

The lines that are shorter than the inch line are the half-inch lines. On smaller scales, the basic unit is not divided into as many divisions. For example, the smallest subdivision on some scales represents two inches.

5.1.1 Types Of Architect's Scales

Architect's scales are available in several types, but the most common include the triangular scale (*Figure 24*) and the flat scale. The quality of architect's scales also vary from cheap plastic scales (costing a dollar or two) to high-quality wooden-laminated tools that are calibrated to precise standards.

The triangular scale (*Figure 24*) is frequently found in drafting and estimating departments or engineering and electrical contracting firms, while the flat scales are more convenient to carry on the job site.

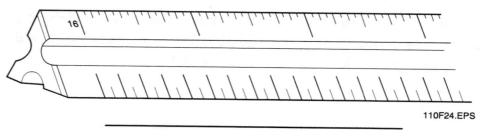

Figure 24. Typical Triangular Architect's Scale

Triangular architect's scales have 12 different scales—two on each edge—as follows:

- Common foot rule (12 inches)
- $\frac{1}{16}$" = 1'–0"
- $\frac{3}{32}$" = 1'–0"
- $\frac{3}{16}$" = 1'–0"
- $\frac{1}{8}$" = 1'–0"
- $\frac{1}{4}$" = 1'–0"
- $\frac{3}{8}$" = 1'–0"
- $\frac{3}{4}$" = 1'–0"
- 1" = 1'–0"
- $\frac{1}{2}$" = 1'–0"
- $1\frac{1}{2}$" = 1'–0"
- 3" = 1'–0"

Two separate scales on one face may seem confusing at first, but after some experience, reading these scales becomes second nature.

In all but one of the scales on the triangular architect's scale, each face has one of the scales placed opposite of the other. For example, on the one-inch face, the one-inch scale is read from left to right, starting from the zero mark. The half-inch scale is read from right to left, again starting from the zero mark.

On the remaining foot-rule scale ($\frac{1}{16}$" = 1'–0"), each $\frac{1}{16}$" mark on the scale represents one foot.

Figure 25 shows all the scales found on the triangular architect's scale.

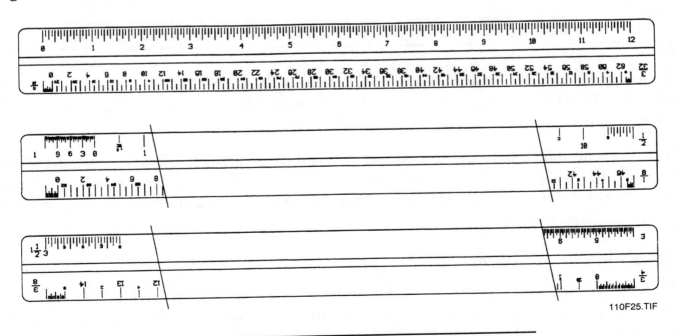

Figure 25. Various Scales On A Triangular Architect's Scale

110F25.TIF

The flat architect's scale shown in *Figure 26* is ideal for workers on most projects. It is easily and conveniently carried in the shirt pocket, and the four scales (⅛", ¼", ½", and 1") are adequate for the majority of projects that will be encountered.

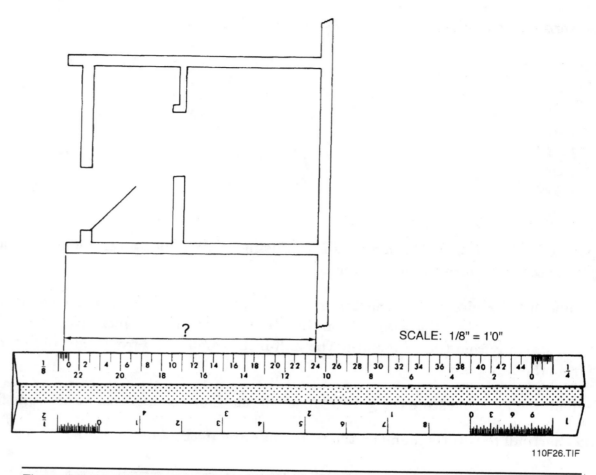

SCALE: 1/8" = 1'0"

110F26.TIF

Figure 26. Using The ⅛" Architect's Scale To Determine The Dimensions On A Drawing

The partial floor plan shown in *Figure 26* is drawn to a scale of ⅛" = 1'–0". The dimension in question is found by placing the ⅛" architect's scale on the drawing and reading the figures. It can be seen that the dimension reads 24'–6".

Every drawing should have the scale to which it is drawn plainly marked on it as part of the drawing title. However, it is not uncommon to have several different drawings on one blueprint sheet—all with different scales. Therefore, always check the scale of each different view found on a drawing sheet.

5.2.0   ENGINEER'S SCALE

The civil engineer's scale is used in basically the same manner as the architect's scale, with the principal difference being that the graduations on the engineer's scale are decimal units rather than feet, as on the architect's scale.

The engineer's scale is used by placing it on the drawing with the working edge away from the user. The scale is then aligned in the direction of the required measurement. Then, by looking down at the scale, the dimension is read.

Civil engineer's scales commonly show the following graduations:

- 1" = 10 units
- 1" = 20 units
- 1" = 30 units
- 1" = 40 units
- 1" = 60 units
- 1" = 80 units
- 1" = 100 units

The purpose of this scale is to transfer the relative dimensions of an object to the drawing or vice versa. It is used mainly on site plans to determine distances between property lines, manholes, duct runs, direct-burial cable runs, and the like.

Site plans are drawn to scale using the engineer's scale rather than the architect's scale. On small lots, a scale of 1 inch = 10 feet or 1 inch = 20 feet is used. For a 1:10 scale, this means that one inch (the actual measurement on the drawing) is equal to 10 feet on the land itself.

On larger drawings, where a large area must be covered, the scale could be 1 inch = 100 feet or 1 inch = 1,000 feet, or any other integral power of 10. On drawings with the scale in multiples of 10, the engineering scale marked 10 is used. If the scale is 1 inch = 200 feet, the engineer's scale marked 20 is used, and so on.

Although site plans appear reduced in scale, depending on the size of the object and the size of the drawing sheet to be used, the actual dimensions must be shown on the drawings at all times. When you are reading the drawing plans to scale, think of each dimension in its full size and not in the reduced scale it happens to be on the drawing (*Figure 27*).

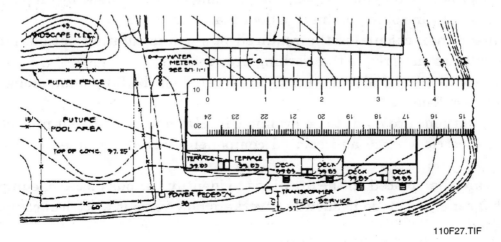

110F27.TIF

Figure 27. Practical Use Of The Engineer's Scale

5.3.0 METRIC SCALE

The metric scale (*Figure 28*) is divided into centimeters (cm), with the centimeters divided into 10 divisions for millimeters (mm) or in 20 divisions for half millimeters. Scales are available with metric divisions on one edge while inch divisions are inscribed on the opposite edge. Many contracting firms that deal in international trade have adopted a dual-dimensioning system expressed in both metric and English symbols. Drawings prepared for government projects may also require metric dimensions.

110F28.TIF

Figure 28. Typical Metric Scale

6.0.0 ANALYZING ELECTRICAL DRAWINGS

The most practical way to learn how to read electrical construction documents is to analyze an existing set of drawings prepared by consulting or industrial engineers.

Engineers or electrical designers are responsible for the complete layout of electrical systems for most projects. Electrical drafters then transform the engineer's designs into working drawings, either using manual drafting instruments or computer-aided design (CAD) systems. The following is a brief outline of what usually takes place in the preparation of electrical design and working drawings:

- The engineer meets with the architect and owner to discuss the electrical needs of the building or project and to discuss various recommendations made by all parties.
- After that, an outline of the architect's floor plan is laid out.
- The engineer then calculates the required power and lighting outlets for the project; these are later transferred to the working drawings.
- All communications and alarm systems are located on the floor plan, along with lighting and power panelboards.
- Circuit calculations are made to determine wire size and overcurrent protection.
- The main electric service and related components are determined and shown on the drawings.
- Schedules are then placed on the drawings to identify various pieces of equipment.
- Wiring diagrams are made to show the workers how various electrical components are to be connected.
- A legend or electrical symbol list is drafted and shown on the drawings to identify all symbols used to indicate electrical outlets or equipment.

- Various large-scale electrical details are included, if necessary, to show exactly what is required of the electricians.
- Written specifications are then made to give a description of the materials and installation methods.

6.1.0 DEVELOPMENT OF SITE PLANS

In general practice, it is usually the owner's responsibility to furnish the architect/engineer with property and topographic surveys, which are made by a certified land surveyor or civil engineer. These surveys show:

- All property lines
- Existing public utilities and their location on or near the property (e.g., electrical lines, sanitary sewer lines, gas lines, water-supply lines, storm sewers, manholes, telephone lines, etc.)

A land surveyor does the property survey from information obtained from a deed description of the property. A property survey shows only the property lines and their lengths, as if the property were perfectly flat.

The topographic survey shows both the property lines and the physical characteristics of the land by using contour lines, notes, and symbols. The physical characteristics may include:

- The direction of the land slope
- Whether the land is flat, hilly, wooded, swampy, high, or low, and other features of its physical nature

All of this information is necessary so that the architect can properly design a building to fit the property. The electrical engineer also needs this information to locate existing electrical utilities and to route the new service to the building, provide outdoor lighting and circuits, etc.

Electrical site work is sometimes shown on the architect's plot plan. However, when site work involves many trades and several utilities (e.g., gas, telephone, electric, television, water, and sewage), it can become confusing if all details are shown on one drawing sheet. In cases like these, it is best to have a separate drawing devoted entirely to the electrical work, as shown in *Figure 29*. This project is an office/warehouse building for Virginia Electric, Inc. The electrical drawings consist of four 24" x 36" drawing sheets, along with a set of written specifications which will be discussed later in this module.

The electrical site or plot plan shown in *Figure 29* has the conventional architect's and engineer's title blocks in the lower right-hand corner of the drawing. These blocks identify the project and project owners, the architect, and the engineer. They also show how this drawing sheet relates to the entire set of drawings. Note the engineer's professional stamp of approval to the left of the engineer's title block. Similar blocks appear on all four of the electrical drawing sheets.

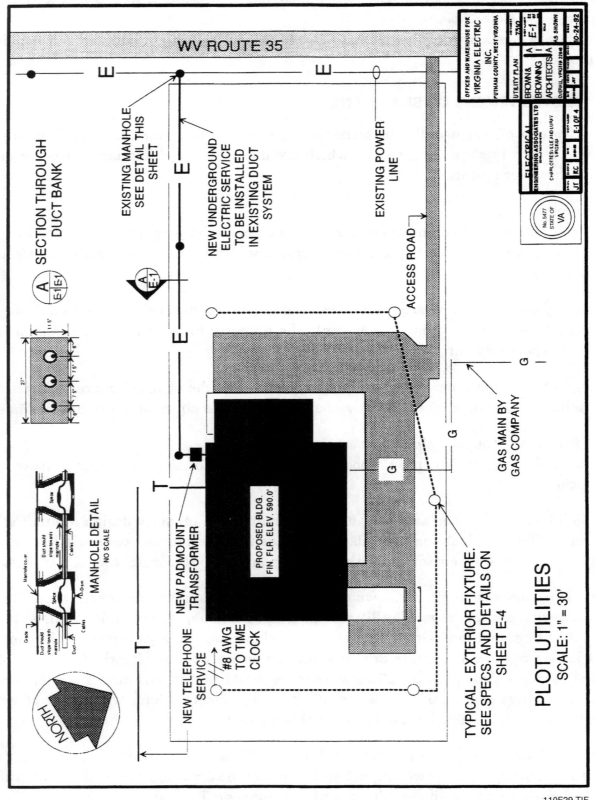

Figure 29. Typical Electrical Site Plan

110F29.TIF

When examining a set of electrical drawings for the first time, always look at the area around the title block. This is where most revision blocks or revision notes are placed. If revisions have been made to the drawings, make certain that you have a clear understanding of what has taken place before proceeding with the work.

Refer again to the drawing in *Figure 29* and note the North Arrow in the upper left corner. A North Arrow shows the direction of true north to help you orient the drawing to the site. Look directly down from the North Arrow to the bottom of the page and notice the drawing title, *Plot Utilities*. Directly beneath the drawing title you can see that the drawing scale of 1" = 30' is shown. This means that each inch on the drawing represents 30 feet on the actual job site. This scale holds true for all drawings on the page unless otherwise noted.

An outline of the proposed building is indicated on the drawing by cross-hatched rectangles along with a callout stating, *Proposed Bldg. Fin. Flr. Elev. 590.0'*. This means that the finished floor level of the building is to be 590 feet above sea level, which in this part of the country will be about two feet above finished grade around the building. This information helps the electrician locate conduit sleeves and stub-ups to the correct height before the finished concrete floor is poured.

The shaded area represents asphalt paving for the access road, drives, and parking lot. Note that the access road leads into a highway which is designated Route 35. This information further helps workers to orient the drawing to the building site.

Existing manholes are indicated by a solid circle, while an open circle is used to show the position of the five new pole-mounted lighting fixtures which are to be installed around the new building. Existing power lines are shown with a light solid line with the letter E placed at intervals along the line. The new underground electric service is shown in the same way except the lines are somewhat wider and darker on the drawing. Note that this new high-voltage cable terminates into a padmount transformer near the proposed building. New telephone lines are similar except the letter *T* is used to identify the telephone lines.

The direct-burial underground cable supplying the exterior lighting fixtures is indicated with dashed lines on the drawing—shown connecting the open circles. A homerun for this circuit is also shown to a time clock.

The manhole detail shown to the right of the North Arrow may seem to serve very little purpose on this drawing since the manholes have already been installed. However, the dimensions and details of their construction will help the electrical contractor or supervisor to better plan the pulling of the high-voltage cable. The same is true of the cross section shown of the duct bank. The electrical contractor knows that three empty ducts are available if it is discovered that one of them is damaged when the work begins.

Although the electrical work will not involve working with gas, the main gas line is shown on the electrical drawing to let the electrical workers know its approximate location while they are installing the direct-burial conductors for the exterior lighting fixtures.

The electrical power plan (*Figure 30*) shows the complete floor plan of the office/warehouse building with all interior partitions drawn to scale. Sometimes, the physical locations of all wiring and outlets are shown on one drawing; that is, outlets for lighting, power, signal and communications, special electrical systems, and related equipment are shown on the same plan. However, on complex installations, the drawing would become cluttered if both lighting and power were shown on the same floor plan. Therefore, most projects will have a separate drawing for power and another for lighting. Riser diagrams and details may be shown on yet another drawing sheet, or if room permits, they may be shown on the lighting or power floor plan sheets.

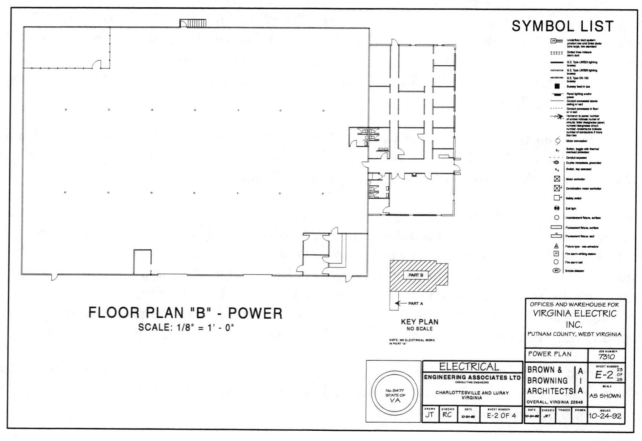

Figure 30. Electrical Power Plan

A closer look at this drawing reveals the title blocks in the lower right corner of the drawing sheet. These blocks list both the architectural and engineering firms, along with information to identify the project and drawing sheet. Also note that the floor plan is titled, *Floor Plan "B" – Power* and is drawn to a scale of ⅛" = 1'–0". There are no revisions shown on this drawing sheet.

7.1.0 KEY PLAN

A key plan appears on the drawing sheet immediately above the engineer's title block (*Figure 31*). The purpose of this key plan is to identify that part of the project to which this sheet applies. In this case, the project involves two buildings: Building A and Building B. Since the outline of Building B is cross-hatched in the key plan, this is the building to which this drawing applies. Note that this key plan is not drawn to scale—only its approximate shape.

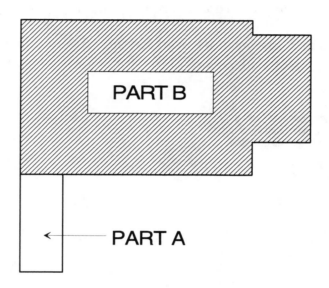

KEY PLAN
NO SCALE

NOTE: NO ELECTRICAL WORK
IN PART "A"

110F31.EPS

Figure 31. Key Plan Appearing On Electrical Power Plan

Although Building A is also shown on this key plan, a note below the key plan title states that there is no electrical work required in Building A.

On some larger installations, the overall project may involve several buildings requiring appropriate key plans on each drawing to help the workers orient the drawings to the appropriate building. In some cases, separate drawing sheets may be used for each room or area in an industrial project—again requiring key plans on each drawing sheet to identify applicable drawings for each room.

7.2.0 SYMBOL LIST

A symbol list appears on the electrical power plan (immediately above the architect's title block) to identify the various symbols used for both power and lighting on this project. In most cases, the only symbols listed are those that apply to the particular project. In other cases, however, a standard list of symbols is used for all projects with the following note:

These are standard symbols and may not all appear on the project drawings; however, wherever the symbol on the project drawings occurs, the item shall be provided and installed.

Only electrical symbols that are actually used for the office/warehouse drawings are shown in the list on the example electrical power plan. A close-up look at these symbols appears in *Figure 32*.

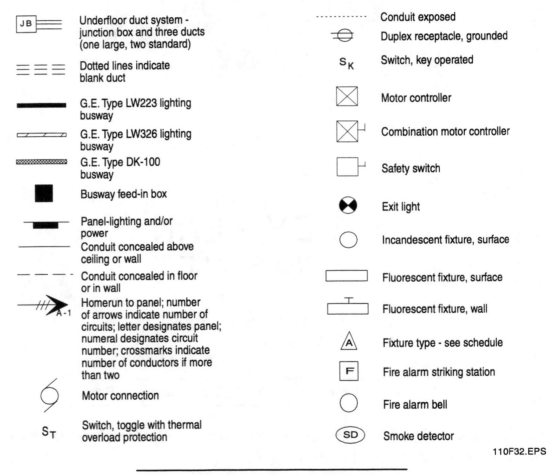

Figure 32. Sample Electrical Symbols List

7.3.0 FLOOR PLAN

A somewhat enlarged view of the electrical floor plan drawing is shown in *Figure 33*. However, due to the size of the drawing in comparison with the size of the pages in this module, it is still difficult to see very much detail. This illustration is meant to show the overall layout of the floor plan and how the symbols and notes are arranged.

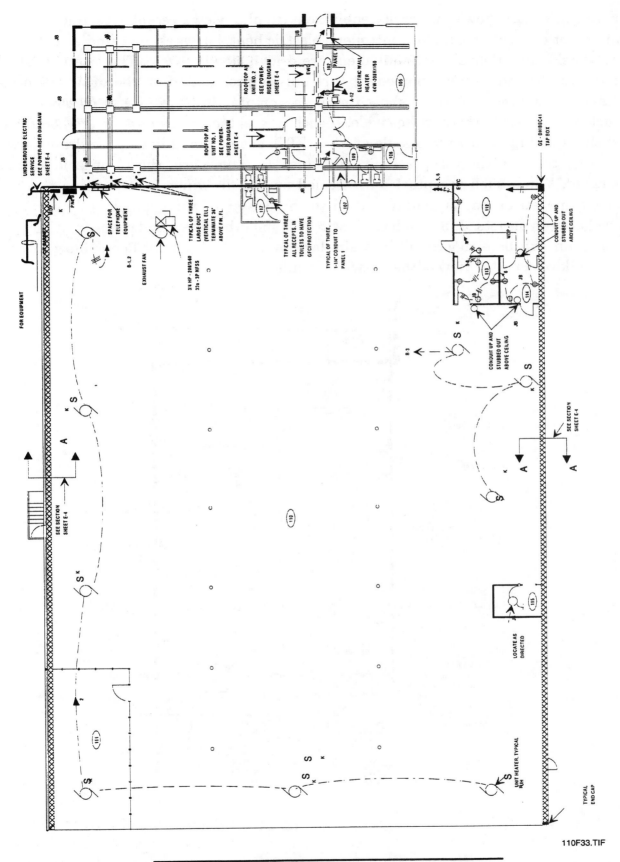

Figure 33. Power Plan For Office/Warehouse Building

110F33.TIF

In general, this plan shows the service equipment (in plan view), receptacles, underfloor duct system, motor connections, motor controllers, electric heat, busways, and similar details. The electric panels and other service equipment are drawn close to scale. The locations of other electrical outlets and similar components are only approximated on the drawings because they have to be exaggerated to show up on the prints. To illustrate, a common duplex receptacle is only about three inches wide. If such a receptacle were to be located on the floor plan of this building (drawn to a scale of ⅛" = 1'–0"), even a small dot on the drawing would be too large to draw the receptacle exactly to scale. Therefore, the receptacle symbol is exaggerated. When such receptacles are scaled on the drawings to determine the proper location, a measurement is usually taken to the center of the symbol to determine the distance between outlets. Junction boxes, switches, and other electrical connections shown on the floor plan will be exaggerated in a similar manner. The partial floor plan drawing in *Figure 34* allows a better view of the drawing details.

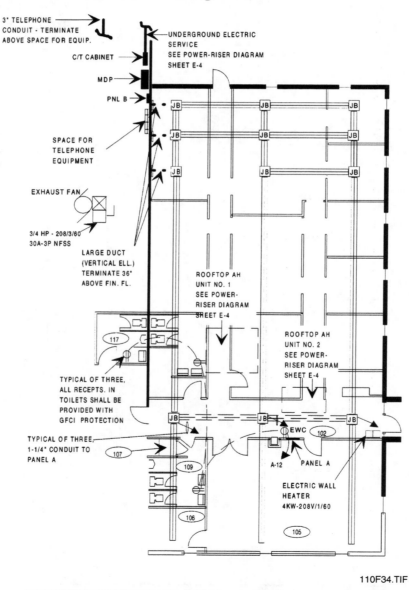

110F34.TIF

Figure 34. Partial Floor Plan For Office/Warehouse Building

ELECTRICAL — TRAINEE TASK MODULE 26110

7.3.1 Notes And Building Symbols

Referring again to *Figure 33*, you will notice numbers placed inside of an oval symbol in each room. These numbered ovals represent the room name or type and correspond to a room schedule in the architectural drawings. For example, room number 112 is designated as the lobby in the room schedule (not shown), room number 113 is designated as office No. 1, etc. On some drawings, these room symbols are omitted and the room names are written out on the drawings.

There are also several notes appearing at various places on the floor plan. These notes offer additional information to clarify certain aspects of the drawing. For example, only one electric heater is to be installed by the electrical contractor; this heater is located in the building's vestibule. Rather than have a symbol in the symbol list for this one heater, a note is used to identify it on the drawing. Other notes on this drawing describe how certain parts of the system are to be installed. For example, in the office area (rooms 112, 113, and 114), you will see the following note: *Conduit up and stubbed out above ceiling*. This empty conduit is for telephone/communications cables that will be installed later by the telephone company.

7.3.2 Busways

The office/warehouse project utilizes three types of busways: two types of lighting busways and one power busway. Only the power busway is shown on the power plan; the lighting busways will appear on the lighting plan.

Figure 33 shows two runs of busways: one running the length of the building on the south end (top wall on drawing), and one running the length of the north wall. The symbol list in *Figure 32* shows this busway to be designated by two parallel lines with a series of X's inside. The symbol list further describes the busway as General Electric Type DK-100. These busways are fed from the main distribution panel (circuits MDP-1 and MDP-2) through GE No. DHIBBC41 tap boxes.

The NEC defines a *busway* as a grounded metal enclosure containing factory-mounted, bare or insulated conductors, which are usually copper or aluminum bars, rods, or tubes.

The relationship of the busway and hangers to the building construction should be checked prior to commencing the installation so that any problems due to space conflicts, inadequate or inappropriate supporting structure, openings through walls, etc. are worked out in advance so as not to incur lost time.

For example, the drawings and specifications may call for the busway to be suspended from brackets clamped or welded to steel columns. However, the spacing of the columns may be such that additional supplementary hanger rods suspended from the ceiling or roof structure may be necessary for the adequate support of the busway. To offer more assistance to workers on the office/warehouse project, the engineer may also provide an additional drawing that shows how the busway is to be mounted.

Other details that appear on the floor plan in *Figure 34* include the general arrangement of the underfloor duct system, junction boxes and feeder conduit for the underfloor duct system, and plan views of the service and telephone equipment, along with duplex receptacle outlets. A note on the drawing requires all receptacles in the toilets to be provided with ground fault circuit interrupter (GFCI) protection. The letters EWC next to the receptacle in the vestibule designates this receptacle for use with an electric water cooler.

8.0.0 LIGHTING FLOOR PLAN

A skeleton view of a lighting floor plan is shown in *Figure 35*. Again, the architect's/engineer's title blocks appear in the lower right corner of the drawing. A key plan, as discussed previously, appears above the engineer's title block. This plan is drawn to the same scale as the power plan; that is, ⅛" = 1' – 0". A lighting fixture schedule appears in the upper right corner of the drawing and some installation notes appear below the schedule.

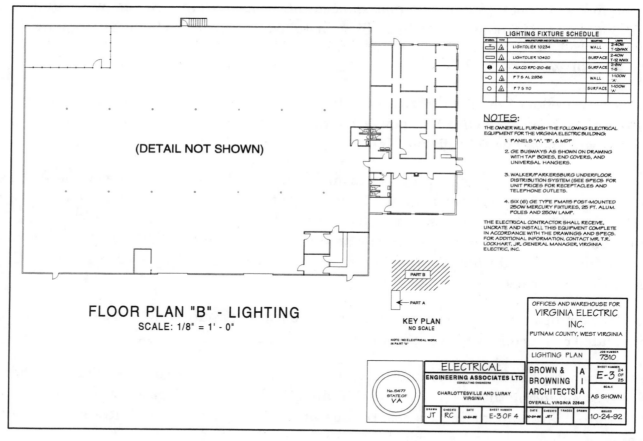

110F35.EPS

Figure 35. Sample Lighting Plan

The lighting outlet symbols found on the drawing for the office/warehouse building represent both incandescent and fluorescent types; a circle on most electrical drawings usually represents an incandescent fixture, and a rectangle represents a fluorescent one. All of these symbols are designed to indicate the physical shape of a particular fixture and are usually drawn to scale.

The type of mounting used for all lighting fixtures is usually indicated in a lighting fixture schedule, which in this case is shown on the drawings. On some projects, the schedule may be found only in the written specifications.

The type of lighting fixture is identified by a numeral placed inside a triangle near each lighting fixture. If one type of fixture is used exclusively in one room or area, the triangular indicator need only appear once with the word *ALL* lettered at the bottom of the triangle.

8.1.0   DRAWING SCHEDULES

A schedule is a systematic method of presenting notes or lists of equipment on a drawing in tabular form. When properly organized and thoroughly understood, schedules are powerful timesaving devices for both those preparing the drawings and workers on the job.

For example, the lighting fixture schedule shown in *Figure 36* lists the fixture and identifies each fixture type on the drawing by number. The manufacturer and catalog number of each type are given along with the number, size, and type of lamp for each.

LIGHTING FIXTURE SCHEDULE

SYMBOL	TYPE	MANUFACTURER AND CATALOG NUMBER	MOUNTING	LAMPS
	A	LIGHTOLIER 10234	WALL	2-40W T-12WWX
	B	LIGHTOLIER 10420	SURFACE	2-40W T-12 WWX
	C	ALKCO RPC-210-6E	SURFACE	2-8W T-5
	D	P 7 S AL 2936	WALL	1-100W 'A'
	E	P 7 S 110	SURFACE	1-100W 'A'

110F36.EPS

Figure 36. Lighting Fixture Schedule

At times, all of the same information found in schedules will be duplicated in the written specifications, but combing through page after page of written specifications can be time consuming, and workers do not always have access to the specifications while working, whereas they usually do have access to the working drawings. Therefore, the schedule is an excellent means of providing essential information in a clear and accurate manner, allowing the workers to carry out their assignments in the least amount of time.

Other schedules that are frequently found on electrical working drawings include:

- Connected load schedule
- Panelboard schedule
- Electric heat schedule
- Kitchen equipment schedule
- Schedule of receptacle types

There are also other schedules found on electrical drawings, depending upon the type of project. However, most will deal with lists of equipment such as motors, motor controllers, and similar items.

9.0.0 ELECTRICAL DETAILS AND DIAGRAMS

Electrical diagrams are drawings that are intended to show electrical components and their related connections. They show the electrical association of the different components, but are seldom, if ever, drawn to scale.

9.1.0 POWER-RISER DIAGRAMS

One-line (single-line) **block diagrams** are used extensively to show the arrangement of electric service equipment. The **power-riser diagram** in *Figure 37*, for example, was used on the office/warehouse building under discussion and is typical of such drawings. The drawing shows all pieces of electrical equipment as well as the connecting lines used to indicate service-entrance conductors and feeders. Notes are used to identify the equipment, indicate the size of conduit necessary for each feeder, and show the number, size, and type of conductors in each conduit.

A panelboard schedule (*Figure 38*) is included with the power-riser diagram to indicate the exact components contained in each panelboard. This panelboard schedule is for the main distribution panel. On the actual drawings, schedules would also be shown for the other two panels (PNL A and PNL B).

In general, panelboard schedules usually indicate the panel number, type of cabinet (either flush- or surface-mounted), panel mains (ampere and voltage rating), phase (single- or three-phase), and number of wires. A four-wire panel, for example, indicates that a solid neutral exists in the panel. Branches indicate the type of overcurrent protection; that is, they indicate the number of poles, trip rating, and frame size. The items fed by each overcurrent device are also indicated.

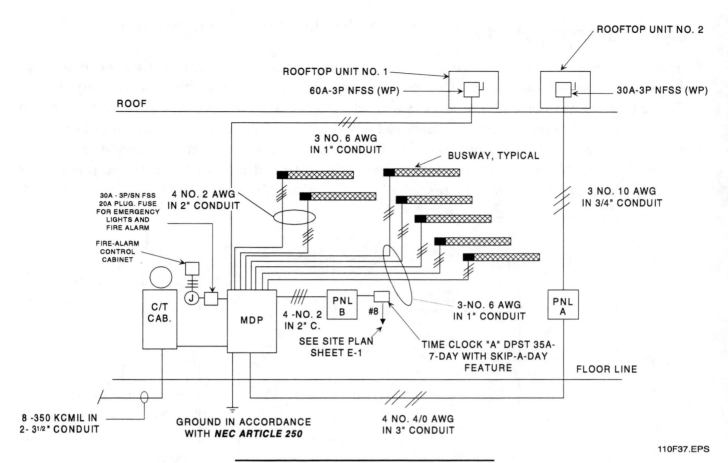

Figure 37. Typical Power-Riser Diagram

110F37.EPS

PANELBOARD SCHEDULE

PANEL No.	CABINET TYPE	PANEL MAINS			BRANCHES					ITEMS FED OR REMARKS
		AMPS	VOLTS	PHASE	1P	2P	3P	PROT.	FRAME	
MDP	SURFACE	600A	120/208	3φ,4-W	-	-	1	225A	25,000	PANEL "A"
					-	-	1	100A	18,000	PANEL "B"
					-	-	1	100A		POWER BUSWAY
					-	-	1	60A		LIGHTING BUSWAY
					-	-	1	70A		ROOFTOP UNIT #1
					-	-	1	70A		SPARE
					-	-	1	600A	42,000	MAIN CIRCUIT BRKR

110F38.EPS

Figure 38. Typical Panelboard Schedule

9.2.0 SCHEMATIC DIAGRAMS

Complete schematic wiring diagrams are normally used only in complicated electrical systems, such as control circuits. Components are represented by symbols, and every wire is either shown by itself or included in an assembly of several wires which appear as one line on the drawing. Each wire should be numbered when it enters an assembly and should keep the same number when it comes out again to be connected to some electrical component in the system. *Figure 39* shows a complete schematic wiring diagram for a three-phase, AC magnetic non-reversing motor starter.

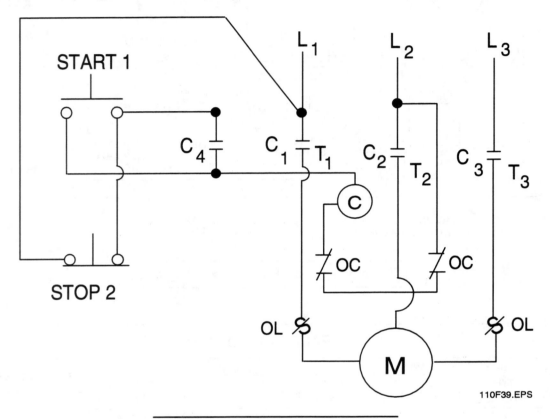

Figure 39. Schematic Wiring Diagram

Note that this diagram shows the various devices in symbol form and indicates the actual connections of all wires between the devices. The three-wire supply lines are indicated by L_1, L_2, and L_3; the motor terminals of motor M are indicated by T_1, T_2, and T_3. Lines L_1 and L_3 each have a thermal overload protection device (OL) connected in series with normally open line contactors C_1 and C_3, respectively, which are both controlled by the magnetic starter coil, C. Each contactor has a pair of contacts that close or open during operation. The control station, consisting of start pushbutton 1 and stop pushbutton 2, is connected across lines L_1 and L_2. An auxiliary contactor (C_4) is connected in series with the stop pushbutton and in parallel with the start pushbutton. The control circuit also has normally closed overload contactors (OC) connected in series with the magnetic starter coil (C).

Any number of additional pushbutton stations may be added to this control circuit similarly to the way in which three-way and four-way switches are added to control a lighting circuit. When adding pushbutton stations, the stop buttons are always connected in series and the start buttons are always connected in parallel. *Figure 40* shows the same motor starter circuit in *Figure 39*, but this time it is controlled by two sets of start/stop buttons.

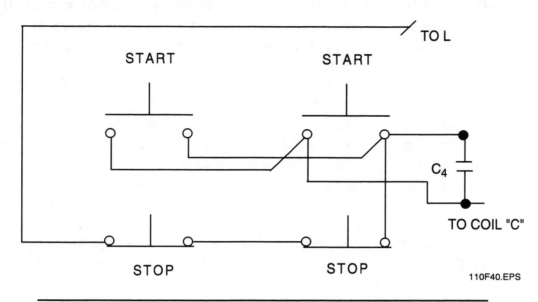

110F40.EPS

Figure 40. Circuit Being Controlled By Two Sets Of Start/Stop Buttons

Schematic wiring diagrams have only been touched upon in this module; there are many other details that you will need to know to perform your work in a proficient manner. Later modules cover wiring diagrams in more detail.

9.3.0   DRAWING DETAILS

A **detail drawing** is a drawing of a separate item or portion of an electrical system, giving a complete and exact description of its use and all the details needed to show the electrician exactly what is required for its installation. For example, the power plan for the office/warehouse has a sectional cut through the busduct. This is a good example of where an extra, detailed drawing is desirable.

A set of electrical drawings will sometimes require large-scale drawings of certain areas that are not indicated with sufficient clarity on the small-scale drawings. For example, the site plan may show exterior pole-mounted lighting fixtures that are to be installed by the contractor.

10.0.0 WRITTEN SPECIFICATIONS

The written specifications for a building or project are the written descriptions of work and duties required of the owner, architect, and consulting engineer. Together with the working drawings, these specifications form the basis of the contract requirements for the construction of the building or project. Those who use the construction drawings and specifications must always be alert to discrepancies between the working drawings and the written specifications. Such discrepancies may occur when:

- Architects or engineers use standard or prototype specifications and attempt to apply them without any modification to specific working drawings.
- Previously-prepared standard drawings are changed or amended by reference in the specifications only and the drawings themselves are not changed.
- Items are duplicated in both the drawings and specifications, but an item is subsequently amended in one and overlooked in the other contract document.

In such instances, the person in charge of the project has the responsibility to ascertain whether the drawings or the specifications take precedence. Such questions must be resolved, preferably before the work begins, to avoid added cost to the owner, architect/engineer, or contractor.

10.1.0 HOW SPECIFICATIONS ARE WRITTEN

Writing accurate and complete specifications for building construction is a serious responsibility for those who design the buildings because the specifications, combined with the working drawings, govern practically all important decisions made during the construction span of every project. Compiling and writing these specifications is not a simple task, even for those who have had considerable experience in preparing such documents. A set of written specifications for a single project will usually contain thousands of products, parts, and components, and the methods of installing them, all of which must be covered in either the drawings and/or specifications. No one can memorize all of the necessary items required to accurately describe the various areas of construction. One must rely upon reference materials such as manufacturer's data, catalogs, checklists, and, most of all, a high-quality master specification.

10.2.0 FORMAT OF SPECIFICATIONS

For convenience in writing, speed in estimating, and ease of reference, the most suitable organization of the specifications is a series of sections dealing successively with the different trades. All the work of each trade should be incorporated into the section devoted to that trade. Those people who use the specifications must be able to find all information needed without spending too much time looking for it.

10.2.1 CSI Format

The Construction Specification Institute (CSI) has developed the Uniform Construction Index that allows all specifications, product information, and cost data to be arranged into a uniform system. This format is followed on most large construction projects in North America. All construction is divided into 16 divisions, and each division has several sections and subsections. The following outline describes the various divisions normally included in a set of specifications for building construction.

Division 1: General Requirements – This division summarizes the work, alternatives, project meetings, submissions, quality control, temporary facilities and controls, products, and the project closeout. Every responsible person involved with the project should become familiar with this division.

Division 2: Site Work – This division outlines work involving such items as paving, sidewalks, outside utility lines (electrical, plumbing, gas, telephone, etc.), landscaping, grading, and other items pertaining to the outside of the building.

Division 3: Concrete – This division covers work involving footings, concrete formwork, expansion and contraction joints, cast-in-place concrete, specially-finished concrete, precast concrete, concrete slabs, and similar items.

Division 4: Masonry – This division covers concrete, mortar, stone, masonry accessories, and similar items.

Division 5: Metals – Metal roofs, structural metal framing, metal joists, metal decking, ornamental metal, and expansion control normally fall under this division.

Division 6: Carpentry – Items falling under this division include rough carpentry, heavy timber construction, trestles, prefabricated structural wood, finish carpentry, wood treatment, architectural woodwork, and the like. Plastic fabrications may also be included in this division.

Division 7: Thermal and Moisture Protection – Waterproofing is the main topic discussed under this division. Other related items such as dampproofing, building insulation, shingles and roofing tiles, preformed roofing and siding, membrane roofing, sheet metal work, wall flashing, roof accessories, and sealants are also included.

Division 8: Doors and Windows – All types of doors and frames are included under this division: metal, plastic, wood, etc. Windows and framing are also included, along with hardware and other window and door accessories.

Division 9: Finishes – Included in this division are the types, quality, and workmanship of lath and plaster, gypsum wallboard, tile, terrazzo, acoustical treatment, ceiling suspension systems, wood flooring, floor treatment, special coatings, painting, and wallcovering.

Division 10: Specialties – Specialty items such as chalkboards and tackboards, compartments and cubicles, louvers and vents that are not connected with the mechanical system, wall and corner guards, access flooring, specialty modules, pest control, fireplaces, flagpoles, identifying devices, lockers, protective covers, postal specialties, partitions, scales, storage shelving, wardrobe specialties, and similar items are covered in this division of the specifications.

Division 11: Equipment – The equipment included in this division could include central vacuum cleaning systems, bank vaults, darkrooms, food service, vending machines, laundry equipment, and many similar items.

Division 12: Furnishing – Items such as cabinets and storage files, fabrics, furniture, rugs and mats, seating, and similar furnishings are included under this division.

Division 13: Special Construction – Such items as air-supported structures, incinerators, and other special items will fall under this division.

Division 14: Conveying Systems – This division covers conveying apparatus such as dumbwaiters, elevators, hoists and cranes, lifts, material-handling systems, turntables, moving stairs and walks, pneumatic tube systems, and power scaffolding.

Division 15: Mechanical – This division includes plumbing, heating, ventilating, and air conditioning and related work. Electric heat is sometimes covered under Division 16, especially if individual baseboard heating units are used in each room or area of the building.

Division 16: Electrical – This division covers all electrical requirements for the building including lighting, power, alarm and communications systems, special electrical systems, and related electrical equipment. This is the division that electricians will use the most.

A sample set of electrical specifications is shown in *Figure 41*.

```
 1    SECTION 16011 - ELECTRICAL OUTLINE OF WORK
 2
 3    PART 1 - GENERAL
 4
 5
 6    The work included in Division 16 electrical includes, but is not necessarily limited to the following
 7    items and systems:
 8
 9    .      Two new 480/277 volt, 3 phase underground services from Duke Power Company pad
10           mounted transformer.
11    .      Connection to power company transformers.
12    .      Concrete pad for power company transformers.
13    .      Installation of new power company meter box.
14    .      All coordination, labor and materials required to connect new chiller service to existing
15           Duke Power Company pad mounted transformer.
16    .      Motor control center.
17    .      Lighting and receptacle panelboard.
18    .      Feeder circuits, including conduits, conductors, troughs and fittings.
19    .      Branch circuits, including conduits, conductors, outlets, boxes, receptacles, switches and
20           fittings.
21    .      Lighting fixtures including lamps.
22    .      Equipment tests.
23    .      Wiring devices
24    .      Grounding systems.
25    .      Safety switches.
26    .      Power connection to all equipment requiring power.
27    .      Provide new lay-in ceiling in corridor of Building 'B'.
28    .      Remove existing ceiling in part of Guidance Area of Building 'A' and provide new lay-in
29           ceiling throughout the entire Guidance Area.
30
31
32    END OF SECTION
33
```

16011-1

Figure 41. Sample Electrical Specifications (1 Of 12)

SECTION 16110 - RACEWAYS

PART 1 - GENERAL:

GENERAL

All wiring shall be in rigid metal conduit, 'RMC', except as otherwise noted.

Electric metallic tubing 'EMT' may be used for concealed and exposed work, except as listed below.

        (1) Exposed to the weather, or in damp location.
        (2) In earth or stone.
        (3) In concrete slabs on grade.
        (4) Where obviously subject to mechanical injury.
        (5) Where specified otherwise.

EMT may be used in lieu of 'RMC' in the following sizes, 1/2 inch through 2-inch for power circuits, 1/2 inch through 4 inch for communications circuits and control wiring as applicable, subject to the use limitations specified above. Intermediate metal conduit 'IMC' may be used for 'RMC'.

Rigid non-metallic conduit (RNMC) may be used in in the following applications only. (All stub-ups shall be 'RMC' or 'IMC' elbows.)

        Below slab, encased all around in not less than three inches of
        concrete and not less than 18 inches of ground cover for power service
        and power feeders, under building. Except, that conduit under
        building foundation, or conduit run through footings or foundations,
        shall be rigid steel conduit.

        Underground, buried not less than 30" below grade for area lighting
        circuits.

        In concrete slab on grade for branch circuits utilizing 3/4 inch or
        smaller conduit when encased by at least 2-inches of concrete all around.

Flexible metallic conduit shall not be used as a wiring method, other than when specifically noted to be used, without prior permission of the Architect/Engineer.

Type AC (BX) armored cable is not permitted in this project.

SLEEVES AND PENETRATIONS:

See Section 16115 for required sleeves and method of achieving raceway penetrations.

APPLICABLE SPECIFICATIONS AND STANDARDS:

The materials specified here shall meet the following specifications and standards in their current edition.

16110-1

110F41B.TIF

Figure 41. Sample Electrical Specifications (2 Of 12)

        (1) UL Standards

                Electric metallic Tubing
                Flexible Metal Conduits-UL-1
                Rigid Metal Conduit UL-6
                Intermediate Metal Conduit UL-1242

        (2) NEMA Standards

                Electric plastic conduit-TC-2

        (3) ANSI Standards

                Specifictions for Rigid Steel Conduit, Zinc Coated,
                ANSI C80.1.

PART 2 - PRODUCTS:

RACEWAYS:

General: Minimum size conduit shall be 1/2 inch.

Rigid Metal Conduit:

Rigid metal conduit shall be schedule 40, of the best quality steel.

The interior and exterior surfaces of the conduit shall be protected with a metallic zinc coating. Rigid steel conduit shall be galvanized by the Hot-Dip process in accordance with ASTM A 123.

Fittings for 'RMC' shall be threaded UL listed.

Electric Metallic Tubing:

Electrical metallic tubing shall be rigid metal conduit of the thin-wall type in straight lengths, elbows or bends for use as raceways for wire or cables in an electrical system.

Electrical metallic tubing shall utilize hexagonal steel type compression threadless fittings of galvanized steel throughout. All fittings shall be UL listed for concrete-tight and rain-tight construction. All EMT entrance fittings shall be provided with insulated throats.

Flexible Metallic Conduit:

Flexible metallic conduit shall conform to UL standard 'Flexible Steel Conduit'. All steel used in the fabrication of the conduit shall be zinc coated.

Liquid-tight flexible steel conduit shall be provided with a protective jacket of polyvinyl chloride extruded over a flexible interlocked galvanized steel core to protect wiring against moisture, oil, chemicals and corrosive fumes.

Flexible conduit connectors shall be UL listed T & B nylon-insulated "Tite-Bite", or equivalent from "Blackhawk.'

16110-2

110F41C.TIF

Figure 41. Sample Electrical Specifications (3 Of 12)

Rigid Non-Metallic Conduit:

Schedule 40 (EPC-40), heavy wall polyvinyl chloride plastic conduit and fittings, UL listed, suitable for 90 degree C. conductors.

Minimum size RNMC conduit shall be 3/4-inch.

Intermediate Metal Conduit:

Intermediate metal conduit 'IMC' shall be zinc coated steel, UL listed and labeled.

Fittings shall be the same type as for 'RMC.'

Wireways and Troughs:

Wireways or troughs shall be of the size noted on the drawings or as required by the NEC of code gauge galvanized steel. Sizes 6 inch x 6 inch and smaller wireways and all troughs shall be of the hinged cover type except as otherwise noted on the drawings. Larger sized wireways shall be of the flangeless screw cover lay-in type. All shall be without knockout, and shall be provided with fittings, supports, and apurtenances as required.

PART 3 - EXECUTION:

INSTALLATION OF CONDUIT AND TUBING:

Metallic raceways shall not be stored exposed to the weather.

Conduits shall be concealed within the walls, ceilings, and floors, where possible, and shall be kept at least 6 inches from parallel runs of flues, steam pipes, or hot water pipes. Exposed runs of conduit or tubing, and conduit or tubing run above suspended ceilings, shall have supports spaced not more than 8 feet apart and shall be installed with runs parallel or perpendicular to walls, structural members, or intersections of vertical planes and ceilings with right-angle turns consisting of cast metal fittings or symmetrical bends. All raceways shall be run in a neat and orderly fashion. Conduits or tubing run in diagonal or disorganized way shall be removed from the premises if so structed by the A/E. Bends and offsets shall be avoided where possible, but where necessary shall be made with an approved hickey or conduit bending machine. Conduit or tubing which has been crushed or deformed in any way shall not be installed.

Conduit and tubing shall be supported on approved types of galvanized wall brackets, ceiling trapezes, strap hangers, or pipe straps, secured by means of toggle bolts on hollow masonry units, expansion bolts in concrete or brick, machine screws on metal surfaces, and wood screws on wood construction. Nails shall not be used as the means of fastening boxes on conduits. Wooden plugs inserted in masonry or concrete shall not be used as a base to secure conduit supports.

Conduits shall be installed in such manner as to insure against trouble from the collection of trapped condensation, and all runs of conduit shall be arranged so as to be devoid of traps where feasible. The contractor shall exercise the necessary precautions to prevent the lodgment of dirt, plaster, or trash in conduit, tubing, fittings, and boxes during the course of installation by the use of T & B pushpennies, appleton pennies, or equal closures. A run of conduit or tubing which has become clogged shall be entirely freed of these accumulations or shall be replaced.

16110-3

Figure 41. Sample Electrical Specifications (4 Of 12)

Conduit shall be securely fastened to all sheet metal enclosures with double galvanized locknuts and insulated bushings, care being observed to see that full number of threads project through to permit the bushing to be drawn tight against the end of conduit, after which the locknuts shall be made up sufficiently tight to insure positive ground continuity between conduit and box.

Double locknuts shall be used on all feeder and motor circuit conduits and where insulated bushings are used. Insulated bushings of molded bakelite shall be used on all conduit entrances, one inch over in size, into junction boxes, panel boxes and motors starters having sheet metal enclosures. Galvanized steel insulated throat fittings shall be used for EMT work.

Rigid metal conduit 'RMC' installed underground 5 feet or more from the building shall have a minimum cover of three feet. 'RMC' directly installed underground shall be coated with two coats of bitumastic or asphalt paint. Conduit installed underground or in concrete on ground shall be made watertight by wrapping the joints with 'Teflon' tape.

<u>EXPANSION FITTINGS:</u>

Conduit crossing expansion joints in concrete slabs shall be provided with suitable expansion fittings, or other suitable means shall be provided to compensate for the building expansion and contraction.

<u>PULL WIRE:</u>

Nylon pull wire not less than 5/32 inches in diameter shall be installed in all empty conduit longer than 10 feet. Pull wire shall be secured at each end and tagged for identification of the use of the conduit.

END OF SECTION

16110-4

Figure 41. Sample Electrical Specifications (5 Of 12)

SECTION 16120 - CONDUCTORS

PART 1 - GENERAL

SCOPE:

This Section applies to secondary conductors for systems rated 600 volts and below.

A complete system of conductors shall be installed in the raceway systems as specified here and shown on drawings.

APPLICABLE SPECIFICATIONS AND STANDARDS:

Compliance:

The materials specified here shall meet the following specifications and standards in their current edition.

    (1) UL Standards:

        Insulation tape
        Wire Connectors and Soldering lugs

    (2) NEMA Standards:

        Thermoplastic - Insulated WC 5 (IPCEA S-61-402)

PART 2 PRODUCTS

CONDUCTORS:

All conductors shall be made of copper.

Conductors, unless otherwise noted, shall be heat and moisture resistant grade, thermoplastic insulated. Conductors No. 8 AWG and larger shall be stranded copper conductors, dual rated, Type THHN-THWN; Conductors No. 10 and smaller shall be solid copper, Type THHN-THWN (dual rated), unless otherwise required below. Branch circuit conductors for all other lighting fixtures shall have a temperature rating of not less than what is required by the UL listing of the fixture with a minimum rating of 90 degrees C.

Conductors for signal and control circuits above 50 volts AC may be THWN-THHN as permitted by NEC, No. 14 AWG. Conductors for signal and control circuits below 50 volts AC may be 300-volt, PVC insulated, No. 14 AWG.

Branch circuit conductors shall be not smaller than No. 12 AWG, except that conductors for branch circuits whose length from panel to center of load exceeds 75 feet for the 280/120 volt system, or 150 feet for 277/480 volt system, shall not be smaller than No. 10 AWG from the panel to the first outlet box in the circuit regardless of what is scheduled on panelboard schedule.

Conductors being connected to transformers and other equipment shall have a temperature rating as required by the transformer or equipment manufacturer.

PART 3 - EXECUTION:

16120-1

110F41F.TIF

Figure 41. Sample Electrical Specifications (6 Of 12)

SPLICES:

Solid Conductor Splices:

Solid conductors namely those sized #10, #12, and #14 AWG copper, and smaller, shall be spliced by twisting securely and by means of hot-dipped solder plus gum rubber tape, plus friction tape, or plastic tape approved as a substitute for friction tape. The contractor shall use Ideal "wire-nuts" for recessed lighting fixture lead splices to branch circuit conductors.

Stranded Conductor Splices:

Namely #8 AWG and larger, shall be spliced by approved mechanical connectors plus gum tape, plus friction or plastic tape. Solderless mechanical connectors, for splices and taps provided with U.L. approval insulating covers, may be used instead of mechanical connectors plus tape.

INSTALLATION OF CONDUCTORS:

Conductors shall be continuous from outlet to outlet, and no splices shall be made except within outlet or junction boxes, troughs and gutters. Junction boxes may be utilized where required. No 'condulet' type fitting shall be used on any service conduits. If other than long radius bends are required, pull boxes sized in accordance with the NEC shall be used.

Home runs may be combined in one conduit, provided all connections are in accordance with National Electrical Code requirements, and the maximum unbalanced current in the neutral does not extend the capacity of the conductor, and the conductors are not required to be derated to below circuit capacity.

Conductors in vertical runs shall be supported as required by NEC 300-19.

COLOR CODING:

Conductors, feeders, and branch circuits shall be color coded by phases as follows:

480/277-volt systems: Phase A-brown; Phase B-orange; Phase C-yellow; Neutral - white with identifiable color stripe, other than green; grounding wire - green.

208/120-volt systems: Phase A-black; Phase B-red; Phase C-blue; neutral-white; Grounding wire-green.

Insulating tape of proper color shall be used to identify the phase conductors No. 6 AWG and largerconductors.

All feeders, sub-feeds to panels, motors, etc., shall be completely phased outas to sequence and rotation. Phase sequence shall be A-B-C from front to rear, top to bottom, left to right when facing equipment.

16120-2

110F41G.TIF

Figure 41. Sample Electrical Specifications (7 Of 12)

```
1    SECTION 16136 - WIRING DEVICES, BOXES & ENCLOSURES
2
3    PART 1 GENERAL
4
5    WORK INCLUDED:
6
7    Work under this Section includes but is not necessarily limited to the following.
8
9    Receptacles, toggle switches, dimmers and photoelectric switching devices.
10   Outlet Boxes
11   Cabinets and Enclosures
12
13   APPLICABLE SPECIFICATIONS AND STANDARDS:
14
15   Equipment specified in this Section shall meet the following specifications and standards.
16
17        (1) UL Standards
18
19             Attachment Plug and Receptacles
20             Electric Cabinet and Boxes
21             Outlet Boxes and Fittings
22             Snap Switches
23
24        (2) NEMA Standards
25
26             Boxes, OS1
27             Wiring Devices, General - Purpose, WD1
28             Wiring Devices, Specific - Purpose, WD5
29
30   PART 2 PRODUCTS
31
32   OUTLET BOXES, PULL BOXES, CABINETS AND ENCLOSURES:
33
34   Boxes:
35
36   Boxes shall have sufficient volume to accomodate the number of conductors entering the box in
37   accordance with the requirements of NFPA 70, Article 370.  Boxes that are exposed to the
38   weather or that are in normally wet locations shall be of the cast-metal type having threaded hubs.
39   Boxes shall be of suitable construction for installation in the environment of their location.  Unless
40   otherwise specifically stated all boxes shall be metallic boxes.
41
42   Zinc-coated or cadium-plated sheet steel boxes, or a class to satisfy the conditions of each outlet,
43   shall be used in concealed work or in exposed work above eight feet from floor.
44
45   Fixture outlet boxes on ceiling shall be not less than 4 inch octagonal.  Fixture outlet boxes in
46   concrete ceiling shall be of the 4 inch octagonal concrete type, set flush with the finished surface.
47   Fixture outlet boxes on plastered ceilings shall be fitted with open covers set to come flush with
48   the finished surface.
49
50   Switch and receptacle outlet boxes in dry walls, plastered walls and pour-in concrete walls shall
51   be not less than 4 inches square cut with appropriate extension to set flush with the finished
52   surface.  One-piece gang or gangable boxes not less than 2 inches deep shall be utilized where the
```

16136-1

Figure 41. Sample Electrical Specifications (8 Of 12)

---

use of 4-inch square boxes is not feasible.

Unless otherwise noted on the drawings, outlet, junction or pull boxes not larger than 5 inches square and within eight feet from floor level in exposed work shall be of cast steel or alloy with threaded hubs and appropriate covers.

Outlet boxes in unplastered masonry and gypsum drywall walls shall be tile type.

Outlet boxes for use with conduit and tubing systems shall be not less than 1-1/2 inches deep.

A device plate or cover shall be provided for each outlet to suit the outlet.

Pull Boxes:

Pull boxes shall be constructed of code-gauge galvanized sheet metal. Boxes shall be of not less than the minimum size required by the National Electrical Code and shall be furnished with screw fastened covers. When several feeders pass through a common pull box they shall be tagged to indicate clearly their electrical characteristics, circuit numbers and panel designations.

Pull boxes shall be furnished and installed where necessary in the raceway system to facilitate conductor installation. Except as otherwise noted for telephone raceways, conduit runs longer than 150 feet, or with more than 360 degrees compound angle bends, shall have a pull box installed at a convenient intermediate location. Normally, when feeder routing is shown on drawings, pull boxes are not acceptable. It is the responsibility of the electrical contractor to provide pull boxes as necessary to meet the stated requirements.

Cabinets:

Cabinet boxes shall be constructed of zinc-coated sheet steel and shall conform with the requirements of Underwriters' Laboratories "Standards for Cabinets and Cutout Boxes". Unless otherwise noted, surface mounted cabinet trims shall have a corrosion inhibiting primer and a lacquer finish. Flush mounted cabinets shall be factory primed ready for finish painting by the General Contractor. Cabinets shall be of suitable construction for installation in the environment of their locations.

Systems cabinet, if shown, shall be provided with interior dimensions not less than those indicated on the drawings. Trims shall provide maximum size openings to the box interiors. Cabinets shall be provided with 5/8 inch fire retardant plywood back-boards having an insulating varnish finish.

Device Plates:

A device plate shall be provided for each outlet (including telephone and computer system outlets) to suit the device installed. Screws shall be of metal with countersunk heads, with finish to match the finish of the plate.

Device plates shall be of the one-piece type, of suitable shape for the devices to be covered. The use of sectional device plates will not be permitted. Plates shall be as follows:

   Plates on surface boxes in unfinished areas shall be of galvanized steel with beveled edges.
   All plates on walls with flush switches or receptacles shall be 302 stainless steel, 0.32"
   nominal thickness.

16136-2

110F41I.TIF

Figure 41. Sample Electrical Specifications (9 Of 12)

Plates on surface boxes in finished areas shall be as for flush outlets.

Switches and Receptacles:

Switches and receptacles shall be as shown on the drawings and in the symbol schedule, and shall meet the latest federal specifications W-S 896 or W-C 596 as verified by UL. Devices shall be the product of one of the followingmanufacturers complying with referenced NEMA Standards - Arrow-Hart Electric Company, Bryant Electric Company, General Electric Company, Harvey Hubbell, Inc., Slater Electric, Inc., Pass & Seymour, Inc., Sierra Electric, or Leviton. 20-ampere and 15 ampere receptacles shall be heavy duty, hospital grade with nylon body. Switches shall be 20-ampere, specification grade.

Color of switches and receptacles shall be gray.

Receptacles with ground fault interrupting (GFI) Protection.

GFI receptacles shall be NEMA 5-15R configuration, UL listed with "noise-suppressed" circuitry to eliminate false trippings. GFI receptacles shall provide protection to all other receptacles installed "downstream" on the same branch circuit. Provide separate neutral conductor for GFI receptacle circuit.

Wiring Device Schedule

For clarity and to identify the class and type of devices required, refer to the following schedule of devices by Pass & Seymour (P&S). Device color shall be as previous specified.

Switches:	Catalogue Number
Single Pole	20 AC 1
Double Pole	20 AC 2
Three Way	20 AC 3

Receptacles:

NEMA Configuration	P&S Catalogue Number
5-15R	5262
5-20R	5362
5-30R	3802
5-50R	3803
6-20R	5862
6-30R	3801
6-50R	3804
10-30R	3860
10-50R	3890
14-30R	3864
15-20R	3821
15-30R	5740
18-20R	3822
L5-20R	L520-R
L5-30R	L530-R
L6-20R	L620-R

16136-3

Figure 41. Sample Electrical Specifications (10 Of 12)

1	L6-30R	L630-R
2	L15-20R	L1520-R
3	L15-30R	L1530-R

4

5 MULTI-OUTLET ASSEMBLIES AND SURFACE METAL RACEWAYS:

6

7 Surface metal raceways shall be provided as indicated on the drawings, complete with all
8 appropriate fittings as required, to provide a safe and complete installation. All components shall
9 be UL listed. Raceways shall be supported on approximately 18" centers with #8 flat lead
10 fasterners. The entire installation shall meet the requirements of Article 352 of the NEC. All field
11 cuts of the raceway shall be made square and shall have no rough edges.

12

13 PART 3 - EXECUTION:

14

15 INSTALLATION OF OUTLETS:

16

17 Location of outlets shown on drawings, other than those dimensioned, are only approximate, the
18 Owner shall have the right to make slight changes in the position of outlets if the Contractor is
19 notified before roughing-in is done. The Contractor shall study the general building plans in relation
20 to the spaces surrounding each outlet in order that his work may fit the other work required by
21 these specifications. When necessary, the Contractor shall relocate outlets of junction boxes so
22 that, when fixtures or other fittings are installed, they will be symmetrically located according to
23 room layout and will not interfere with other work or equipment. Do not install outlets back to
24 back. Outlets flush mounted in fire or sound wall shall be mounted not closer than 24" apart
25 horizontally when on same face of wall and 12" apart horizontally when mounted on opposite sides
26 of wall. For those outlets in fire or sound rated walls that must be closer than 24" apart, provide
27 Nelson Firestop system 'putty pads' all around outlets to maintain fire rating.

28

29 Minimum length of conduit connecting to adjacent flush in sound rated walls outlets sound rated
30 walls shall be 18".

31

32 Boxes shall be installed in a rigid and satisfactory manner, either by wood screws on wood,
33 expansion shields on masonry, or machine screws on steel work.

34

35 Recessed boxes in dry wall type construction shall be supported from both adjacent studs, or by
36 the use of metal stud brackets as manufactured by E-Z Mount Bracket Co., or equivalent.

37

38 PLATES:

39

40 Plates for receptacles and switches shall be installed with all four edges in continuous contact with
41 the finished wall surfaces without the use of mats or similar devices. Plaster fillings will not be
42 permitted. Plates shall be installed vertically and with an alignment tolerance of 1/16 inch.

43

44 SURFACE METAL RACEWAYS

45

46 Surface metal raceways shall be installed where noted on the drawings. Support on approximately
47 30" centers with #8 flat head screws. In addition, couplings, fittings and boxes shall be supported
48 independent of the raceway.

49

50 INSTALLATION OF PULL BOXES

51

52 Pull boxes shall be installed overhead or on walls at locations free of interference with equipment,

16136-4

110F41K.TIF

Figure 41. Sample Electrical Specifications (11 Of 12)

---

1   ducts, piping and activities being carried out at the premises. Pull boxes with covers larger than
2   4 square feet shall be provided with two handles.
3
4
5   END OF SECTION
6

16136-5

Figure 41. Sample Electrical Specifications (12 Of 12)

Note: Line 30 of the above specifications refers to color coding. Please refer
to ***NEC Section 210-4(d)*** for specific requirements.

SUMMARY

In this module, you learned the symbols and conventions used on architectural and engineering drawings. As an electrician, you need to know how to recognize the basic symbols used on electrical drawings and other drawings used in the building construction industry. You should also know where to find the meaning of symbols that you do not immediately recognize. Schedules, diagrams, and specifications often provide detailed information that is not included on the working drawings.

Reading architectural and engineering drawings takes practice and study. Now that you have the basic skills, take the time to master them.

References

For advanced study of topics covered in this task module, the following books are suggested:

American Electrician's Handbook, Latest Edition, McGraw-Hill, New York, NY.
Illustrated Guide to the NEC, Latest Edition, Craftsman Book Company, Carlsbad, CA.
National Electrical Code Handbook, Latest Edition, National Fire Protection Association, Quincy, MA.

REVIEW QUESTIONS

Questions 1 through 7 refer to the seven electrical symbols shown below. In the spaces provided, place the letter corresponding to the correct answer found in the list.

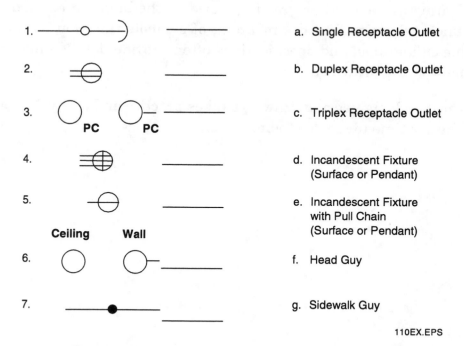

1. _____   a. Single Receptacle Outlet

2. _____   b. Duplex Receptacle Outlet

3. _____   c. Triplex Receptacle Outlet

4. _____   d. Incandescent Fixture (Surface or Pendant)

5. _____   e. Incandescent Fixture with Pull Chain (Surface or Pendant)

6. _____   f. Head Guy

7. _____   g. Sidewalk Guy

110EX.EPS

Questions 8 through 10 refer to the sample electrical specifications shown in *Figure 41*.

8. Which of the following tasks is *not* covered under Division 16?
   a. Lighting fixtures
   b. New lay-in ceiling in the corridor of Building B
   c. Concrete pad for power company transformers
   d. New chiller

9. What is the minimum size RNMC conduit allowed by the specifications?
   a. ½ inch
   b. ¾ inch
   c. 1 inch
   d. ⁵⁄₁₆ inch

10. The color code specified for a 208/120V system is _____.
    a. phase A – black; phase B – red; phase C – blue
    b. neutral – white; grounding wire – green
    c. phase A – brown; phase B – orange; phase C – yellow
    d. both a and b

notes

ANSWERS TO REVIEW QUESTIONS

Answer		**Section**
1.	g	4.0.0
2.	b	4.0.0
3.	e	4.0.0
4.	c	4.0.0
5.	a	4.0.0
6.	d	4.0.0
7.	f	4.0.0
8.	d	10.2.1/Figure 41
9.	b	10.2.1/Figure 41
10.	d	10.2.1/Figure 41

METRIC CONVERSION CHART

inches (fractions)	inches (decimals)	mm	inches (fractions)	inches (decimals)	mm	inches (fractions)	inches (decimals)	mm	inches (fractions)	inches (decimals)	mm
—	.0004	.01	25/32	.781	19.844	2-3/16	2.165	55.	3-11/16	3.6875	93.663
—	.004	.10	—	.7874	20.	—	2.1875	55.563	—	3.7008	94.
r	.01	.25	51/64	.797	20.241	—	2.2047	56.	3-23/32	3.719	94.456
1/64	.0156	.397	13/16	.8125	20.638	2-7/32	2.219	56.356	—	3.7401	95.
—	.0197	.50	—	.8268	21.	—	2.244	57.	3-3/4	3.750	95.250
—	.0295	.75	53/64	.828	21.034	2-1/4	2.250	57.150	—	3.7795	96.
1/32	.03125	.794	27/32	.844	21.431	2-9/32	2.281	57.944	3-25/32	3.781	96.044
—	.0394	1.	55/64	.859	21.828	—	2.2835	58.	3-13/16	3.8125	96.838
3/64	.0469	1.191	—	.8661	22.	2-5/16	2.312	58.738	—	3.8189	97.
—	.059	1.5	7/8	.875	22.225	—	2.3228	59.	3-27/32	3.844	97.631
1/16	.062	1.588	57/64	.8906	22.622	2-11/32	2.344	59.531	—	3.8583	98.
5/64	.0781	1.984	—	.9055	23.	—	2.3622	60.	3-7/8	3.875	98.425
—	.0787	2.	29/32	.9062	23.019	2-3/8	2.375	60.325	—	3.8976	99.
3/32	.094	2.381	59/64	.922	23.416	—	2.4016	61.	3-29/32	3.9062	99.219
—	.0984	2.5	15/16	.9375	23.813	2-13/32	2.406	61.119	—	3.9370	100.
7/64	.109	2.778	—	.9449	24.	2-7/16	2.438	61.913	3-15/16	3.9375	100.013
—	.1181	3.	61/64	.953	24.209	—	2.4409	62.	3-31/32	3.969	100.806
1/8	.125	3.175	31/32	.969	24.606	2-15/32	2.469	62.706	—	3.9764	101.
—	.1378	3.5	—	.9843	25.	—	2.4803	63.	4	4.000	101.600
9/64	.141	3.572	63/64	.9844	25.003	2-1/2	2.500	63.500	4-1/16	4.062	103.188
5/32	.156	3.969	1	1.000	25.400	—	2.5197	64.	4-1/8	4.125	104.775
—	.1575	4.	—	1.0236	26.	2-17/32	2.531	64.294	—	4.1338	105.
11/64	.172	4.366	1-1/32	1.0312	26.194	—	2.559	65.	4-3/16	4.1875	106.363
—	.177	4.5	1-1/16	1.062	26.988	2-9/16	2.562	65.088	4-1/4	4.250	107.950
3/16	.1875	4.763	—	1.063	27.	2-19/32	2.594	65.881	4-5/16	4.312	109.538
—	.1969	5.	1-3/32	1.094	27.781	—	2.5984	66.	—	4.3307	110.
13/64	.203	5.159	—	1.1024	28.	2-5/8	2.625	66.675	4-3/8	4.375	111.125
—	.2165	5.5	1-1/8	1.125	28.575	—	2.638	67.	4-7/16	4.438	112.713
7/32	.219	5.556	—	1.1417	29.	2-21/32	2.656	67.469	4-1/2	4.500	114.300
15/64	.234	5.953	1-5/32	1.156	29.369	—	2.6772	68.	—	4.5275	115.
—	.2362	6.	—	1.1811	30.	2-11/16	2.6875	68.263	4-9/16	4.562	115.888
1/4	.250	6.350	1-3/16	1.1875	30.163	—	2.7165	69.	4-5/8	4.625	117.475
—	.2559	6.5	1-7/32	1.219	30.956	2-23/32	2.719	69.056	4-11/16	4.6875	119.063
17/64	.2656	6.747	—	1.2205	31.	2-3/4	2.750	69.850	—	4.7244	120.
—	.2756	7.	1-1/4	1.250	31.750	—	2.7559	70.	4-3/4	4.750	120.650
9/32	.281	7.144	—	1.2598	32.	2-25/32	2.781	70.6439	4-13/16	4.8125	122.238
—	.2953	7.5	1-9/32	1.281	32.544	—	2.7953	71.	4-7/8	4.875	123.825
19/64	.297	7.541	—	1.2992	33.	2-13/16	2.8125	71.4376	—	4.9212	125.
5/16	.312	7.938	1-5/16	1.312	33.338	—	2.8346	72.	4-15/16	4.9375	125.413
—	.315	8.	—	1.3386	34.	2-27/32	2.844	72.2314	5	5.000	127.000
21/64	.328	8.334	1-11/32	1.344	34.131	—	2.8740	73.	—	5.1181	130.
—	.335	8.5	1-3/8	1.375	34.925	2-7/8	2.875	73.025	5-1/4	5.250	133.350
11/32	.344	8.731	—	1.3779	35.	2-29/32	2.9062	73.819	5-1/2	5.500	139.700
—	.3543	9.	1-13/32	1.406	35.719	—	2.9134	74.	—	5.5118	140.
23/64	.359	9.128	—	1.4173	36.	2-15/16	2.9375	74.613	5-3/4	5.750	146.050
—	.374	9.5	1-7/16	1.438	36.513	—	2.9527	75.	—	5.9055	150.
3/8	.375	9.525	—	1.4567	37.	2-31/32	2.969	75.406	6	6.000	152.400
25/64	.391	9.922	1-15/32	1.469	37.306	—	2.9921	76.	6-1/4	6.250	158.750
—	.3937	10.	—	1.4961	38.	3	3.000	76.200	—	6.2992	160.
13/32	.406	10.319	1-1/2	1.500	38.100	3-1/32	3.0312	76.994	6-1/2	6.500	165.100
—	.413	10.5	1-17/32	1.531	38.894	—	3.0315	77.	—	6.6929	170.
27/64	.422	10.716	—	1.5354	39.	3-1/16	3.062	77.788	6-3/4	6.750	171.450
—	.4331	11.	1-9/16	1.562	39.688	—	3.0709	78.	7	7.000	177.800
7/16	.438	11.113	—	1.5748	40.	3-3/32	3.094	78.581	—	7.0866	180.
29/64	.453	11.509	1-19/32	1.594	40.481	—	3.1102	79.	—	7.4803	190.
15/32	.469	11.906	—	1.6142	41.	3-1/8	3.125	79.375	7-1/2	7.500	190.500
—	.4724	12.	1-5/8	1.625	41.275	—	3.1496	80.	—	7.8740	200.
31/64	.484	12.303	—	1.6535	42.	3-5/32	3.156	80.169	8	8.000	203.200
—	.492	12.5	1-21/32	1.6562	42.069	3-3/16	3.1875	80.963	—	8.2677	210.
1/2	.500	12.700	1-11/16	1.6875	42.863	—	3.1890	81.	8-1/2	8.500	215.900
—	.5118	13.	—	1.6929	43.	3-7/32	3.219	81.756	—	8.6614	220.
33/64	.5156	13.097	1-23/32	1.719	43.656	—	3.2283	82.	9	9.000	228.600
17/32	.531	13.494	—	1.7323	44.	3-1/4	3.250	82.550	—	9.0551	230.
35/64	.547	13.891	1-3/4	1.750	44.450	—	3.2677	83.	—	9.4488	240.
—	.5512	14.	—	1.7717	45.	3-9/32	3.281	83.344	9-1/2	9.500	241.300
9/16	.563	14.288	1-25/32	1.781	45.244	—	3.3071	84.	—	9.8425	250.
—	.571	14.5	—	1.8110	46.	3-5/16	3.312	84.1377	10	10.000	254.001
37/64	.578	14.684	1-13/16	1.8125	46.038	3-11/32	3.344	84.9314	—	10.2362	260.
—	.5906	15.	1-27/32	1.844	46.831	—	3.3464	85.	—	10.6299	270.
19/32	.594	15.081	—	1.8504	47.	3-3/8	3.375	85.725	11	11.000	279.401
39/64	.609	15.478	1-7/8	1.875	47.625	—	3.3858	86.	—	11.0236	280.
5/8	.625	15.875	—	1.8898	48.	3-13/32	3.406	86.519	—	11.4173	290.
—	.6299	16.	1-29/32	1.9062	48.419	—	3.4252	87.	—	11.8110	300.
41/64	.6406	16.272	—	1.9291	49.	3-7/16	3.438	87.313	12	12.000	304.801
—	.6496	16.5	1-15/16	1.9375	49.213	—	3.4646	88.	13	13.000	330.201
21/32	.656	16.669	—	1.9685	50.	3-15/32	3.469	88.106	—	13.7795	350.
—	.6693	17.	1-31/32	1.969	50.006	3-1/2	3.500	88.900	14	14.000	355.601
43/64	.672	17.066	2	2.000	50.800	—	3.5039	89.	15	15.000	381.001
11/16	.6875	17.463	—	2.0079	51.	3-17/32	3.531	89.694	—	15.7480	400.
45/64	.703	17.859	2-1/32	2.03125	51.594	—	3.5433	90.	16	16.000	406.401
—	.7087	18.	—	2.0472	52.	3-9/16	3.562	90.4877	17	17.000	431.801
23/32	.719	18.256	2-1/16	2.062	52.388	—	3.5827	91.	—	17.7165	450.
—	.7283	18.5	—	2.0866	53.	3-19/32	3.594	91.281	18	18.000	457.201
47/64	.734	18.653	2-3/32	2.094	53.181	—	3.622	92.	19	19.000	482.601
—	.7480	19.	2-1/8	2.125	53.975	3-5/8	3.625	92.075	—	19.6850	500.
3/4	.750	19.050	—	2.126	54.	3-21/32	3.656	92.869	20	20.000	508.001
49/64	.7656	19.447	2-5/32	2.156	54.769	—	3.6614	93.			

The NCCER makes every effort to keep these manuals up-to-date and free of technical errors. We appreciate your help in this process. If you have an idea for improving this manual, or if you find an error, a typographical mistake, or an inaccuracy in the NCCER's Craft Training Manuals, please write us, using this form or a photocopy. Be sure to include the exact module number, page number, a description of the problem, and the correction, if possible. Your input will be brought to the attention of the Technical Review Committee. Thank you for your assistance.

Instructors – If you found that additional materials were necessary in order to teach this module effectively, please let us know so that we may include them in the Equipment/Materials list in the Instructor's Guide.

Write: Curriculum Development and Revision Department
National Center for Construction Education and Research
P.O. Box 141104
Gainesville, FL 32614-1104

Fax: 352-334-0932

Craft _____ Module Name _____

Copyright Date _____ Module Number _____ Page Number(s) _____

Description of Problem _____

_____

_____

_____

_____

(Optional) Correction of Problem _____

_____

_____

(Optional) Your Name and Address _____

_____

_____

Wiring: Commercial and Industrial

Module 26111

**NATIONAL
CENTER FOR
CONSTRUCTION
EDUCATION AND
RESEARCH**

WIRING: COMMERCIAL & INDUSTRIAL

OBJECTIVES

Upon completion of this module, the trainee will be able to:

1. Identify and state the functions and ratings of single-pole, double-pole, three-way, four-way, dimmer, special, and safety switches.
2. Explain NEMA classifications as they relate to switches and enclosures.
3. Explain the NEC requirements concerning wiring devices.
4. Identify and state the functions and ratings of straight blade, twist lock, and pin and sleeve receptacles.
5. Identify and define receptacle terminals and disconnects.
6. Identify and define ground fault circuit interrupters.
7. Explain the box mounting requirements in the NEC.
8. Use a wire stripper to strip insulation from a wire.
9. Use a solderless connector to splice wires together.
10. Identify and state the functions of limit switches and relays.
11. Identify and state the function of switchgear.

Prerequisites

Successful completion of the following Task Modules is required before beginning study of this Task Module: Core Curricula; Electrical Level 1, Modules 26101 through 26110.

Required Trainee Materials

1. Trainee Task Module
2. Copy of the latest edition of the *National Electrical Code*
3. Appropriate Personal Protective Equipment

Note: The designations "National Electrical Code," "NE Code," and "NEC," where used in this document, refer to the *National Electrical Code®*, which is a registered trademark of the National Fire Protection Association, Quincy, MA. *All National Electrical Code (NEC) references in this module refer to the 1999 edition of the NEC.*

Course Map

This course map shows all the task modules in the first level of the Electrical curricula. The suggested training order begins at the bottom and proceeds up. Skill levels increase as a trainee advances on the course map. The training order may be adjusted by the local Training Program Sponsor.

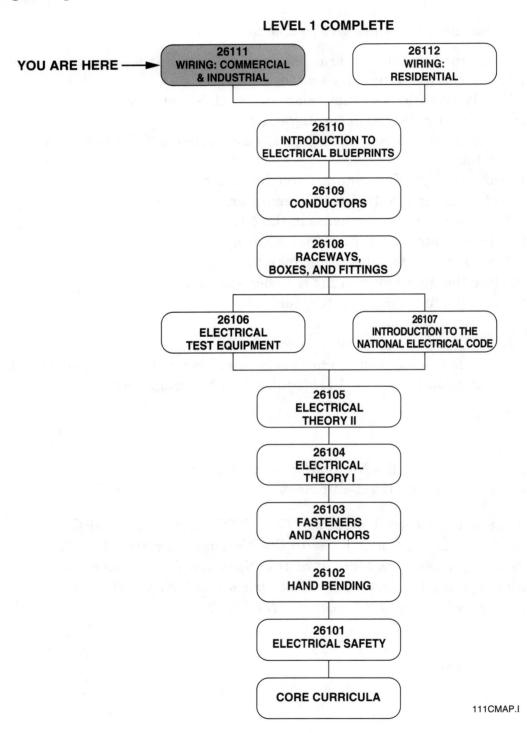

LEVEL 1 COMPLETE

YOU ARE HERE →

26111
WIRING: COMMERCIAL & INDUSTRIAL

26112
WIRING: RESIDENTIAL

26110
INTRODUCTION TO ELECTRICAL BLUEPRINTS

26109
CONDUCTORS

26108
RACEWAYS, BOXES, AND FITTINGS

26106
ELECTRICAL TEST EQUIPMENT

26107
INTRODUCTION TO THE NATIONAL ELECTRICAL CODE

26105
ELECTRICAL THEORY II

26104
ELECTRICAL THEORY I

26103
FASTENERS AND ANCHORS

26102
HAND BENDING

26101
ELECTRICAL SAFETY

CORE CURRICULA

111CMAP.I

TABLE OF CONTENTS

TABLE OF CONTENTS (Continued)

Trade Terms Introduced In This Module

Attachment plug (plug cap): A device which, by insertion in a receptacle, establishes a connection between the conductors of the attached flexible cord and the conductors connected permanently to the receptacle.

Branch circuit: The circuit conductors between the final overcurrent device protecting the circuit and the outlet(s).

Disconnecting means: A device, group of devices, or other means by which the conductors of a circuit can be disconnected from their source of supply.

Grounding: The process of connecting noncurrent-carrying metal parts of equipment to earth to provide an intentional path for fault current.

Limit switch: A type of pilot device that uses a mechanical movement to open or close electrical contacts.

Outlet: A point on a wiring system at which current is taken to supply utilization equipment.

Pilot device: Any of several types of electrical switches that are actuated by air pressure, temperature, level, or other mechanical or electrical means. These devices are wired to energize relays, which in turn connect power to larger loads.

Polarized: A physical difference between the grounded and positive and negative poles of a plug or receptacle. The difference is accomplished by making the grounded (neutral) blade of a plug larger than the ungrounded blade. The purpose of polarizing is to ensure that the grounded lead of a plug-and-cord assembly cannot be connected to the ungrounded terminal of a receptacle.

Receptacle: A contact device installed at the outlet for the connection of a single attachment plug.

Receptacle outlet: An outlet in which one or more receptacles are installed.

Relay: An electromagnetic switch for controlling one or more separate circuits remotely by applying voltage to the coil. Contacts attached to a moving plunger are connected to, or disconnected from, stationary contacts as the plunger is pulled by the magnetic field of the energized coil.

Switchgear: A term used for a wide variety of enclosed circuit breakers, disconnects, busways, and associated items that distribute power from a common point.

1.0.0 INTRODUCTION

As an electrician, a large part of your job will be involved with wiring and wiring devices. There are many different types of wiring devices and methods of installing them. This module will help you become familiar with some common devices and the methods for installing them. As with almost all aspects of an electrician's job, the NEC is an important reference for use when installing wiring devices.

2.0.0 SWITCHES

The purpose of a switch is to make and break an electrical circuit. In that way, the switch can control the operation of lighting and various kinds of equipment connected in the circuit. *NEC Article 380* is one of the areas in the NEC that covers the installation and use of switches.

In the past, all switches were mechanical devices that were manually operated to open or close a circuit. Modern technology has greatly expanded the role and design of early switches. Today, switches can be activated by light, heat, chemicals, and electrical energy.

2.1.0 COMMON TERMS

Two general terms are used to define switches. The term *pole* refers to the number of conducting paths that the switch will control in the circuit. A single-pole switch breaks the connection on only one conducting path in a circuit. A double-pole switch is used to break the connection on both conducting paths in a two-conductor circuit, usually a 240V, two-wire circuit.

The term *throw* refers to the number of internal operations that a switch can perform. For example, a single-pole, single-throw (SPST) switch will complete a single circuit only when it is thrown in one direction (the ON position). This circuit will be opened when the switch is thrown in the opposite direction (the OFF position). The common ON/OFF snap switch is an SPST switch.

A double-pole, single-throw (DPST) switch opens or closes two conducting paths at the same time. In this case, both circuits are either open or closed. A DPST switch is common to 240V circuits. In these circuits, both conductors are considered to be energized.

The single-pole, double-throw (SPDT) switch, also known as a *three-way switch*, is often used to control a single load, such as a lamp, from two different locations. SPDT switches are discussed in more detail later in this module.

A double-pole, double-throw (DPDT) switch usually directs a 240V, two-wire circuit through one of two different paths. One example of a DPDT switch is an electrical transfer switch used to energize certain circuits from either the main electric service or from an emergency standby generator. The DPDT switch prevents the circuits from being energized from both sources at once.

A four-way switch cannot be classified as an example of a double-pole, double-throw switch. While a four-way switch does throw the current path in two different directions by moving the switch toggle, only one pole on each side of the switch is allowing current to pass.

Figure 1 summarizes the basic switches and shows their symbols.

SWITCH TYPE	ABBREVIATION	SYMBOL
SINGLE-POLE SINGLE-THROW	S1	
DOUBLE-POLE SINGLE-THROW	S2	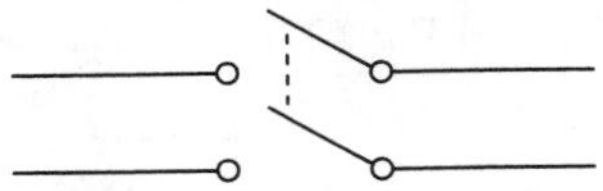
SINGLE-POLE DOUBLE-THROW (THREE-WAY)	S3	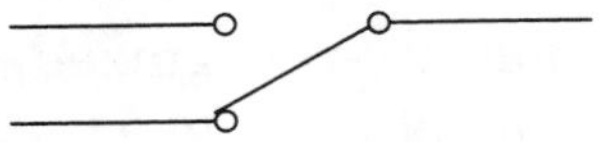
DOUBLE-POLE DOUBLE-THROW	N/A	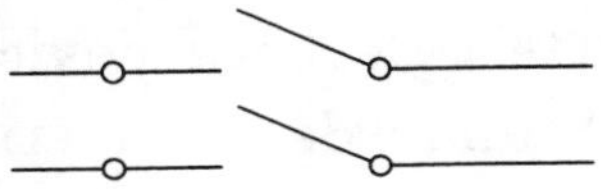
FOUR-WAY	S4	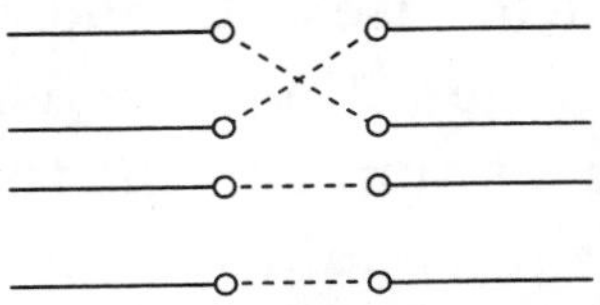

111F01.EPS

Figure 1. Switch Symbols

2.2.0   IDENTIFYING SWITCHES

Switches vary in their grades, capacities, and purposes. For example, a switch that is appropriate for activating a tungsten-filament lamp would not meet the NEC requirements for controlling motor starting. It is very important that switches be used for their designated purpose.

There is a great deal of information given on the switch itself which can help you determine its function and rating (see *Figure 2*). Most information related to the switch rating will be found on the front of the switch.

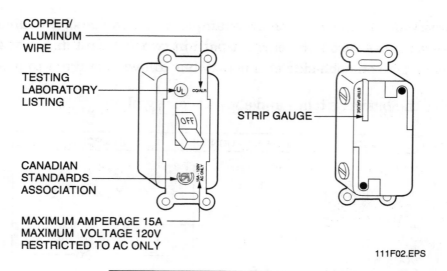

COPPER/
ALUMINUM
WIRE

TESTING
LABORATORY
LISTING

CANADIAN
STANDARDS
ASSOCIATION

MAXIMUM AMPERAGE 15A
MAXIMUM VOLTAGE 120V
RESTRICTED TO AC ONLY

STRIP GAUGE

111F02.EPS

Figure 2. Typical Single-Pole Switch

- The testing laboratory label is an indication that the device has undergone extensive testing by a nationally-recognized testing laboratory and has met the minimum safety requirements. In *Figure 2*, the switch is marked with a "UL" which stands for Underwriters' Laboratories, which was created by the National Board of Fire Underwriters to test electrical devices and materials.

- The Canadian Standards Association (CSA) label is an indication that the material or device has undergone a similar testing procedure by the Canadian Standards Association and is acceptable for use in Canada.

- Current and voltage ratings are listed by maximum amperage and maximum voltage. In *Figure 2*, the maximum current through the device should not exceed 15 amps and the maximum voltage should not exceed 120V. This switch is to be used on alternating current (AC) only. Failure to follow these recommendations could result in damage to the unit or a possible fire hazard.

- The CO/ALR symbol indicates that a switch may be used with copper, aluminum, or copper-clad aluminum wire. The letters ALR stand for aluminum revised. The CO/ALR mark replaces an earlier mark, CU/AL. It was discovered that the earlier CU/AL-rated switches were not suitable for aluminum wire in the 15- to 20-amp range.

Note: Older switches marked CU/AL should be used for copper only.

- A tungsten (T) rating indicates that the switch can be used with lamps which use tungsten. Tungsten is a metal that is used as the filament in standard incandescent lamps. When it is cold, tungsten has a very low resistance. This low resistance causes the initial current flow to be six to ten times that of the normal current flow. Once the tungsten heats up, which takes about $\frac{1}{250}$th of a second, the current flow drops back to normal. This severe overloading can reduce the life expectancy of the switch.

Another way to identify the function and rating of a switch is by color coding. Switches are typically constructed with a ground screw attached to the metallic strap of the switch. The ground screw is usually a hex-head screw and is indicated by the color green.

Note: The color green is used only for ground connections in wiring.

On a three-way switch, the color black or bronze (brass) is used to show the common pivot point for the switch. Current will always enter or leave through the pivot point.

3.0.0   TYPES OF SWITCHES

There are many types of switches available. This section covers the most common types.

3.1.0   SINGLE-POLE SWITCHES

Single-pole switches are easy to identify because they only have two terminals for conductor connections (*Figure 3*).

Note: *Figure 3* shows the white wire to the switch as the hot wire. This allows the use of standard black/white pair cable to be used for switch legs, while retaining a black wire for the hot wire to the load.

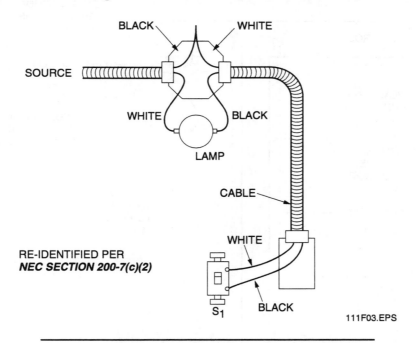

Figure 3. Single-Pole Pictorial For Cable Installation

Many single-pole switches have a third or ground terminal. This type of switch is probably the most widely used switch in residential applications. It is used for the on and off control of lights, fans, and numerous other appliances from a single point.

Note: According to **NEC Section 200-7(c)(2)**, single-pole, three-way, and four-way switch loops can use the white wire (usually reserved for the grounded conductor) for supply to the switch, but the conductor must be permanently re-identified for this use by painting or other effective means.

Figure 4 shows how a single-pole switch controlling a single light would be represented on a set of blueprints. *Figure 5* shows a schematic representation of a single-pole switch.

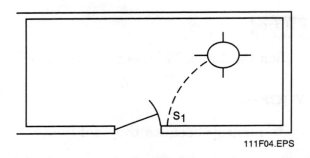

111F04.EPS

Figure 4. Single-Pole Switch As Shown On A Blueprint

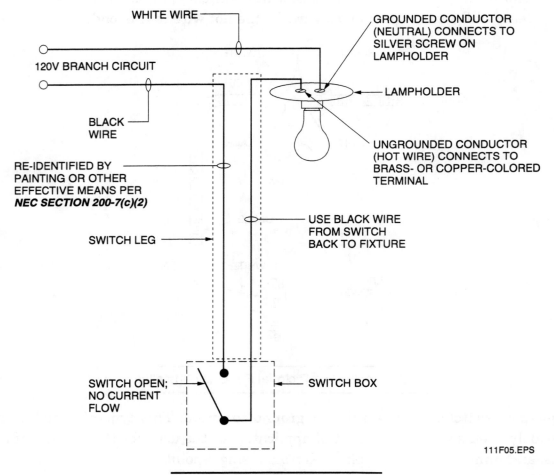

111F05.EPS

Figure 5. Single-Pole Schematic

3.2.0 DOUBLE-POLE SWITCHES

Double-pole switches are really two switches in one. They are used primarily when it is necessary to completely isolate the load. Double-pole switches are connected so that both circuit conductors can be opened or closed at the same time. The double-pole switch has four terminals for conductor connections. These terminals are identified according to polarity. One set of terminals will usually be brass in color and the other set of terminals will be white or silver-colored. Like the single-pole switch, the double-pole switch has an identified on or off position.

3.3.0 THREE-WAY SWITCHES

A three-way switch does not perform three separate operations, as its name might suggest. The three-way switch is a single-pole, double-throw switch without a center off position. A common use for a three-way switch is to control a lamp from two separate locations, such as from the top and bottom of a stairway. In this case, a pair of three-way switches would be used. *Figure 6* shows a pictorial drawing of such a circuit.

Note: The switch loop between the two three-way switches requires four conductors when the **grounding** wire is included.

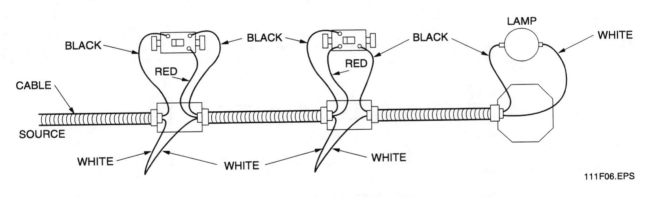

Figure 6.   Three-Way Pictorial For Cable Installation

The three-way switch has three terminal connections (four if a ground or green terminal is present). The single terminal located at one end of the switch is called the *common*. This terminal is darker in color (either black or brass-colored) than the other two terminals. The hot leg is always connected to this darker terminal. The remaining two terminals are lighter in color and are referred to as the *traveler terminals*. *Figure 7* shows how the actual connection of the three-way switching circuit is made.

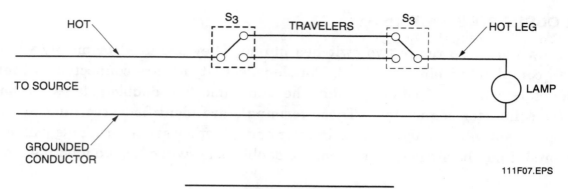

Figure 7. Three-Way Schematic

With the hot leg connected to the common of the first switch, the current is given a choice of paths to follow depending upon the relative position of the switch. The current then proceeds along one of the travelers (or alternating switch loops) to the second switch. At the second switch, the circuit will either be completed or opened, depending upon the position of this second switch. There are four possible combinations of the switch contact positions, as shown in *Figure 8*.

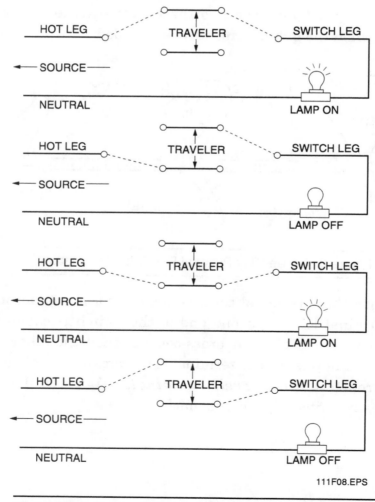

Figure 8. Possible Combinations Of A Three-Way Switch

Of the four possible combinations of switch contact positions, only two will result in a complete circuit. Changing the position of either switch will always change the state of the lamp.

CAUTION: Wiring a three-way switching situation in any other fashion will result in a polarity change at the light fixture. This is prohibited by **NEC Section 200-11**. If the polarity is changed on the light fixture, the shell of the light socket will then be hot in relation to ground. This is a personal safety hazard that could result in someone receiving a shock.

Figure 9 illustrates how a three-way switch is shown on a blueprint.

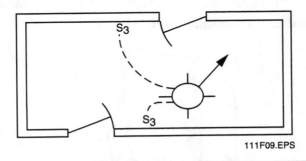

111F09.EPS

Figure 9. Three-Way Switch As Shown On A Blueprint

3.4.0   FOUR-WAY SWITCHES

Four-way switches can be used if it is necessary to control a light or group of lights from more than two locations. The four-way switch (or switches) must always be connected in the traveler conductors between the three-way switches. *Figure 10* shows how a four-way switch could be added to the circuit previously shown in *Figure 6*. The path for the current to follow is relative to the positions of the switch handle.

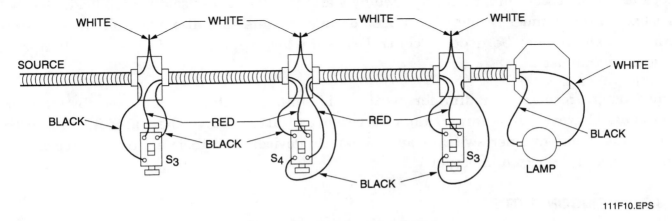

111F10.EPS

Figure 10. Four-Way Pictorial

Note that the only difference between this circuit and the three-way switch circuit is the inclusion of the four-way switch in the traveler conductors between the two three-way switches. If it is necessary to control a lamp (or lamps) from *more* than three positions, more four-way switches can be installed in the travelers or carrier conductors.

3.5.0 SPECIAL SWITCHES

Another common switch is the dimmer switch shown in *Figure 11*. This switch not only turns the light on and off, but also provides a means of controlling the light's brilliance. Dimmer switches are used where full brightness is not always desired. The standard models are available in single-pole and three-way types. They mount in a standard switch box. Single-pole dimmers can replace single-pole switches. However, three-way dimmers must be used with a three-way toggle switch at the other switch location. In other words, two three-way dimmers cannot be used to control the same load.

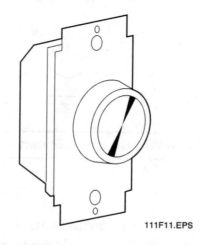

111F11.EPS

Figure 11. Dimmer Switch

Incandescent dimmers are normally rated at 600, 1,000, 1,500, or 2,000 watts, 120 volts AC. Because incandescent dimmers are rated by wattage, it is easy to select lamp loads that will not exceed the dimmer rating. Specialized dimmer systems are also available. Incandescent dimmers should not be used to control receptacles, appliances, fluorescent lamps, or fluorescent ballasts.

The wiring procedure for dimmer switches is similar to the method used for standard switches. However, it is important to give special attention to the connections and the grounding. Some dimmer switches are solid-state devices, and their electronic circuitry can be damaged if connected improperly.

3.6.0 DISCONNECTS

NEC Article 100 defines a disconnect as a device, group of devices, or other means by which the conductors of a circuit can be disconnected from their source of supply. All disconnect devices (switches or circuit breakers) for load devices and circuits must be clearly and

ELECTRICAL — TRAINEE TASK MODULE 26111

permanently marked to show the purpose of the disconnect. This is a must according to the NEC. Under OSHA, this rule applies to all existing electrical systems, no matter how old, and also to new, modernized, expanded, or altered electrical systems. **NEC Section 110-22** covers the identification of **disconnecting means**.

CAUTION: Effectively-identified disconnect devices are critically important to safety when a switch or circuit breaker has to be opened to quickly deenergize a circuit.

3.6.1 SAFETY SWITCHES

In the early 1900s, an open knife switch was commonly used for manual switching and disconnecting. It consisted of a stationary jaw and a movable blade pivoting in a hinge post. These early devices carried current adequately, but were insufficient from a safety standpoint. The load-make and load-break manual switching relied on an operator to open or close the blade quickly, and there were also exposed live parts.

Because of the danger of exposed live parts, the switches were eventually mounted in service boxes to increase operator safety. In the early 1920s, an externally-operated handle was added to make the switch safe. A quick-make, quick-break mechanism was connected to the handle and installed inside the enclosure in the 1930s. This made opening and closing the breaker contacts independent of the operator's hand speed. These devices became known as *safety switches*.

Further safety features have been added over the years:

* Arc shields for electrical protection and maximum horsepower ratings
* Interiors which could be removed easily for wire pulling
* Coilproof dual-defeatable interlocks which increase safety and ease of inspection by authorized personnel
* Bold on/off and handle markings which clearly indicate the contact position from a distance
* Faceplates which cover the front of the switch and protect exposed wires

Figure 12 shows a safety switch that is typically used in commercial and industrial applications.

Safety switches are individually enclosed, manually-operated disconnects that must be installed and operated in accordance with the NEC. Safety switches are voltage-rated and must be used with the appropriate voltage level. Unless otherwise noted, all safety switches are listed under UL 98 (Enclosed Switches). Safety switches must provide several features that are required by the NEC, including the following:

* Protection from live parts by a metallic enclosure that completely surrounds the disconnecting means
* Highly visible indication of the contact position
* Provisions for padlocking the external, manually-operated handle in the OFF position

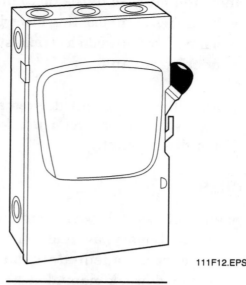

Figure 12. Safety Switch

4.0.0 NEMA CLASSIFICATIONS

The National Electrical Manufacturers' Association (NEMA) is a nonprofit organization supported by the manufacturers of electrical equipment and supplies. NEMA develops standards that are used when purchasing equipment. The classifications help to match the product to the application.

4.1.0 SWITCH CLASSIFICATIONS

There are two NEMA classifications for enclosed switches: general-duty switches and heavy-duty, motor-rated switches.

4.1.1 General-Duty Switches

A general-duty switch is designed for use in residential and commercial applications. It is used where the service factor is not high (e.g., in lighting, air conditioning, and appliance loads). There are several kinds of general-duty, single-throw switches. They range from a simple two-wire switch to a solid neutral switch, with one blade and one fuse, which is rated for 120 volts AC. The other end of the range includes four-wire switches, with solid neutral, three-blade, three-fuse construction, which are rated for 240 volts AC. General-duty two- and three-pole switches are designed for industrial use where the service demands are not high. They are rated from 30 to 1,200 amps, up to 600 volts AC or DC, and are available in fusible and nonfusible types. The safety switch shown in *Figure 12* is a general-duty switch.

4.1.2 Heavy-Duty, Motor-Rated Switches

Heavy-duty, motor-rated switches are designed to be used in production industries. Heavy-duty switches are rated from 30 to 1,200 amps, up to 600 volts AC or DC. They are available in both fusible and nonfusible types.

When a heavy-duty switch is in the OFF position, the blades are visible for safety. Most of these switches also feature quick-make, quick-break operating mechanisms and a full-cover interlock. The quick-make, quick-break operation prevents an operator from closing or opening the switch too slowly. Slow operation of the switch can cause the blades and clips to arc when a load is connected. Arcing is more pronounced on larger switches. In time, the blades and clips burn away. *Figure 13* shows a heavy-duty switch.

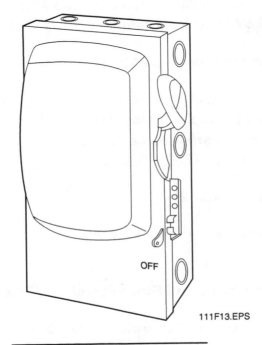

OFF

111F13.EPS

Figure 13. Heavy-Duty Switch

When the switch is closed, the door is locked. The door can be opened only when the switch handle is in the open position. In addition, the switch cannot be closed with the door open. However, for maintenance purposes, the interlock device can be tripped so that the switch can be closed with the door open. The switch handle can be locked with a padlock to prevent unauthorized operation. When you need to work on the circuit or the switch controls, pull the handle to open the circuit. Next, open the switch door and remove the fuses. Padlock the handle in the open position; that way, the switch cannot be closed by anyone while you are working. Tie a safety tag with your name onto the padlock. When you are finished working on the circuit, replace the fuses, and remove your lock and tag. After verifying that it is safe to close the switch, do so.

> **WARNING!**   Make sure to always follow your site-specific lockout/tagout procedure.

The manufacturer's list shows the switch's current rating, voltage rating, and the types of enclosures available. If the switch is intended for interrupting a motor circuit, the switch must also be horsepower-rated.

4.2.0 ENCLOSURE CLASSIFICATIONS

NEMA also classifies the enclosures designed for specific conditions. When choosing devices for use in hazardous locations, the proper NEMA-type enclosure must be specified. The NEMA numbers are not marked on the enclosures, but are included in the manufacturers' catalogs. *NEC Table 430-91* provides a complete list of enclosures.

4.2.1 General Purpose Indoor (Type 1)

The Type 1 general purpose indoor enclosure is made of sheet metal. This enclosure is primarily intended to protect people against accidental contact with equipment. It is not dust-tight. This enclosure often has pre-punched knockouts that are easily removed for connection to a raceway. *Figure 13* shows a NEMA Type 1 enclosure.

4.2.2 Drip-Proof Indoor (Type 2)

The Type 2 general purpose indoor enclosure is similar to Type 1, but it also has drip shields or their equivalent.

4.2.3 Dust-Tight, Rain-Tight, And Sleet-Resistant (Type 3)

The Type 3 enclosure is intended for outdoor use to protect against windblown dust and water.

4.2.4 Rainproof And Sleet-Resistant (Type 3R)

The Type 3R enclosure (*Figure 14*) is intended for outdoor use. This enclosure meets the requirements of UL 508. This enclosure often contains knockouts below the level of the lowest live component.

4.2.5 Dust-Tight, Rain-Tight, And Sleet-Proof (Type 3S)

Similar to Type 3, the Type 3S enclosure is intended for use outdoors to protect equipment from windblown dust and water. However, the enclosure can operate even when covered by external snow and sleet.

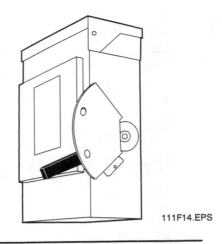

Figure 14. NEMA Type 3R Enclosure

Note: Types 3, 3R, and 3S are not considered watertight but are intended to keep water at a safe distance from live internal parts. These enclosures may have weep holes built in to allow any water that may enter to drain quickly.

4.2.6   Watertight And Dust-Tight (Type 4)

The Type 4 enclosure is made of rigid cast metal or sheet metal. It is designed to withstand a standard hose-stream test.

4.2.7   Submersible, Watertight, And Dust-Tight (Type 6)

The Type 6 enclosure can be submerged in water. It is also resistant to dust and sleet (ice), and can be installed either indoors or outdoors.

4.2.8   Hazardous Locations, Class I (Type 7)

The Type 7 enclosure is used for air-break devices. It meets the NEC requirements for Class I, Groups A, B, C, or D hazardous locations. These locations may have flammable gases or vapors present. See *Figure 15* for an example of a Type 7 enclosure.

4.2.9   Hazardous Locations, Class I (Type 8)

A Type 8 enclosure is similar to the Type 7 enclosure. However, the Type 8 enclosure is designed for oil-immersed equipment.

4.2.10   Hazardous Locations, Class II (Type 9)

The Type 9 enclosure meets the NEC requirements for Class II, Groups E, F, and G hazardous locations. These locations have combustible dust in the air.

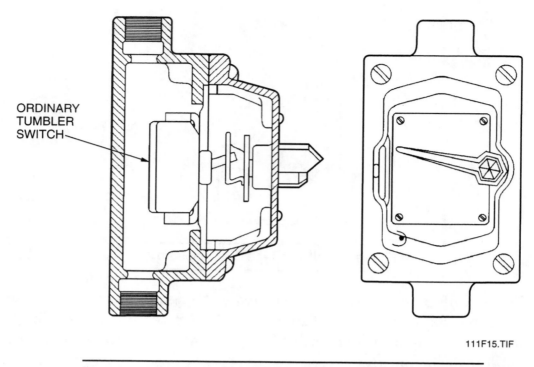

ORDINARY
TUMBLER
SWITCH

111F15.TIF

Figure 15. Snap Switch In Explosion-Proof (Type 7) Enclosure

4.2.11 Bureau Of Mines (Type 10)

The Type 10 enclosure meets the requirements of the U.S. Bureau of Mines.

4.2.12 Corrosion-Resistant, Drip-Proof Indoor (Type 11)

The Type 11 enclosure is designed for oil-immersed equipment and is suitable for use in locations subject to corrosive acid or fumes.

4.2.13 Dust-Tight And Drip-Tight Industrial Use (Type 12)

The Type 12 enclosure keeps out dust, lint, fibers, and other airborne particles. A gasket fits between the case and the cover. There are no holes in the enclosure and no conduit knockouts or conduit openings.

4.2.14 Oil-Tight And Dust-Tight (Type 13)

The Type 13 enclosure is used indoors mainly to house **pilot devices**, including limit switches, foot switches, and pushbuttons. This enclosure protects the devices from lint, dust, and seepage.

Note: Hazardous locations are described in ***NEC Article 500***.

4.2.15 Ingress Protection (IP) Classifications

The Ingress Protection (IP) Classification System uses a two-digit number to classify the degree of protection provided by an enclosure (*Table 1*). The first digit (0 through 6) represents the degree of protection against solid objects, while the second digit (0 through 8) represents the degree of protection against water.

Note: The IP Classification System does not indicate corrosion resistance.

FIRST NUMBER Degree of protection against solid objects	SECOND NUMBER Degree of protection against water
0. Not protected.	0. Not protected.
1. Protected against a solid object greater than 50mm, such as a hand.	1. Protected against water dripping vertically, such as condensation.
2. Protected against a solid object greater than 12mm, such as a finger.	2. Protected against dripping water when tilted up to 15°.
3. Protected against a solid object greater than 2.5mm, such as a wire or tool.	3. Protected against water when spraying at an angle of up to 60°.
4. Protected against a solid object greater than 1.0mm, such as wire or thin strips of metal.	4. Protected against water splashing from any direction.
5. Dust-protected. Prevents ingress of dust sufficient to cause harm.	5. Protected against jets of water from any direction.
6. Dust-tight. No dust ingress.	6. Protected against heavy seas or powerful jets of water. Prevents ingress sufficient to cause harm.
	7. Protected against harmful ingress of water when immersed between a depth of 150mm to 1m.
	8. Protected against submersion. Suitable for continuous immersion in water.

Table 1. IP Classification System

5.0.0   RECEPTACLES

A **receptacle** is a contact device which is installed at an **outlet** so that a single **attachment plug** can be connected. According to *NEC Article 100,* a **receptacle outlet** is an outlet where one or more receptacles are installed. Receptacles are rated according to their voltage and amperage capacity. This rating determines the number and configuration of the contacts. *Figure 16* shows a portion of the NEMA configurations for general purpose nonlocking plugs and receptacles, as well as other common receptacles. In most cases, a distinction is made between the system grounded conductor (W) and the grounding conductor (G). You will note that plugs are often **polarized** (one blade is wider than the other). This prevents the user from plugging the device in the wrong way.

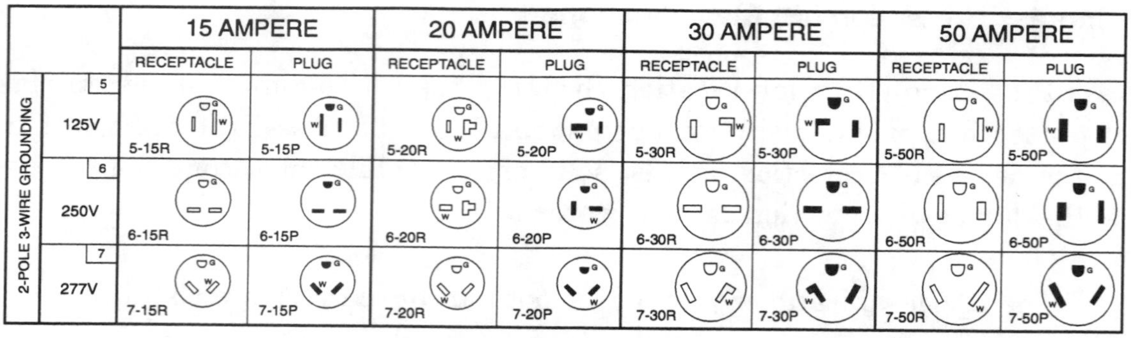

		15 AMPERE		20 AMPERE		30 AMPERE		50 AMPERE	
		RECEPTACLE	PLUG	RECEPTACLE	PLUG	RECEPTACLE	PLUG	RECEPTACLE	PLUG
2-POLE 3-WIRE GROUNDING	5 125V	5-15R	5-15P	5-20R	5-20P	5-30R	5-30P	5-50R	5-50P
	6 250V	6-15R	6-15P	6-20R	6-20P	6-30R	6-30P	6-50R	6-50P
	7 277V	7-15R	7-15P	7-20R	7-20P	7-30R	7-30P	7-50R	7-50P

111F16.EPS

Figure 16. Typical NEMA Plug Configurations

Grounded conductors, which are labeled W, may carry current in normal operation. **NEC Article 100** defines a grounded conductor as a system or circuit conductor that is intentionally grounded. Grounded conductors are commonly referred to as *neutrals*.

Grounding conductors (equipment grounds) are labeled G. They carry current only when part of the equipment, which was never intended to be energized, becomes energized through a fault. These are commonly referred to as *grounding conductors*. **NEC Article 100** defines a grounding conductor as a conductor used to connect equipment or the grounded circuit of a wiring system to a grounding electrode or electrodes.

NEC Section 210-7 states that receptacles installed on 15A and 20A **branch circuits** must be of the grounding type. *Figure 17* shows a grounding-type duplex receptacle. On all new installations, a grounding-type device must be installed.

The receptacle current rating must be at least equal to that of the conductors feeding it. If there is more than one receptacle on the circuit, the total load plugged into these receptacles cannot exceed the rating of the conductors. **NEC Tables 210-21(b)(2) and (3)** list these requirements.

Note: **NEC Section 210-21** allows the installation of 15A receptacles on a 20A circuit if the circuit feeds more than one receptacle.

5.1.0   IDENTIFYING RECEPTACLES

Receptacles, like switches, have various symbols and information printed on them that help to determine their proper use and ratings (*Figure 18*). In addition, the shapes and positions of the openings also determine the use of the receptacle.

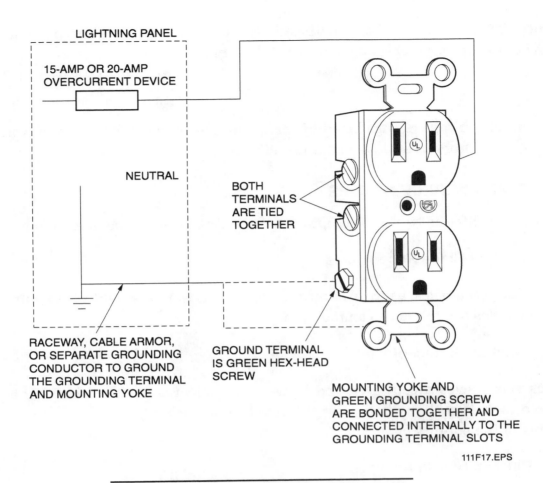

LIGHTNING PANEL

15-AMP OR 20-AMP
OVERCURRENT DEVICE

NEUTRAL

BOTH
TERMINALS
ARE TIED
TOGETHER

RACEWAY, CABLE ARMOR,
OR SEPARATE GROUNDING
CONDUCTOR TO GROUND
THE GROUNDING TERMINAL
AND MOUNTING YOKE

GROUND TERMINAL
IS GREEN HEX-HEAD
SCREW

MOUNTING YOKE AND
GREEN GROUNDING SCREW
ARE BONDED TOGETHER AND
CONNECTED INTERNALLY TO THE
GROUNDING TERMINAL SLOTS

111F17.EPS

Figure 17. Grounding-Type Duplex Receptacle

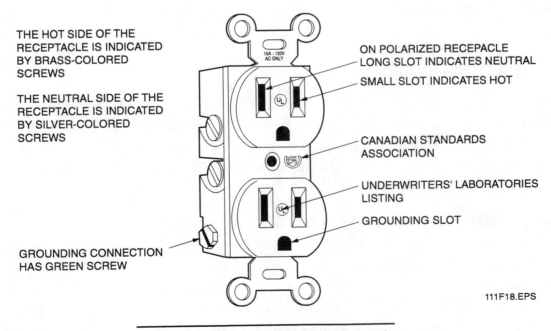

THE HOT SIDE OF THE
RECEPTACLE IS INDICATED
BY BRASS-COLORED
SCREWS

THE NEUTRAL SIDE OF THE
RECEPTACLE IS INDICATED
BY SILVER-COLORED
SCREWS

15A - 120V
AC ONLY

ON POLARIZED RECEPACLE
LONG SLOT INDICATES NEUTRAL

SMALL SLOT INDICATES HOT

CANADIAN STANDARDS
ASSOCIATION

UNDERWRITERS' LABORATORIES
LISTING

GROUNDING SLOT

GROUNDING CONNECTION
HAS GREEN SCREW

111F18.EPS

Figure 18. Standard Duplex Receptacle

The printed information on the receptacle is similar to the information printed on switches. The information on receptacles includes the following:

- The UL label
- The CSA label
- Current and voltage ratings listed by maximum amperage and maximum voltage (may be shown on the front or back, depending on the switch)

5.2.0 TYPES OF RECEPTACLES

There are many different types of receptacles. This section covers the most common types.

5.2.1 Straight Blade Receptacles

Straight blade receptacles, as their name implies, will accept a straight blade connector or plug. These receptacles are the most common type.

5.2.2 Twist Lock Receptacles

Twist lock receptacles accept a curved blade connector or plug. The plug/connector and the receptacle lock together with a slight twist. The locking action prevents accidental unplugging of the equipment.

5.2.3 Pin And Sleeve Receptacles

Pin and sleeve devices have a unique locking feature. These receptacles are made with an extremely heavy-duty plastic housing that makes them virtually indestructible. They come with long brass pins for long life and are color coded according to voltage for easy identification.

5.3.0 RECEPTACLE TERMINALS

The terminals on a receptacle are color coded. The green terminal is the equipment ground connection and is associated with the U-shaped slot. The silver-colored terminal is associated with the neutral (white) conductor and the larger of the two vertical slots on the receptacle. The brass-colored terminal is associated with the hot (black) conductor and the smaller vertical slot on the receptacle.

Similar to switches, receptacles have either screw terminals for attaching wires or quick-wire terminals located on the back. Although UL-approved, the push-in quick-wire terminals on receptacles and switches do not provide as good a connection as a wire that is properly terminated on the screw terminal. Use quick-wire terminals only for lightly loaded circuits. Receptacle terminals must be listed for the type of wire that is used in the system. If copper or copper-clad wire is used, all types of terminals are acceptable. If aluminum wire is used, the terminal must be rated for aluminum.

5.4.0 GROUND FAULT CIRCUIT INTERRUPTERS

The NEC defines a ground fault circuit interrupter (GFCI) as a protective device that functions to deenergize a circuit or portion thereof within an established period of time when a current to ground exceeds some predetermined value that is less than that required to operate the overcurrent protective device of the supply circuit. ***NEC Section 210-8*** defines locations for GFCIs.

Electricity normally flows into a tool or appliance through an energized wire and returns through a neutral wire. A GFCI compares the amount of current going out of the hot wire to the amount of current returning on the neutral wire. Any current that goes out of the hot wire but does not return on the neutral wire must be taking an unintended path or ground fault. Ground faults may be caused by worn or damp insulation or faulty wiring. If someone is in that current path, that person can complete the circuit necessary for the flow of current. Shock, injury, or electrocution can result, depending on the age and health of the person involved, how well grounded the person is, and the path the current takes through the person's body.

When a ground fault leak of some predetermined value (usually 5mA) is detected, the sensitive electronic circuitry within the GFCI will automatically trip a switch. This interrupts the circuit and cuts off the power. The interruption takes place within 1/40th of a second or less and prevents possible injury. *Figure 19* shows a typical GFCI. GFCIs are available as circuit breakers, receptacles, or on extension cords and adapters that plug into standard receptacles.

111F19.EPS

Figure 19. Typical GFCI

Note: When installed properly, one GFCI receptacle can protect all of the receptacles on the same circuit, as shown in *Figure 20*.

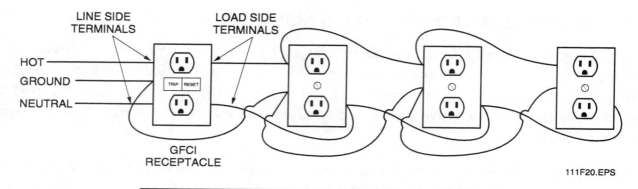

LINE SIDE TERMINALS

LOAD SIDE TERMINALS

HOT

GROUND

NEUTRAL

TRIP RESET

GFCI RECEPTACLE

111F20.EPS

Figure 20. GFCI Protecting Several Standard Receptacles

NEC Section 305-6 covers the rules that concern GFCI protection for all 15A, 20A, and 30A receptacle outlets on construction sites. The NEC states that ground fault circuit interrupters (either GFCI circuit breakers or GFCI receptacles) must be used to provide personnel protection for all receptacles of the designated rating that are in use by employees and are not part of the permanent wiring of the building or structure.

6.0.0 WIRING TECHNIQUES

When you terminate conductors at devices, there is one simple rule to follow: use the proper procedure and the correct tool.

6.1.0 BENDING WIRE

Suppose you need to wrap the stripped end of a conductor around a screw lug. If you used a pair of needle-nose pliers to grip the conductor, you would probably scar the conductor and weaken it in the process. A conductor can overheat if it is scarred or scratched at its point of termination. Overheating creates a very real fire hazard.

Instead, the right tool to use for bending wire is the wire stripper. Many wire strippers have a hole in each of the jaws. When the stripped end of the conductor is inserted into one of these holes, you can safely bend the conductor into the proper shape for terminating it. Care must be taken to wrap the U-shaped conductor around the screw with the short end of the U wrapped around the screw in a clockwise direction. Before the screw is tightened, the stripped end of the conductor should be bent against the screw. To do this, use the screw shank with the wire stripper and press the end of the conductor against the screw. Then, use the jaws of the wire stripper to bend the conductor, as shown in *Figure 21.*

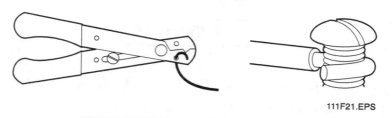

111F21.EPS

Figure 21. Bending The Conductor

ELECTRICAL — TRAINEE TASK MODULE 26111

After the conductor is bent, the screw can be tightened to make a good electrical connection. The wire is actually twisted around the screw as it is tightened. For this reason, the wire must be wrapped around the screw in a clockwise direction.

CAUTION: The torque applied to the terminal screw must be just right. Too much torque can strip the terminal. Too little torque can cause the termination to come apart.

Note: Certain receptacles are specially-rated and designed for stranded wire.

6.2.0   BRANCH CIRCUIT WIRING

According to *NEC Article 100*, a branch circuit is the circuit conductors between the final overcurrent device protecting the circuit and the outlet(s). That means that all outlets are supplied by branch circuits.

6.2.1   Planning

The first step in branch circuit wiring is to check the drawings to determine the exact location of all outlet boxes for general purpose receptacles, switches, lighting fixtures, and special purpose receptacles. The NEC requirements for the number of circuits, protective measures, and special circuits are determined at this time.

Note: The drawings may not show the actual conductor runs or homeruns for the wiring installation. Prior to planning the branch circuit wiring for an electrical installation, you should review the applicable articles of the NEC.

6.2.2   Installation

After you have studied the electrical drawings and determined the general layout of the electrical system, the general rough-in stage begins. The rough-in stage involves the following:

- Installing outlet boxes and the service-entrance panel
- Establishing the conductor pathway
- Pulling or installing the cable or raceway

Roughing-in and the following step (installing switches, receptacles, etc.) require that you have an overall knowledge of both tasks. For example, it is difficult to install the proper conductors for a four-way switch in a particular location without knowing the working characteristics of the switch or the various configurations in which the switch can be used.

Note: Before you begin the roughing-in stage of the wiring, review the applicable articles of the NEC.

Before any actual installation begins, study the electrical floor plan and consult with the builder or architect to ensure that no last-minute changes have been made in the electrical system. At this time, you should select the type of boxes that will be used at particular locations. The sizes of the required boxes must be determined in accordance with *NEC Article 370*. Pay particular attention to *NEC Section 370-16(a) and (b)*.

The size of the box is rated according to its cubic-inch capacity (volume). Although the NEC states what practices must be followed, the NEC does not explain why such a practice is required except that all provisions are designed to promote safety and prevent damage. Minimum volume capacities are required for different reasons. For example, too many wires too close together can cause overheating. In addition, adequate volume allows for greater air circulation and reduces crowding, which could cause damage to the insulation or connections of the wires.

WARNING! Strict observance of the NEC is mandatory and all required calculations must be taken very seriously.

According to the NEC, an outlet box or junction box has to be installed at every point in the electrical wiring system where the wiring cable is going to be spliced or terminated. Each of these points must also be accessible for future revisions or repairs. Failure to make all boxes accessible is a violation of the NEC.

Many industrial/commercial outlet boxes are surface-mounted and connect directly to the electrical raceway (*Figure 22*). These boxes are often mounted to the building steel or to a support channel or strut which is welded or clamped to the steel. The installation of outlet boxes or raceways should be coordinated so as not to interfere with piping, ductwork, fireproofing, and other items which may or may not already be in place.

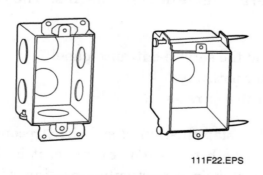

111F22.EPS

Figure 22. Outlet Boxes

Commercial wiring may involve flush wiring similar to residential work. *NEC Sections 370-20 and 370-21* require that recessed switch, receptacle, and fixture boxes be mounted flush with the finished face of the wall surface when combustible material, such as plywood or paneling, is being used. If noncombustible material is used as the finish surface, the box may be recessed by as much as ¼ inch. The distance that the box is extended past the wall stud or building member is referred to as the *mounting depth* or *setout*. Many boxes have several setout measurements inscribed on the box (*Figure 23*).

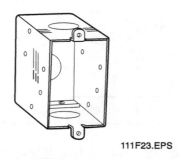

111F23.EPS

Figure 23. Box With Setout Measurements Inscribed

These marks provide a quick and easy method of measuring the setout required for a particular wall finish. This saves time and effort when installing the box.

Note: Front-mounting bracket boxes will have the setout determined by the bracket assembly.

Install all boxes in accordance with the electrical drawings. The spacing should be as even as possible. The box center is the midpoint on the vertical dimension of the box. Make sure that you check the door swing direction so that the switches are not installed behind a door. Measure the height of the switch boxes from the floor so that they will be at the proper height when installed.

6.2.3   Splicing

Wires are spliced when they are joined or connected. Splices are required when installing switches, lighting, and other electrical equipment. There are three major steps for making a splice: removing the insulation, making the splice, and insulating the splice.

Wire strippers should be used to remove the insulation. Knives should not be used because of the possibility of nicking the wire. Use the following procedure to strip wire using a wire stripper:

Step 1   Insert the wire into the proper slot on the side of the stripper marked for solid wire.

Step 2   Twist the stripper back and forth a few times to cut through the insulation.

Step 3   Pull the stripper back to pull the insulation off.

Most industrial and commercial lighting and receptacle circuits are spliced using solderless connectors. Solderless connectors, commonly referred to as *Wirenuts®*, are plastic caps that contain a tapered and threaded metal insert, as shown in *Figure 24*.

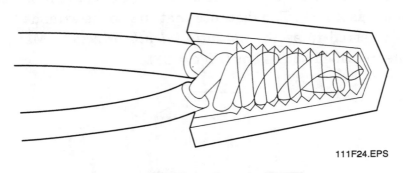

111F24.EPS

Figure 24. Solderless Connector

When using a solderless connector, screw the wire nut over the twisted ends to hold the conductors together. (Some manufacturers recommend first twisting the conductors together to ensure good electrical contact.) Solderless connectors are rated for size, temperature, and number of conductors. Make sure you use a properly-rated connector, and always follow the manufacturer's instructions when using any connector.

Most industrial and commercial control circuits and signal wires (for devices such as transmitters, motor stop/start circuits, etc.) are spliced on fixed terminal blocks with crimp-on lugs. Examples of lugs are shown in *Figure 25*. Lugs help to keep the conductors organized and greatly aid in troubleshooting. When installing lugs, make sure your crimping tool is designed for the particular lugs you are using.

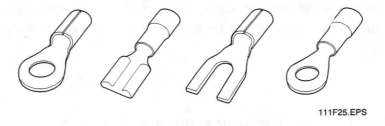

111F25.EPS

Figure 25. Crimp-On Wire Lugs

7.0.0 CONTROL DEVICES

One of the major differences between industrial/commercial wiring and residential wiring is the complexity of controls for electrical machinery. Whereas most loads in a home are turned on manually using a wall switch, industrial machinery is controlled by a variety of devices, including relays and limit switches.

7.1.0 RELAYS

A **relay** acts as a remote-controlled switch that opens and closes to supply power to a load such as a motor, heater, or light. The electromagnetic coil of a relay acts as a load for the control circuit. This allows switches and other controlling devices to control a large load while only carrying a very small current. For this reason, these controlling devices are often referred to as *pilot devices*. The pilot devices work together in a control circuit to ultimately close a relay that has contacts that are rated for the full load current. These relay contacts are what connects the load to its source voltage. *Figure 26* shows a typical relay along with its internal electrical arrangement.

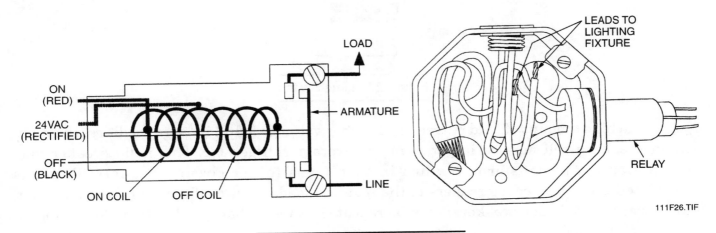

Figure 26. Typical Plug-In Relay

The pilot devices that control a relay may be activated by a variety of methods, such as air or hydraulic pressure for pressure switches, heat for temperature switches, or light for photoelectric switches, to name a few. One of the most popular industrial pilot devices is the mechanical **limit switch**.

7.2.0 LIMIT SWITCHES

Limit switches use mechanical movement to open or close electrical contacts. Limit switches are frequently used on manufacturing equipment, conveyor belts, automatic doors, and many other devices that have repeated motion or travel limits. A typical limit switch is shown in *Figure 27*. Limit switches can be fitted with numerous types of mechanical arms to suit specific equipment configurations.

8.0.0 POWER DISTRIBUTION EQUIPMENT

In residential wiring, power distribution equipment is usually limited to a breaker panel or fuse panel. Modern panels contain the main and branch circuit breakers in one relatively small enclosure. Industrial and commercial wiring calls for distribution equipment that is usually much larger.

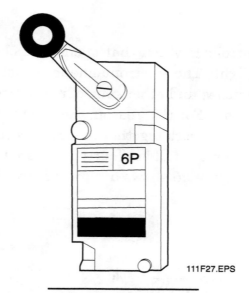

Figure 27. Limit Switch

Due to the high amount of load current and higher voltages found in large facilities, the main circuit breaker itself may be larger than the average refrigerator. This main breaker may feed a step-down transformer which in turn feeds feeder circuit breakers. Power is distributed from the feeder breakers to the required loads. This configuration (main breaker, transformer, and feeder breakers) may be repeated at more than one location, depending on the size of the facility.

Distribution equipment, commonly referred to as **switchgear,** may be comprised of individually-enclosed and mounted devices that are wired together, or it may all be included in one enclosure, as shown in *Figure 28.* Despite the many configurations of switchgear available, the purpose is always the same: to provide a safe system of controlling a large power source and supplying this power to the remote loads where it is needed.

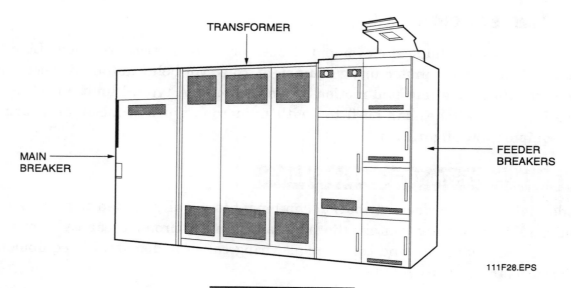

Figure 28. Switchgear

SUMMARY

Wiring and installing wiring devices will make up a large portion of your job as an electrician. Knowing how to relate the rules and regulations in the NEC to your job will improve your skills and ensure that the devices are installed properly.

References

For advanced study of topics covered in this task module, the following book is suggested:

National Electrical Code Handbook, Latest Edition, National Fire Protection Association, Quincy, MA.

REVIEW QUESTIONS

1. The designation for a double-pole, single-throw switch is _____.
   a. S1
   b. S2
   c. S3
   d. S4

2. A three-way switch is a type of _____ switch.
   a. SPST
   b. DPDT
   c. DPST
   d. SPDT

3. Older switches marked CU/AL should be used for _____.
   a. copper wire only
   b. aluminum wire only
   c. either copper or aluminum wire
   d. aluminum-clad copper wire only

4. Four-way switches can be used to control lights from more than _____ location(s).
   a. one
   b. two
   c. three
   d. four

5. Dimmer switches are normally rated at _____ watts.
   a. 600
   b. 1,000
   c. 1,500
   d. all of the above

6. Which of the following is *not* true regarding modern safety switches?
   a. They include arc shields.
   b. They have removable interiors.
   c. They have coilproof dual-defeatable interlocks.
   d. They are called *open knife switches.*

7. NEMA Type 1 enclosures _____.
   a. can be used in all locations
   b. can only be used indoors
   c. are dust-tight
   d. both b and c

8. For a Class II, Group E hazardous location, you would use a _____ enclosure.
   a. Type 7
   b. Type 8
   c. Type 9
   d. Type 10

9. An IP Classification of 6-8 means that _____.
   a. the enclosure offers the highest degree of protection
   b. the enclosure is dust-tight
   c. the enclosure is suitable for continuous immersion
   d. all of the above

10. Outlet boxes are covered extensively in _____.
   a. *NEC Article 500*
   b. *NEC Article 310*
   c. *NEC Article 370*
   d. *NEC Article 422*

ANSWERS TO REVIEW QUESTIONS

Answers		**Section**
1.	b	2.1.0
2.	d	2.1.0
3.	a	2.2.0
4.	b	3.4.0
5.	d	3.5.0
6.	d	3.6.1
7.	b	4.2.1
8.	c	4.2.10
9.	d	4.2.15
10.	c	6.2.2

The NCCER makes every effort to keep these manuals up-to-date and free of technical errors. We appreciate your help in this process. If you have an idea for improving this manual, or if you find an error, a typographical mistake, or an inaccuracy in the NCCER's Craft Training Manuals, please write us, using this form or a photocopy. Be sure to include the exact module number, page number, a description of the problem, and the correction, if possible. Your input will be brought to the attention of the Technical Review Committee. Thank you for your assistance.

Instructors – If you found that additional materials were necessary in order to teach this module effectively, please let us know so that we may include them in the Equipment/Materials list in the Instructor's Guide.

Write: Curriculum Development and Revision Department
National Center for Construction Education and Research
P.O. Box 141104
Gainesville, FL 32614-1104
Fax: 352-334-0932

Craft _____ Module Name _____

Copyright Date _____ Module Number _____ Page Number(s) _____

Description of Problem _____

_____

_____

_____

(Optional) Correction of Problem _____

_____

_____

(Optional) Your Name and Address _____

_____

_____

_____

notes

Wiring: Residential

Module 26112

Electrical Trainee Task Module 26112

NATIONAL
CENTER FOR
CONSTRUCTION
EDUCATION AND
RESEARCH

WIRING: RESIDENTIAL

OBJECTIVES

Upon completion of this module, the trainee will be able to:

1. Describe how to determine electric service requirements for dwellings.
2. Explain the grounding requirements of a residential electric service.
3. Calculate and select service-entrance equipment.
4. Select the proper wiring methods for various types of residences.
5. Explain the role of the NEC in residential wiring.
6. Compute branch circuit loads and explain their installation requirements.
7. Explain the types and purposes of equipment grounding conductors.
8. Explain the purpose of ground fault circuit interrupters and tell where they must be installed.
9. Size outlet boxes and select the proper type for different wiring methods.
10. Describe rules for installing electric space heating and HVAC equipment.
11. Describe the installation rules for electrical systems around swimming pools, spas, and hot tubs.
12. Explain how wiring devices are selected and installed.
13. Describe the installation and control of lighting fixtures.

Prerequisites

Successful completion of the following Task Modules is required before beginning study of this Task Module: Core Curricula; Electrical Level 1, Task Modules 26101 through 26111.

Required Trainee Materials

1. Trainee Task Module
2. Copy of the latest edition of the *National Electrical Code*
3. Appropriate Personal Protective Equipment

Note: The designations "National Electrical Code," "NE Code," and "NEC," where used in this document, refer to the *National Electrical Code®*, which is a registered trademark of the National Fire Protection Association, Quincy, MA. *All National Electrical Code (NEC) references in this module refer to the 1999 edition of the NEC.*

Course Map

This course map shows all of the task modules in the first level of the Electrical curricula. The suggested training order begins at the bottom and proceeds up. Skill levels increase as a trainee advances on the course map. The training order may be adjusted by the local Training Program Sponsor.

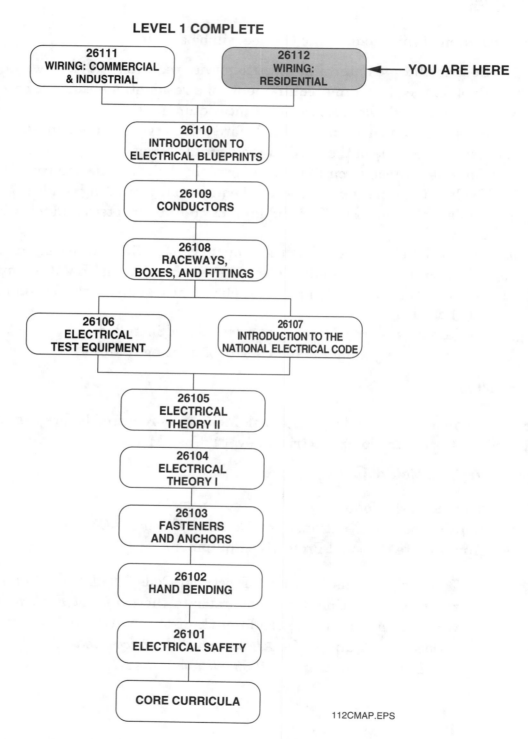

LEVEL 1 COMPLETE

26111
WIRING: COMMERCIAL & INDUSTRIAL

26112
WIRING: RESIDENTIAL ← YOU ARE HERE

26110
INTRODUCTION TO ELECTRICAL BLUEPRINTS

26109
CONDUCTORS

26108
RACEWAYS, BOXES, AND FITTINGS

26106
ELECTRICAL TEST EQUIPMENT

26107
INTRODUCTION TO THE NATIONAL ELECTRICAL CODE

26105
ELECTRICAL THEORY II

26104
ELECTRICAL THEORY I

26103
FASTENERS AND ANCHORS

26102
HAND BENDING

26101
ELECTRICAL SAFETY

CORE CURRICULA

112CMAP.EPS

TABLE OF CONTENTS

TABLE OF CONTENTS (Continued)

Trade Terms Introduced In This Module

Appliance: Equipment designed for a particular purpose (i.e., using electricity to produce heat, light, mechanical motion, etc.). Appliances are usually self-contained, are generally available for applications other than industrial use, and are normally produced in standard sizes or types.

Armored (Type AC) cable: Cable that consists of wires with a spiral-wound, flexible steel outer jacketing. See *BX®*.

Bonding bushing: A special conduit bushing equipped with a conductor terminal to take a bonding jumper. It also has a screw or other sharp device to bite into the enclosure wall to bond the conduit to the enclosure without a jumper when there are no concentric knockouts left in the wall of the enclosure.

Bonding jumper: A bare or green insulated conductor used to ensure the required electrical conductivity between metal parts required to be electrically connected. Bonding jumpers are frequently used from a bonding bushing to the service-equipment enclosure to provide a path around concentric knockouts in an enclosure wall, and they may also be used to bond one raceway to another.

Branch circuit: The portion of a wiring system extending beyond the final overcurrent device protecting a circuit.

BX®: A name for armored cable; although used generically, BX® is a registered trademark of the General Electric Company.

Feeder: A circuit, such as conductors in conduit or a cable run, that carries current from the service equipment to a subpanel or a branch circuit panel or to some point in the wiring system.

Load center: A type of panelboard that is normally located at the service entrance of a residential installation. It usually contains the main disconnect.

Nonmetallic-sheathed (Type NM) cable: A type of cable that is popular for use in residential and small commercial wiring systems. In general, it may be used for both exposed and concealed work in normally dry locations. See *Romex®*.

Romex®: General Cable's trade name for Type NM cable; however, it is often used generically to refer to any nonmetallic-sheathed cable.

Roughing in: The first stage of an electrical installation, when the raceway, cable, wires, boxes, and other equipment are installed. This is the electrical work that must be done before any finishing work can be done.

Service drop: The overhead conductors, through which electrical service is supplied, between the last power company pole and the point of their connection to the service facilities located at the building.

Service entrance: The point where power is supplied to a building (including the equipment used for this purpose). The service entrance includes the service main switch or panelboard, metering devices, overcurrent protective devices, and conductors/raceways for connecting to the power company's conductors.

Service-entrance conductors: The conductors between the point of termination of the overhead service drop or underground service lateral and the main disconnecting device in the building.

Service-entrance equipment: Equipment that provides overcurrent protection to the feeder and service conductors, a means of disconnecting the feeders from energized service conductors, and a means of measuring the energy used.

Service lateral: The underground conductors through which service is supplied between the power company's distribution facilities and the first point of their connection to the building or area service facilities located at the building.

Switch: A mechanical device used for turning an electrical circuit on and off.

Switch leg: A circuit routed to a switch box for controlling electric lights.

1.0.0 INTRODUCTION

The use of electricity in houses began shortly after the opening of the California Electric Light Company in 1879 and Thomas Edison's Pearl Street Station in New York City in 1882. These two companies were the first to enter the business of producing and selling electric service to the public. In 1886, the Westinghouse Electric Company secured patents which resulted in the development and introduction of alternating current; this paved the way for rapid acceleration in the use of electricity.

The primary use of early home electrical systems was to provide interior lighting, but today's uses of electricity include:

- Heating and air conditioning
- Electrical **appliances**
- Interior and exterior lighting
- Communications systems
- Alarm systems

In planning any electrical system, there are certain general steps to be followed, regardless of the type of construction. In planning a residential electrical system, the electrician must take certain factors into consideration. These include:

- Wiring method
- Overhead or underground electrical service

- Type of building construction
- Type of service entrance and equipment
- Grade of wiring devices and lighting fixtures
- Selection of lighting fixtures
- Type of heating and cooling system
- Control wiring for the heating and cooling system
- Signal and alarm systems

The experienced electrician readily recognizes, within certain limits, the type of system that will be required. However, always check the local code requirements when selecting a wiring method. The NEC provides minimum requirements for the practical safeguarding of persons and property from hazards arising from the use of electricity. These minimum requirements are not necessarily efficient, convenient, or adequate for good service or future expansion of electrical use. Some local building codes require electrical installations that surpass the requirements of the NEC. For example, **NEC Section 230-51(a)** requires that service cable be secured by means of cable straps placed every 30 inches. The electrical inspection department in one area requires these cable straps to be placed at a minimum distance of 18 inches.

If more than one wiring method may be practical, a decision as to which type of system to use should be made prior to beginning the installation.

In a residential occupancy, the electrician should know that a 120/240-volt (V), single-phase **service entrance** will invariably be provided by the utility company. The electrician knows that the service and **feeders** will be three-wire, that the **branch circuits** will be either two- or three-wire, and that the safety **switches**, service equipment, and panelboards will be three-wire, solid neutral. On each project, however, the electrician must determine where the point of service-drop attachment to the building will be located and how much of the service is to be provided as part of the electrical contract, with the remaining portion being installed by the utility company.

Note: See *Appendix A* for a general reference to other codes and electrical standards that apply to residential electrical installations.

2.0.0   SIZING THE ELECTRICAL SERVICE

It may be difficult to decide at times which comes first, the layout of the outlets or the sizing of the electric service. In many cases, the service (main disconnect, panelboard, service conductors, etc.) can be sized using the NEC before the outlets are actually located. In other cases, the outlets will have to be laid out first. However, in either case, the service entrance and panelboard locations will have to be determined before the circuits can be installed—so the electrician will know in which direction (and to what points) the circuit homeruns will terminate. In this module, an actual residence will be used as a model to size the electric service according to the latest edition of the NEC.

2.1.0 FLOOR PLANS

A floor plan is a drawing that shows the length and width of a building and the rooms that it contains. A separate plan is made for each floor.

Figure 1 shows how a floor plan is developed. An imaginary cut is made through the building as shown in the view on the left. The top half of this cut is removed (bottom view), and the resulting floor plan is what the remaining structure looks like when viewed directly from above.

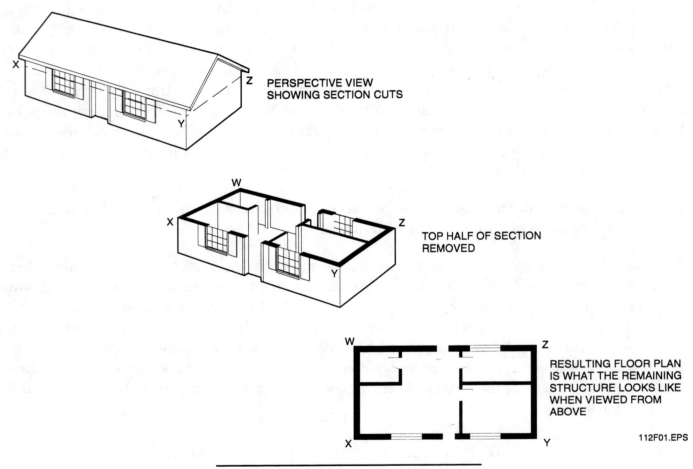

PERSPECTIVE VIEW
SHOWING SECTION CUTS

TOP HALF OF SECTION
REMOVED

RESULTING FLOOR PLAN
IS WHAT THE REMAINING
STRUCTURE LOOKS LIKE
WHEN VIEWED FROM
ABOVE

112F01.EPS

Figure 1. Principles Of Floor Plan Layout

The floor plan for a small residence is shown in *Figure 2*. This building is constructed on a concrete slab with no basement or crawl space. There is an unfinished attic above the living area and an open carport just outside the kitchen entrance. Appliances include a 12 kilovolt-ampere (kVA) electric range and a 4.5kVA water heater.

There is also a washer/dryer (rated at 5.5kVA) in the utility room. Gas heaters are installed in each room with no electrical requirements. In this module, the electrical requirements of this example building will be computed.

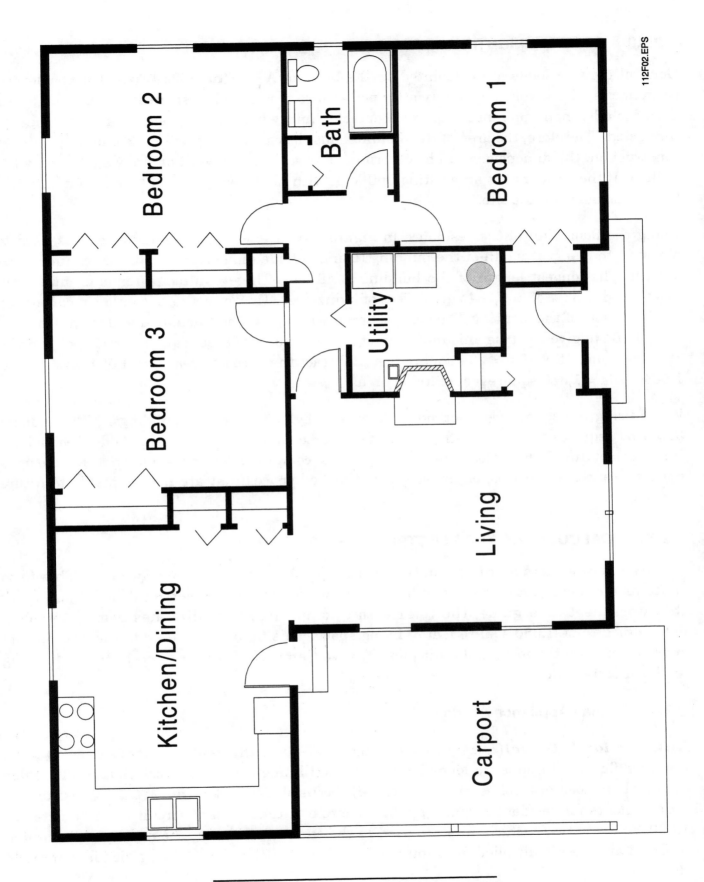

Figure 2. Floor Plan Of A Typical Residence

2.2.0 GENERAL LIGHTING LOADS

General lighting loads are calculated on the basis of **NEC Table 220-3(a)**. For residential occupancies, three volt-amperes (watts) per square foot of living space is the figure to use. This includes non-appliance duplex receptacles into which lamps, televisions, etc. may be connected. Therefore, the area of the building must be calculated first. If the building is under construction, the dimensions can be determined by scaling the working drawings used by the builder. If the residence is an existing building, with no drawings, actual measurements will have to be made on the site.

Using the floor plan of the residence in *Figure 2* as a guide, an architect's scale is used to measure the longest width of the building (using outside dimensions). It is determined to be 33 feet. The longest length of the building is 48 feet. These two measurements multiplied together give 33 x 48 = 1,584 square feet of living area. However, there is an open carport on the lower left of the drawing. This carport area will have to be calculated and then deducted from 1,584 to give the true amount of living space. This open area (carport) is 12 feet wide by 19.5 feet long. 12 x 19.5 = 234 square feet. Subtract the carport area from 1,584 square feet: 1,584 – 234 = 1,350 square feet of living area.

When using the square-foot method to determine lighting loads for buildings, **NEC Section 220-3(a)** requires the floor area for each floor to be computed from the outside dimensions. When calculating lighting loads for residences, the computed floor area must not include open porches, carports, garages, or unused or unfinished spaces that are not adaptable to future use.

2.3.0 CALCULATING THE ELECTRIC SERVICE LOAD

Figure 3 shows a standard calculation worksheet for a single-family dwelling. This form contains numbered blank spaces to be filled in while making the service calculation. Using this worksheet as a guide, the total area of our sample dwelling has been previously determined to be 1,350 square feet of living space. This figure is entered in the appropriate space (Box 1) on the form and multiplied by 3 volt-amperes (VA) for a total general lighting load of 4,050VA (Box 2).

2.3.1 Small Appliance Loads

NEC Section 210-11(c)(1) requires at least two 120V, 20A small appliance circuits to be installed for small appliance loads in the kitchen, dining area, breakfast nook, and similar areas where toasters, coffee makers, etc. will be used. **NEC Section 220-16** gives further requirements for residential small appliance circuits; that is, each circuit must be rated at 1,500VA. Since two such circuits are used in the sample residence, the number 2 is entered in Box 3 and then multiplied for a total small appliance load of 3,000VA, which is entered in Box 4.

I. GENERAL LIGHTING LOADS

TYPE OF LOAD	CALCULATION	TOTAL VA	NEC REFERENCE
Lighting load	(1) 1,350 sq. ft. x 3VA =	(2) 4,050 VA	Table 220-3(a)
Small appliance loads	(3) circuits x 1,500VA=	(4) VA	Section 220-16(a)
Laundry load	(5) circuits x 1,500VA=	(6) VA	Section 220-16(b)
Lighting, small appliance, laundry	Total VA =	(7) VA	

LOAD	CALCULATION	DEMAND FACTOR	TOTAL VA
Lighting, small appliance, laundry	First 3,000VA x	100% =	(8) 3,000VA
Lighting, small appliance, laundry	(9) Remaining VA x	35% =	(10) VA
Lighting, small appliance, laundry	Add #8 & #10 above	=	(11) VA

TOTAL CALCULATED LOAD FOR LIGHTING, SMALL APPLIANCES, & LAUNDRY CIRCUIT(S) (Enter Item #11 above in box #12)

(12)

II. LARGE APPLIANCE LOADS

TYPE OF LOAD	NAMEPLATE RATING	DEMAND FACTOR	TOTAL VA	NEC REFERENCE
Electric range	Not over 12kVA	Use 8kVA	(13) 8,000VA	Table 220-19
Clothes dryer	(14) VA	100%	(15) VA	Table 220-18
Water heater	(16) VA	100%	(17) VA	
Other appliances	(18) VA	100%	(19) VA	

TOTAL CALCULATED LOAD FOR LARGE APPLIANCES (Add items # 13, #15, #17, and #19 above). ENTER TOTAL IN BOX #20

(20)

TOTAL CALCULATED LOAD (Add boxes #12 & #20) ENTER TOTAL IN BOX #21

III. CONVERT VA TO AMPERES

(21)

$$\frac{Total\ VA\ in\ Box\ \#21}{240\ (volts)} = amperes \qquad \frac{VA}{240\ (volts)} = (22)\ \underline{\hspace{2cm}} amperes$$

IV. FEEDER CONDUCTOR SIZE—NEC SECTION 310-15(b)(6)

FEEDER CONDUCTOR SIZE		
COPPER AWG SIZE	ALUMINUM OR COPPER-CLAD AL	CONDUCTOR SERVICE SIZE
4	2	100
3	1	110
2	1/0	125
1	2/0	150
1/0	3/0	175
2/0	4/0	200

112F03.TIF

Figure 3. Calculation Worksheet For Residential Requirements

2.3.2 Laundry Circuit

NEC Section 210-11(c)(2) requires an additional 20A branch circuit to be provided for the exclusive use of the laundry area (Box 5). This circuit must not have any other outlets connected except for the laundry receptacle(s). Therefore, enter 1,500VA in Box 6 on the form.

So far, there is enough information to complete the first portion of the service calculation form:

- General lighting   4,050VA (Box 2)
- Small appliance load   3,000VA (Box 4)
- Laundry load   1,500VA (Box 6)
- Total general lighting
  and appliance loads   8,550VA (Box 7)

2.3.3 Demand Factors

All residential electrical outlets are never used at one time. There may be a rare instance where all the lighting may be on for a short time every night, but even so, all the small appliances, all burners on the electric range, water heater, furnace, dryer, washer, and the numerous receptacles throughout the house will never be used simultaneously. Knowing this, the NEC allows a diversity or *demand factor* to be used when sizing electric services. Our calculation continues as follows:

- The first 3,000VA is rated at 100%   3,000VA (Box 8)
- The remaining 5,550VA (Box 9)
  may be rated at 35% (the
  allowable demand factor).
  Therefore, 5,550 x 0.35 =   1942.5VA (Box 10)
- Net general lighting and
  small appliance load   4,942.5VA (Box 11)

Also enter this number in Box 12 on the form.

The electric range, water heater, and clothes dryer must now be considered in the service calculation. Although it was previously learned that the nameplate rating of the electric range is 12kVA, seldom will every burner be on the maximum setting at one time. Nor will the oven remain on all the time during cooking. When the oven reaches the temperature set on the oven controls, the thermostat shuts off the power until it cools down. Again, the NEC allows a diversity or demand factor to be used for this appliance.

When one electric range is installed and the nameplate rating is not over 12kVA, **NEC Table 220-19** allows a demand factor resulting in a total rating of 8kVA. Therefore, 8kVA may be used in the service calculation instead of the nameplate rating of 12kVA. The electric clothes dryer and water heater, however, must be calculated at 100% when using this method to calculate residential electric services. The total large appliance load is entered in Box 20 on the form. Here is the service calculation thus far:

- Net general lighting and
  small appliance load   4,942.5VA (Box 11) and (Box 12)
- Electric range
  (using demand factor)   8,000VA (Box 13)
- Clothes dryer   5,500VA (Box 14) and (Box 15)
- Water heater   4,500VA (Box 16) and (Box 17)
- Total large appliance load   18,000VA (Box 20)
- Total load   22,942.5VA (Box 21)

2.3.4 Required Service Size

The conventional electric service for residential use is 120/240V, three-wire, single-phase. Services are sized in amperes, and when the volt-amperes are known on single-phase services, amperes may be found by dividing the highest voltage into the total volt-amperes. For example:

$$22,942.5VA \div 240V = 95.6A \text{ (Box 22)}$$

NEC Section 230-79(c) requires a minimum of 100 amps for the service disconnecting means for a one-family dwelling. **NEC Section 230-42(b)** requires ungrounded conductors to have an ampacity of not less than the service disconnecting means specified in **NEC Section 230-79(c)**. Therefore, the minimum service size for a single-family dwelling is 100 amps.

2.4.0 DEMAND FACTORS

NEC Section 220 Part B provides the rules regarding the application of demand factors to certain types of loads. Recall that a demand factor is the maximum amount of volt-amp load expected at any given time compared to the total connected load of the circuit. The maximum demand of a feeder circuit is equal to the connected load times the demand factor. The loads to which demand factors apply can be found in the NEC as listed below:

- Lighting loads   **NEC Table 220-11**
- Receptacle loads   **NEC Table 220-13**
- Dryer loads   **NEC Table 220-18**
- Range loads   **NEC Table 220-19**
- Kitchen equipment loads   **NEC Table 220-20**
- Dwelling unit loads   **NEC Table 220-30**
- Existing dwelling unit loads   **NEC Section 220-31**
- Multi-family dwelling unit loads   **NEC Section 220-32**

2.5.0 GENERAL LIGHTING AND RECEPTACLE LOAD DEMAND FACTORS

NEC Table 220-11 provides the demand factors allowed for various types of lighting situations.

2.6.0 SMALL APPLIANCE AND LAUNDRY LOADS FOR DWELLING UNITS

Each two-wire small appliance branch circuit required by *NEC Section 210-11(c)* for small appliances supplied by 15A or 20A receptacles on 20A branch circuits in the kitchen, pantry, dining room, and breakfast room is calculated at 1,500VA.

These loads should be included with the general lighting load and subjected to the same demand factors permitted in *NEC Table 220-11*.

2.6.1 Laundry Circuit Load

A 1,500VA feeder load is added to load calculations for each two-wire laundry branch circuit installed in a home. The branch circuit is required by *NEC Section 210-11(c)*. This load may also be added to the general lighting load and subjected to the same demand factors provided in *NEC Section 220-11*.

2.6.2 Dryer Load

The dryer load for each electric clothes dryer is 5,000VA or the actual nameplate value of the dryer, whichever is larger. Demand factors listed in *NEC Table 220-18* may be applied for more than one dryer in the same dwelling.

2.6.3 Range Load

Range loads and other cooking appliances are covered under *NEC Section 210-19(c)*. The feeder demand loads for household electric ranges, wall-mounted ovens, countertop cooking units, and other similar household appliances individually rated above 1¾kW are permitted to be computed in accordance with *NEC Table 220-19*. If two or more single-phase ranges are supplied by a three-phase, four-wire feeder, the total load is computed by using twice the maximum number connected between any two phases.

2.6.4 Demand Loads For Electric Ranges

- Dwelling unit loads   *NEC Section 220-30*
- Existing dwelling unit loads   *NEC Section 220-31*
- Multi-family dwelling unit loads   *NEC Table 220-32*
- School loads   *NEC Table 220-34*
- Restaurant loads   *NEC Table 220-36*
- Farm loads (individual loads)   *NEC Table 220-40*
- Farm load (total of loads)   *NEC Table 220-41*

2.6.5 Calculating The Demand Load

To illustrate the use of demand factors, consider the following example: the total connected load of a noncontinuous receptacle load in an office building is 26,000VA.

Refer to **NEC Table 220-13** and find that the first 10 kVA or less is calculated at 100%:

10kVA x 100% = 10,000VA

26,000VA – 10,000VA = 16,000VA

The remainder over 10kVA is calculated at 50%:

16,000VA x 50% = 8,000VA

10,000VA + 8,000VA = 18,000VA

The calculated receptacle load in this example would be 18,000VA.

2.7.0 DEMAND FACTORS FOR NEUTRAL CONDUCTORS

The neutral conductor of electrical systems generally carries only the current imbalance of the phase conductors. For example, in a single-phase feeder circuit with one phase conductor carrying 50 amps and the other carrying 40 amps, the neutral conductor would carry 10 amps. Since the neutral in many cases will never be required to carry as much current as the phase conductors, the NEC allows us to apply a demand factor. (See **NEC Section 220-22** for details.) Note that in certain circumstances such as electrical discharge lighting, data processing equipment, and other similar equipment, a demand factor cannot be applied to the neutral conductors because these types of equipment produce harmonic currents that increase the heating effect in the neutral conductor.

3.0.0 SIZING RESIDENTIAL NEUTRAL CONDUCTORS

The neutral conductor in a three-wire, single-phase service carries only the unbalanced load between the two ungrounded (hot) wires or legs. Since there are several 240V loads in the above calculations, these 240V loads will be balanced and therefore reduce the load on the service neutral conductor. Consequently, in most cases, the service neutral does not have to be as large as the ungrounded (hot) conductors.

In the previous example, the water heater does not have to be included in the neutral conductor calculation, since it is strictly 240V with no 120V loads. The clothes dryer and electric range, however, have 120V lights that will cause a current imbalance between phases. The NEC allows a demand factor of 70% for these two appliances. Using this information, the neutral conductor may be sized accordingly:

- General lighting and appliance load   4,942.5VA
- Electric range (8,000VA x 0.70)   5,600VA
- Clothes dryer (5,500VA x 0.70)   3,850VA
- Total   14,392.5VA

To find the total phase-to-phase amperes, divide the total volt-amperes by the voltage between phases:

14,392.5VA ÷ 240V = 59.96A or 60A

The **service-entrance conductors** have now been calculated and must be rated at 100A with a neutral conductor rated for at least 60A. See *Figure 4* for a completed calculation form for the residence in question.

SERVICE LOAD CALCULATION
ONE-FAMILY DWELLING—STANDARD CALCULATION

I. GENERAL LIGHTING LOADS

TYPE OF LOAD	CALCULATION	TOTAL VA	NEC REFERENCE
Lighting load	(1) 1,350 sq. ft. x 3VA =	(2) 4,050 VA	Table 220-3(a)
Small appliance loads	(3) 2 circuits x 1,500VA=	(4) 3,000 VA	Section 220-16(a)
Laundry load	(5) 1 circuits x 1,500VA=	(6) 1,500 VA	Section 220-16(b)
Lighting, small appliance, laundry	Total VA =	(7) 8,550 VA	

LOAD	CALCULATION	DEMAND FACTOR	TOTAL VA
Lighting, small appliance, laundry	First 3,000VA x	100% =	(8) 3,000VA
Lighting, small appliance, laundry	(9) Remaining VA 5,550 x	35% =	(10) 1,942.5 VA
Lighting, small appliance, laundry	Add #8 & #10 above	=	(11) 4,942.5 VA

TOTAL CALCULATED LOAD FOR LIGHTING, SMALL APPLIANCES, & LAUNDRY CIRCUIT(S) (Enter Item #11 above in box #12)

(12) 4,942.5VA

II. LARGE APPLIANCE LOADS

TYPE OF LOAD	NAMEPLATE RATING	DEMAND FACTOR	TOTAL VA	NEC REFERENCE
Electric range	Not over 12kVA	Use 8kVA	(13) 8,000VA	Table 220-19
Clothes dryer	(14) 5,500 VA	100%	(15) 5,500 VA	Table 220-18
Water heater	(16) 4,500 VA	100%	(17) 4,500 VA	
Other appliances	(18) VA	100%	(19) VA	

TOTAL CALCULATED LOAD FOR LARGE APPLIANCES (Add items #15, #17, and #19 above).
ENTER TOTAL IN BOX #20

(20) 18,000VA

TOTAL CALCULATED LOAD (Add boxes #12 & #20)
ENTER TOTAL IN BOX #21

(21) 22,942.5VA

III. CONVERT VA TO AMPERES

$$\frac{Total\ VA\ in\ Box\ \#21}{240\ (volts)} = amperes \qquad \frac{22,942.5VA}{240\ (volts)} = \boxed{(22)\ 95.6\ amperes}$$

112F04.TIF

Figure 4. Completed Calculation Form

In *NEC Section 310-15(b)(6)*, special consideration is given to 120/240V, single-phase residential services. Conductor sizes are shown in *NEC Table 310-15(b)(6)*. Reference to this table shows that the NEC allows a No. 4 AWG copper or No. 2 AWG aluminum or copper-clad aluminum conductor for a 100A service.

When sizing the grounded conductor for services, the provisions stated in *NEC Sections 215-2, 220-22, and 230-42* must be met, along with other applicable sections.

4.0.0   SIZING THE LOAD CENTER

Each ungrounded conductor in all circuits must be provided with overcurrent protection either in the form of fuses or circuit breakers. If more than six such devices are used, a means of disconnecting the entire service must be provided using either a main disconnect switch or a main circuit breaker.

To calculate the number of fuse holders or circuit breakers required in the sample residence, look at the general lighting load first. The total general lighting load of 4,050VA can be divided by 120V to find the amperage:

4,050VA ÷ 120V = 33.75A

Either 15A or 20A circuits may be used for the lighting load. Two 20A circuits (2 x 20) equal 40A, so two 20A circuits would be adequate for the lighting. However, two 15A circuits total only 30A and 33.75A are needed. Therefore, if 15A circuits are used, three will be required for the total lighting load. In this example, three 15A circuits will be used.

In addition to the lighting circuits, the sample residence will require a minimum of two 20A for the small appliance load and one 20A circuit for the laundry. So far, the following branch circuits can be counted:

• General lighting load	Three 15A circuits
• Small appliance load	Two 20A circuits
• Laundry load	One 20A circuit
• Total	Six branch circuits

Most **load centers** and panelboards are provided with an even number of circuit breaker spaces or fuse holders (e.g., four, six, eight, ten, etc.). But before the panelboard can be selected, space must be provided for the 240V loads—each requiring a two-pole circuit breaker or fuse holder. The ratings of overcurrent protection devices are sized by dividing the demand volt-amperes by the voltage. For example, since the demand load of the electric range is 8kVA, 8,000VA divided by 240V equals 33.33A. The closest standard overcurrent protection device is 40A; this will be the size used—a 40A, two-pole circuit breaker. In some existing installations, you might find a two-pole fuse block containing two 40A cartridge fuses being used to feed a residential electric range. The remaining 240V circuits are calculated in a similar fashion, which results in the following:

- Electric range   One 40A, two-pole circuit breaker
- Clothes dryer   One 30A, two-pole circuit breaker
- Water heater   One 30A, two-pole circuit breaker

These three appliances will therefore require an additional six spaces (two poles x three appliances) in the load center or panelboard. Adding these six spaces to the six required for the general lighting and small appliance loads requires at least a 12-space load center to handle the circuits in the sample residence.

4.1.0 GROUND FAULT CIRCUIT INTERRUPTERS

Under certain conditions, the time it takes to open an overcurrent protection device can be critical. Because of this fact, the NEC requires ground fault circuit interrupters (GFCIs) to be installed on the following:

- Circuits feeding impedance heating units operating at voltages greater than 30V; such units are frequently used for de-icing and snow-melting equipment
- Circuits for electrically-operated pool covers
- Power or lighting circuits for swimming pools, fountains, and similar locations
- Receptacles in both commercial and residential garages
- Receptacles in residential bathrooms
- Receptacles installed outdoors where there is direct grade-level access
- Receptacles installed in residential crawl spaces or unfinished basements
- Countertop receptacles mounted within six feet of a wet bar sink
- Receptacles installed in bathhouses
- Receptacles installed in bathrooms of commercial, industrial, or any other buildings
- Receptacles installed on roofs of any building except dwellings
- Kitchen countertop receptacles
- Branch circuits derived from autotransformers

GFCI circuit breakers require the same mounting space as standard single-pole circuit breakers and provide the same branch circuit wiring protection as standard circuit breakers. They also provide Class A ground fault protection.

GFCI breakers that are UL-listed are available in single- and two-pole construction; 15A, 20A, 25A, and 30A ratings; and have a 10,000A interrupting capacity. Single-pole units are rated at 120VAC; two-pole units are rated at 120/240VAC.

GFCI breakers can be used not only in load centers and panelboards, but they are also available factory-installed in meter pedestals and power outlet panels for recreational vehicle (RV) parks and construction sites.

The GFCI sensor continuously monitors the current balance in the ungrounded or energized (hot) load conductor and the neutral load conductor. If the current in the neutral load wire becomes less than the current in the hot load wire, then a ground fault exists, since a portion of the current is returning to the source by some means other than the neutral load wire. When a current imbalance occurs, the sensor, which is a differential current transformer, sends a signal to the solid-state circuit, which activates the ground trip solenoid mechanism and breaks the hot load connection (*Figure 5*). A current imbalance as low as six milliamps (6mA) will cause the circuit breaker to interrupt the circuit. This is indicated by the trip indicator on the front of the device.

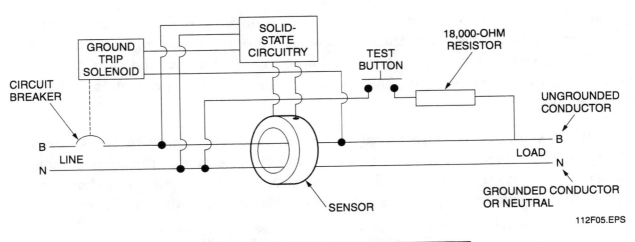

112F05.EPS

Figure 5. Operating Circuitry Of A Typical GFCI

The two-pole GFCI breaker (*Figure 6*) continuously monitors the current balance between the two hot conductors and the neutral conductor. As long as the sum of these three currents is zero, the device will not trip; that is, if there are ten amps of current in the A load wire, five amps in the neutral, and five amps in the B load wire, then the sensor is balanced and will not produce a signal. A current imbalance from a ground fault condition as low as 6mA will cause the sensor to produce a signal of sufficient magnitude to trip the device.

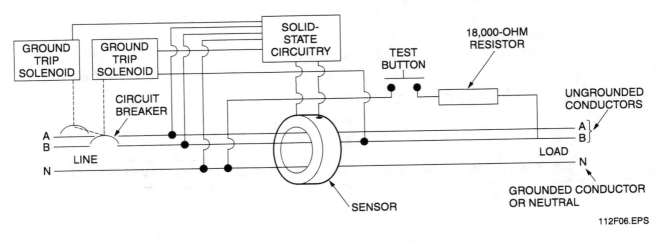

112F06.EPS

Figure 6. Operating Characteristics Of A Two-Pole GFCI

4.1.1   Single-Pole GFCI Circuit Breakers

The single-pole GFCI breaker has two load lugs and a white wire pigtail in addition to the line side plug-on or bolt-on connector. The line side hot connection is made by installing the GFCI breaker in the panel just like any other circuit breaker is installed. The white wire pigtail is attached to the panel neutral (S/N) assembly. Both the neutral and hot wires of the branch circuit being protected are terminated in the GFCI breaker. These two load lugs are clearly marked LOAD POWER and LOAD NEUTRAL in the breaker case. Also in the case is the identifying marking for the pigtail, PANEL NEUTRAL.

Single-pole GFCI circuit breakers must be installed on independent circuits. Circuits that employ a neutral common to more than one hot conductor cannot be protected against ground faults by a single-pole breaker because a common neutral cannot be split and retain the necessary hot wire to neutral wire balance under normal use to prevent the GFCI breaker from tripping.

Care should be exercised when installing GFCI breakers in existing panels. Be sure that the neutral wire for the branch circuit corresponds with the hot wire of the same circuit. Always remember that unless the current in the neutral wire is equal to that in the hot wire (within 6mA), the GFCI breaker senses this as being a possible ground fault (see *Figure 7*).

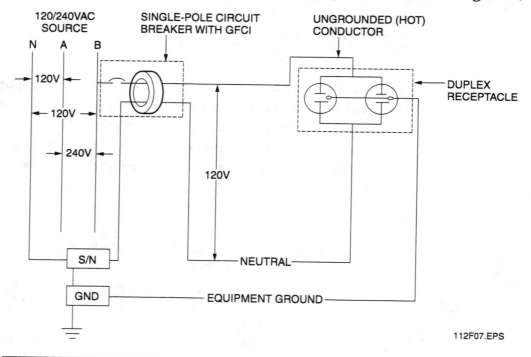

112F07.EPS

Figure 7. Operating Characteristics Of A Single-Pole Circuit Breaker With A GFCI

4.1.2   Two-Pole GFCI Circuit Breakers

A two-pole GFCI circuit breaker can be installed on a 120/240VAC single-phase, three-wire system; the 120/240VAC portion of a 120/240VAC three-phase, four-wire system; or the two phases and neutral of a 120/208VAC three-phase, four-wire system. Regardless of the application, the installation of the breaker is the same—connections are made to two hot buses and the panel neutral assembly. When installed on these systems, protection is provided for two-wire 240VAC or 208VAC circuits, three-wire 120/240VAC or 120/208VAC circuits, and 120VAC multiwire circuits.

The circuit in *Figure 8* illustrates the problems that are encountered when a common load neutral is used for two single-pole GFCI breakers. Either or both breakers will trip when a load is applied at the #2 duplex receptacle. The neutral current from the #2 duplex receptacle flows through breaker #1; this increase in neutral current through breaker #1 causes an imbalance in its sensor, thus causing it to produce a fault signal. At the same time, there is no neutral current flowing through breaker #2; therefore, it also senses a current imbalance. If a load is applied at the #1 duplex receptacle, and there is no load at the #2 duplex receptacle, then neither breaker will trip because neither breaker will sense a current imbalance.

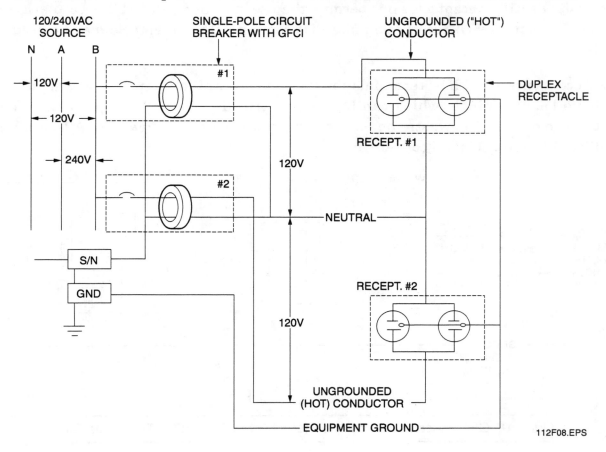

Figure 8. Circuit Depicting Common Load Neutral

Junction boxes can also present problems when they are used to provide taps for more than one branch circuit. Even though the circuits are not wired using a common neutral, sometimes all neutral conductors are connected together. Thus, parallel neutral paths are established, producing an imbalance in each GFCI breaker sensor, causing them to trip.

The two-pole GFCI breaker eliminates the problems encountered when trying to use two single-pole GFCI breakers with a common neutral. Because both hot currents and the neutral current pass through the same sensor, no imbalance occurs between the three currents, and the breaker will not trip.

4.1.3   Direct-Wired GFCI Receptacles

Direct-wired GFCI receptacles provide Class A ground fault protection on 120VAC circuits.

They are available in both 15A and 20A arrangements. The 15A unit has a NEMA 5-15R receptacle configuration for use with 15A plugs only. The 20A device has a NEMA 5-20R receptacle configuration for use with 15A or 20A plugs. Both 15A and 20A units have a 120VAC, 20A circuit rating. This is to comply with **NEC Table 210-24**, which requires that 15A circuits use 15A receptacles but permits the use of either 15A or 20A receptacles on 20A circuits. Therefore, GFCI receptacle units that contain a 15A receptacle may be used on 20A circuits.

These receptacles have line terminals for the hot, neutral, and ground wires. In addition, they have load terminals which can be used to provide ground fault protection for other receptacles electrically downstream on the same branch circuit (*Figure 9*). All terminals will accept No. 14 to No. 10 AWG copper wire.

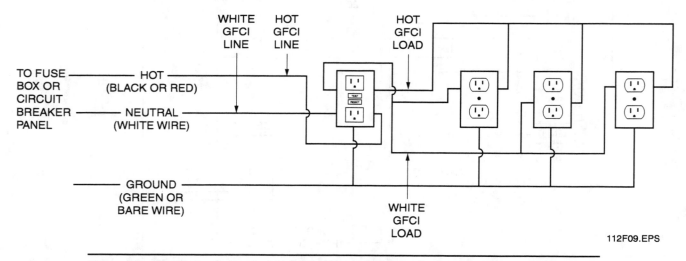

112F09.EPS

Figure 9.   GFCI Receptacle Used To Protect Other Outlets On The Same Circuit

GFCI receptacles have a two-pole tripping mechanism that breaks both the hot and the neutral load connections.

When tripped, the RESET button pops out. The unit is reset by pushing the button back in.

GFCI receptacles have the additional benefit of noise suppression. Noise suppression minimizes false tripping due to spurious line voltages or radio frequency (RF) signals between 10 and 500 megahertz (Mhz).

GFCI receptacles can be mounted without adapters in wall outlet boxes that are at least 1.5 inches deep.

5.0.0   GROUNDING ELECTRIC SERVICES

The grounding system is a major part of the electrical system. Its purpose is to protect people and equipment against the various electrical faults that can occur. It is sometimes possible for higher-than-normal voltages to appear at certain points in an electrical system or in the electrical equipment connected to the system. Proper grounding ensures that the electrical charges that cause these high voltages are channeled to the earth or ground before damaging equipment or endangering people.

When referring to *ground*, we are talking about ground potential or earth ground. If a conductor is connected to the earth or to some conducting body that serves in place of the earth, such as a driven ground rod (electrode) or cold water pipe, the conductor is said to be *grounded*. The neutral conductor in a three- or four-wire service, for example, is intentionally grounded and therefore becomes a *grounded conductor*. However, a wire used to connect this neutral conductor to a grounding electrode or electrodes is referred to as a *grounding electrode conductor*. Note the difference in the two meanings; one is grounded, while the other provides a means of grounding.

There are two general classifications of protective grounding:

• System grounding
• Equipment grounding

The system ground relates to the **service-entrance equipment** and its interrelated and bonded components. That is, system and circuit conductors are grounded to limit voltages due to lightning, line surges, or unintentional contact with higher voltage lines, and to stabilize the voltage to ground during normal operation.

Equipment grounding conductors are used to connect the noncurrent-carrying metal parts of equipment, conduit, outlet boxes, and other enclosures to the system grounded conductor, the grounding electrode conductor, or both, at the service equipment or at the source of a separately-derived system. Equipment grounding conductors are bonded to the system grounded electrode conductor to provide a low-impedance path for fault current that will facilitate the operation of overcurrent devices under ground fault conditions.

Equipment grounding is discussed later in this module. *NEC Article 250* covers general requirements for grounding and bonding. Read this article carefully.

To better understand a complete grounding system, a conventional residential system will be examined, beginning at the power company's high-voltage lines and transformer, as shown in *Figure 10*. The pole-mounted transformer is fed with a two-wire, single-phase 7,200V system which is transformed and stepped down to a three-wire, 120/240V, single-phase electric service suitable for residential use. Note that the voltage between line A and line B is 240V. However, by connecting a third wire (neutral) on the secondary winding of the transformer—between the other two—the 240V is split in half, providing 120V between either line A or line B and the neutral conductor. Consequently, 240V is available for household appliances such as ranges, hot water heaters, clothes dryers, etc., while 120V is available for lights, small appliances, televisions, etc.

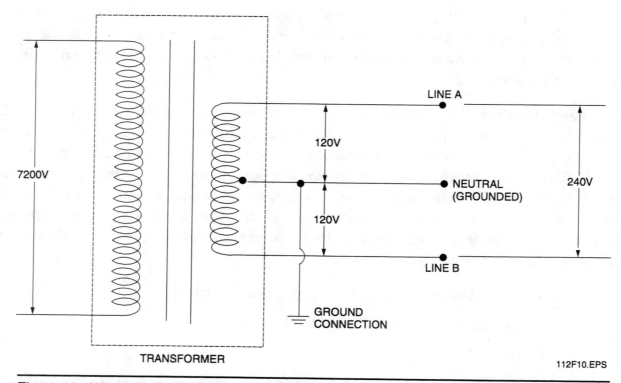

112F10.EPS

Figure 10. Wiring Diagram Of A 7,200V To 120/240V, Single-Phase Transformer Connection

Referring again to *Figure 10*, conductors A and B are ungrounded conductors, while the neutral is a grounded conductor. If only 240V loads were connected, the neutral (grounded conductor) would carry no current. However, since 120V loads are present, the neutral will carry the unbalanced load and become a current-carrying conductor. For example, if line A carries 60A and line B carries 50A, the neutral conductor would carry only 10A (60 – 50 = 10). This is why the NEC allows the neutral conductor in an electric service to be smaller than the ungrounded conductors. The typical pole-mounted service entrance is normally routed by messenger cable from a point on the pole to a point on the building being served, terminating in a meter housing. Another service conductor is installed between the meter housing and the main service switch or #2 panelboard. This is the point where most systems are grounded—the neutral bus in the main panelboard. See *Figure 11*.

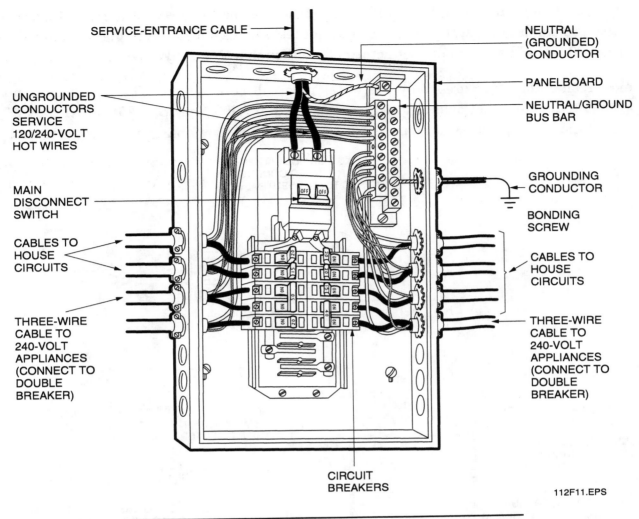

SERVICE-ENTRANCE CABLE

NEUTRAL (GROUNDED) CONDUCTOR

PANELBOARD

NEUTRAL/GROUND BUS BAR

UNGROUNDED CONDUCTORS SERVICE 120/240-VOLT HOT WIRES

MAIN DISCONNECT SWITCH

GROUNDING CONDUCTOR

BONDING SCREW

CABLES TO HOUSE CIRCUITS

CABLES TO HOUSE CIRCUITS

THREE-WIRE CABLE TO 240-VOLT APPLIANCES (CONNECT TO DOUBLE BREAKER)

THREE-WIRE CABLE TO 240-VOLT APPLIANCES (CONNECT TO DOUBLE BREAKER)

CIRCUIT BREAKERS

112F11.EPS

Figure 11. Interior View Of Panelboard Showing Connections

5.1.0 GROUNDING METHODS

Methods of grounding an electric service are covered in **NEC Section 250-50**. In general, all of the following (if available) and any made electrodes must be bonded together to form the grounding electrode system:

- An underground water pipe may be used if it is in direct contact with the earth for not less than 10 feet.
- The metal frame of a building may be used if it is effectively grounded.
- An electrode encased by at least two inches of concrete may be used if it is located within and near the bottom of a concrete foundation or footing that is in direct contact with the earth. The electrode must also be at least 20 feet long and must be made of electrically-conductive coated steel reinforcing bars or rods of not less than ½ inch in diameter, or consisting of at least 20 feet of bare copper conductor not smaller than No. 4 AWG wire size.

- A ground ring encircling the building or structure may be used if it is in direct contact with the earth at a depth below grade of not less than 2½ feet. This ring must consist of at least 20 feet of bare copper conductor not smaller than No. 2 AWG wire size.

In most residential structures, only the water pipe will be available. This water pipe must be supplemented by an additional electrode, as specified in **NEC Sections 250-50(a) and 250-52**. *Figure 12* shows a typical residential electric service and the available grounding electrodes.

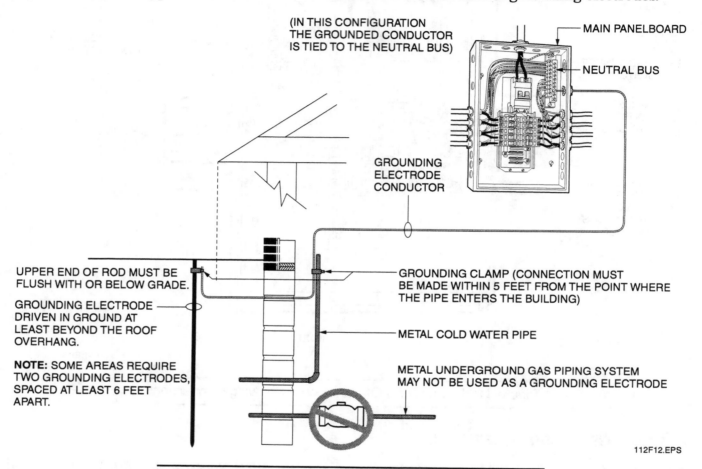

(IN THIS CONFIGURATION THE GROUNDED CONDUCTOR IS TIED TO THE NEUTRAL BUS)

MAIN PANELBOARD

NEUTRAL BUS

GROUNDING ELECTRODE CONDUCTOR

UPPER END OF ROD MUST BE FLUSH WITH OR BELOW GRADE.

GROUNDING ELECTRODE DRIVEN IN GROUND AT LEAST BEYOND THE ROOF OVERHANG.

NOTE: SOME AREAS REQUIRE TWO GROUNDING ELECTRODES, SPACED AT LEAST 6 FEET APART.

GROUNDING CLAMP (CONNECTION MUST BE MADE WITHIN 5 FEET FROM THE POINT WHERE THE PIPE ENTERS THE BUILDING)

METAL COLD WATER PIPE

METAL UNDERGROUND GAS PIPING SYSTEM MAY NOT BE USED AS A GROUNDING ELECTRODE

112F12.EPS

Figure 12. Components Of A Residential Grounding System

The sample residence has a metal underground water pipe that is in direct contact with the earth for more than 10 feet, so this is one valid grounding source. The house also has a metal underground gas piping system, but this may not be used as a grounding electrode [**NEC Section 250-52(a)**]. **NEC Section 250-50(a)** further states that the underground water pipe must be supplemented by an additional electrode of a type specified in **NEC Section 250-50** or in **NEC Section 250-52**. Since the other grounding methods (a grounded metal building frame, concrete-encased electrode, or a ground ring) are not normally available for most residential applications, **NEC Section 250-52** must be used to determine the supplemental electrode.

In most cases, this supplemental electrode will consist of either a driven rod or pipe electrode, the specifications for which are shown in *Figure 13*.

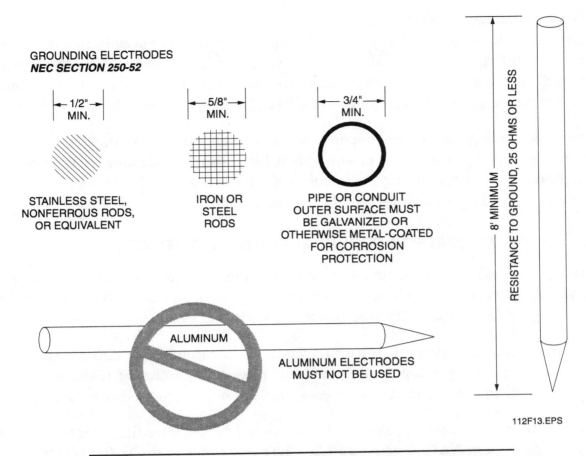

Figure 13. Specifications For Rod And Pipe Grounding Electrodes

An alternate method to a pipe or rod is a plate electrode. Each plate electrode must expose not less than two square feet of surface to the surrounding earth. Plates made of iron or steel must be at least ¼-inch thick, while plates of nonferrous metal such as copper need only be .06-inch thick.

Either type of electrode must have a resistance to ground of 25 ohms or less. If not, they must be augmented by an additional electrode, with the two electrodes spaced not less than six feet from each other. In fact, many locations require two electrodes regardless of the resistance to ground. This is not a NEC requirement, but is required by certain power companies and local ordinances in some areas. Always check with the local inspection authority for such rules that surpass the requirements of the NEC.

5.1.1 Grounding Conductors

The grounding conductor connecting the panelboard neutral bus to the water pipe and grounding electrodes must be made of copper, aluminum, or copper-clad aluminum. The material selected must be resistant to any corrosive condition existing at the installation, or it must be suitably protected against corrosion. The grounding conductor may be either solid or stranded, covered or bare, but it must be in one continuous length without a splice or joint, except for the following conditions:

- Splices in busbars are permitted.
- Where a service consists of more than a single enclosure, it is permissible to connect taps to the grounding electrode conductor, provided the taps are made within the enclosures.
- Grounding electrode conductors may also be spliced at any location by means of irreversible compression-type connectors listed for the purpose or by exowelding.

The size of grounding conductors depends on service-entrance size, that is, the size of the largest service-entrance conductor or equivalent for parallel conductors. *NEC Table 250-66* lists the proper sizes of grounding conductors for various sizes of electric services. Equipment grounding is covered in detail later in this module.

5.2.0 GROUNDING PATH TO GROUND ELECTRODE AT SERVICE

A grounding electrode conductor is required to be used to connect the equipment grounding conductors, the service-entrance enclosures, and the grounded (neutral) service conductor to the grounding electrode. See *NEC Section 250-24(a)*.

An unspliced main **bonding jumper** is required to connect the equipment grounding conductor and the service-equipment enclosure to the system grounded (neutral) conductor of the electric utility system inside the service-equipment or service-conductor enclosure.

Where an alternating current system is connected to a grounding electrode in or at a building as specified in *NEC Section 250-24*, the same common grounding electrode must be used to ground conductor enclosures and equipment in or on that building. See *NEC Section 250-58*.

Where separate services supply a building and are required to be connected to a grounding electrode, the same grounding electrode is required to be used. See *NEC Section 250-58*. Two or more grounding electrodes that are effectively bonded together are considered to be a single grounding electrode system.

5.2.1 Use Of A Grounded (Neutral) Circuit Conductor For Grounding Equipment

For supply side equipment, a grounded (neutral) circuit conductor is permitted to be used to ground noncurrent-carrying metal parts of equipment, raceways, and other enclosures on the supply (line) side of the service disconnecting means. See *NEC Section 250-58* for additional information on this subject.

5.2.2 Method Of Bonding Service Equipment

Electrical continuity at service equipment is required to be assured by one of the methods specified below:

- The service equipment must be bonded to the grounded (neutral) service conductor using a listed pressure connector, listed clamp, or other listed means.

Note: Connection devices or fittings that depend solely upon solder must not
be used for this connection.

- Threaded couplings and connectors on enclosures with joints are required to be made up wrench-tight where rigid metal conduit and intermediate metal conduit are installed.
- Threadless couplings and connectors are required to be made up tight for rigid metal conduit, intermediate metal conduit, and electrical metallic tubing. Standard locknuts or bushings cannot be used for the bonding required by this action.
- Bonding jumpers must be used around concentric or eccentric knockouts in enclosures that are punched or otherwise formed so as to impair the electrical connection to ground.
- Other approved devices may be used, such as bonding-type locknuts and **bonding bushings**.

5.2.3   Main And Equipment Bonding Jumpers

The material of the main and equipment bonding jumpers must be of copper or other corrosion-resistant material. A main bonding jumper may be in the form of a wire, bus, screw, or similar suitable conductor.

Where a main bonding jumper is in the form of a screw, it is required to be identified with a green finish so that the head of the screw is visible for inspection. A main and equipment bonding jumper must be attached using a listed pressure connector, listed clamp, or other listed means.

The bonding jumper cannot be smaller than the sizes given in *NEC Table 250-66* for grounding electrode conductors. A properly-sized main bonding jumper will usually be included as a part of a listed panelboard.

5.2.4   Bonding And Grounding Requirements For Other Systems

An accessible means external to the service-equipment enclosure is required for connecting intersystem conducting and grounding conductors and connections for the communications, radio and television (TV), and community antenna television (CATV) circuits. This can be accomplished using at least one of the following:

- An exposed metallic service raceway
- An exposed grounding electrode conductor
- A No. 6 copper conductor that is not shorter than six inches and is bonded to the metallic service raceway or service-equipment enclosure using a listed and identified fitting

If a copper conductor is used, the other end of this conductor must be accessible and must be located on the exterior outside wall so that an external connection of the intersystem bonding or grounding conductor can be made, as required by *NEC Sections 800-40, 810-21, and 820-40* for communications, radio and TV equipment, and CATV circuits.

Note: All of the above requirements preclude the use of screw(s) to secure the equipment cover(s), screw on connectors, etc.

Provide a permanent sign at this intersystem connection that reads as follows: *CONNECTION FOR GROUNDING OR BONDING OF COMMUNICATIONS, TV, AND CATV SYSTEMS.*

5.2.5 Grounding Electrode Conductor Connections

The grounding electrode conductor is required to be connected to the grounding fitting using listed pressure connectors, listed clamps, or other listed means. Connections that depend on solder must never be used. To prevent corrosion, the ground clamp must be listed for the material of the grounding electrode and the grounding electrode conductor.

Where used on a pipe, ground rod, or other buried electrodes, the ground clamp must be listed for direct soil burial. More than one conductor is not permitted to be connected to the grounding electrode using a single clamp or fitting unless the clamp or fitting is specifically listed for the connection of more than one conductor.

For the connection to an electrode, you must use one of the following:

- A listed, bolted clamp of cast bronze or brass, or plain or malleable iron
- A pipe fitting, pipe plug, or other approved device that is screwed into a pipe or pipe fitting
- A listed sheet metal strap-type ground clamp with a rigid metal base that seats on the electrode with a strap that will not stretch during or after installation

Note: The UL listing states that "strap-type ground clamps are not suitable for attachment of the grounding electrode conductor of an interior wiring system to a grounding electrode."

5.2.6 Bonding Of Piping Systems

The interior metal water piping system must be bonded to the service-equipment enclosure, the grounded (neutral) conductor at the service, the grounding electrode conductor where of sufficient size, or to the grounding electrode(s) used.

The bonding jumper must be sized in accordance with *NEC Table 250-66* and installed in accordance with *NEC Sections 250-64(b) and (e)*. The points of attachment of the bonding jumper to the metal piping must always be accessible per *NEC Section 250-104*.

An exception for two-family dwellings permits the interior metal water piping system for each occupancy to be bonded to the panelboard enclosure (other than at the service equipment) supplying that occupancy where the interior metal water piping system for each unit is metallically isolated by the use of a nonmetallic water piping system. The size of this bonding jumper must be in accordance with *NEC Table 250-122*.

5.2.7 Other Metal Piping

Interior metal piping that may become energized is required to be bonded to the service-equipment enclosure, the grounded conductor at the service, the grounding electrode conductor where of sufficient size, or the grounding electrode(s) used. This bonding jumper must be sized in accordance with *NEC Table 250-122* using the rating of the circuit that may energize the piping. The equipment grounding conductor for the circuit that may energize the piping is permitted to serve as the bonding means.

Note: Bonding all piping and metal air ducts within the premises will provide additional safety.

5.2.8 Grounding Electrode System

A bonding jumper is required to be installed in accordance with *NEC Section 250-64(b)*. The bonding jumper must be sized in accordance with *NEC Table 250-66* and be connected using a listed lug, listed pressure connector, or listed clamp. See *NEC Section 250-70*.

The unspliced grounding electrode conductor is permitted to be run to any convenient grounding electrode that is available in the grounding electrode system. It will be sized for the largest grounding electrode conductor required among all the available electrodes.

If available on the premises at each building or structure served, each item below, and any made electrodes used in accordance with *NEC Sections 250-52(c) and (d)*, are required to be bonded together to form the grounding electrode system.

- A metal underground water pipe may be used, provided that it is in direct contact with the earth for 10 feet or more (including any metal well casing effectively bonded to the pipe) and electrically continuous (or made electrically continuous by bonding around insulating joints or sections or insulating pipe) to the points of connection of the grounding electrode conductor and the bonding conductors. Continuity of the grounding path or the bonding connection to interior piping must never rely upon water meters. A metal underground water pipe must also be supplemented by an additional electrode of a type specified in *NEC Sections 250-50 or 250-52*. The supplemental electrode is permitted to be bonded to the grounding electrode conductor, grounded service-entrance conductor, grounded service raceway, and grounded service enclosure, or the interior metal water piping at any convenient point. Where the supplemental electrode is a made electrode (such as a rod, pipe, or plate), that portion of the bonding jumper which is the sole connection to the supplemental grounding electrode is not required to be larger than No. 6 copper wire or No. 4 aluminum wire. Where used outside, aluminum or copper-clad aluminum grounding electrode conductors are not permitted to be installed within 18 inches of the earth [*NEC Section 250-64(a)*].
- The metal frame of the building may be used, provided that it is effectively grounded.

Note: *Effectively grounded* means intentionally connected to earth through one or more ground connection(s) of sufficiently low impedance and having sufficient current-carrying capacity to prevent the buildup of voltages which may result in a hazard to people or connected equipment.

- Concrete-encased electrodes may be used when they are encased by at least two inches of concrete. This type of electrode is located within and near the bottom of a concrete foundation or footing that is in direct contact with the earth. It consists of a total of at least 20 feet of one or more lengths of steel reinforcing bars or rods, with each being not less than ½ inch in diameter. It may also consist of a continuous unbroken length of at least 20 feet of No. 4 or larger bare copper conductor.
- A ground ring encircling the building or structure may be used if it is in direct contact with the earth at a depth below the surface of not less than 1½ feet. It must consist of an unbroken length of at least 20 feet of a No. 2 bare copper conductor.

5.2.9 Made Electrodes And Other Electrodes

Where none of the electrodes specified above is available, one or more of the electrodes specified below must be used.

Where practical, made electrodes are required to be embedded below the permanent moisture level. Any nonconductive coatings, such as paint or enamel, must be removed. Where more than one electrode is used, each electrode of one grounding system (including that used for lightning rods) must be at least six feet from any other electrode of another grounding system.

CAUTION: Metal underground gas piping systems must never be used as a grounding electrode.

Where two or more electrodes are effectively bonded together, they are treated as a single electrode system.

Permissible made electrodes are:

- Other local metal underground systems or structures, such as piping systems and underground tanks
- Rod and pipe electrodes

Rod and pipe electrodes are required to be not less than eight feet in length and must consist of the following material:

- Electrodes of pipe or conduit cannot be smaller than ¾ inch in diameter. Pipe or conduit electrodes of iron or steel must have their outer surface galvanized or be otherwise coated for corrosion protection.

- Rods of iron or steel must be at least ⅝ inch in diameter. Nonferrous or stainless steel rods or their equivalent that are less than ⅝ inch in diameter are required to be listed and cannot be less than ½ inch in diameter.

Rod and pipe electrodes must be installed in the following manner:

- The electrode must be driven to a depth of not less than eight feet, unless rock bottom is encountered. In this situation, the electrode is permitted to be driven at an angle not to exceed 45° from the vertical, or it must be buried in a trench that is at least 2½-feet deep.
- The upper end of the electrode is required to be flush with or below ground level unless the above-ground end and the grounding electrode conductor attachment are protected against physical damage. See *NEC Section 250-10*.
- Plate electrodes must expose not less than two square feet of surface to exterior soil. Electrodes of iron or steel plates must be at least ¼-inch thick. Electrodes of nonferrous metal must be at least 0.06-inch thick.
- Aluminum electrodes are not permitted.

5.2.10 Resistance Of Made Electrodes

An electrode consisting of a single rod, pipe, or plate that does not register a resistance to ground of 25 ohms or less must be augmented by one additional electrode of any of the types specified in *NEC Sections 250-50 or 250-52*. Where multiple rod, pipe, or plate electrodes are installed to meet the requirements of *NEC Section 250-56*, they must be at least six feet apart. The paralleling efficiency of rods longer than eight feet is improved by spacing greater than six feet.

5.2.11 Use Of Lightning Rods

Lightning rod conductors and driven pipes, rods, or other made electrodes used for grounding lightning rods are not to be used in lieu of the made grounding electrodes required by *NEC Section 250-52* for the grounding of wiring systems and equipment.

See *NEC Section 250-106* for more information on lightning protection systems.

5.2.12 Grounding Conductor Material

The materials for grounding conductors must meet the following specifications:

- The grounding electrode conductor must be of copper, aluminum, or copper-clad aluminum. The material selected must be resistant to any corrosive condition existing at the installation or must otherwise be suitably protected against corrosion.
- The grounding electrode conductor may be solid or stranded, and it may be insulated, covered, or bare.

- The grounding electrode conductor must be installed in one continuous length without a splice or joint. An exception permits the connection of taps to the grounding electrode conductor. Each such tap conductor will be required to extend to the inside of each such enclosure. The size for the grounding electrode conductor must be in accordance with *NEC Section 250-66*. Tap conductors are permitted to be sized for the largest conductor serving the respective enclosure(s). Tap conductors must be connected to the grounding electrode conductor in such a manner that the grounding electrode conductor remains without any splice or joint.

5.2.13   Installation Of Grounding Electrode Conductors

A grounding electrode conductor or its enclosure is required to be securely fastened to the surface on which it is carried. A No. 4 or larger copper or aluminum grounding electrode conductor is required to be protected where it will be exposed to severe physical damage. A No. 6 or larger grounding electrode conductor that is free from exposure to physical damage is permitted to be run along the surface of the building without metal covering or protection where it is securely fastened to the building. Otherwise, it must be installed in rigid metal conduit, intermediate metal conduit, rigid nonmetallic conduit, electrical metallic tubing, or cable armor.

Grounding electrode conductors smaller than No. 6 must be protected by rigid metal conduit, intermediate metal conduit, rigid nonmetallic conduit, electrical metallic tubing, or cable armor.

Aluminum or copper-clad aluminum grounding electrode conductors cannot be used where they are in direct contact with masonry or the earth or where subject to corrosive conditions. Where used outside, aluminum or copper-clad aluminum grounding electrode conductors must be installed at least 18 inches from the earth.

Metal enclosures for grounding electrode conductors are required to be electrically continuous from the point of attachment to metal cabinets or metallic equipment enclosures to the grounding electrode. They must also be securely fastened to the ground clamp or fitting.

A metal enclosure that is not physically continuous from a metal cabinet or metallic equipment enclosure to the grounding electrode must be made electrically continuous by bonding each end to the enclosed grounding electrode conductor.

5.2.14   Size Of Grounding Electrode Conductor

The size of the grounding electrode conductor for an alternating current system cannot be less than that given in *NEC Table 250-66*.

The exceptions to this rule are as follows:

- Where connected to made electrodes as in **NEC Sections 250-52(c) or (d)**, that portion of the grounding electrode conductor that is the sole connection to the grounding electrode is not required to be larger than No. 6 copper wire or No. 4 aluminum wire.
- Where connected to a concrete-encased electrode as in **NEC Section 250-50(c)**, that portion of the grounding electrode conductor that is the sole connection to the grounding electrode is not required to be larger than No. 4 copper wire.
- Where connected to a ground ring as in **NEC Section 250-50(a)(2)**, that portion of the grounding electrode conductor that is the sole connection to the grounding electrode is not required to be larger than No. 6 copper.

Where multiple sets of service-entrance conductors are used as permitted in **NEC Section 230-40, Exception No. 2**, the equivalent size of the largest service-entrance conductor is required to be determined by the largest sum of the areas of the corresponding conductors of each set.

Where there are no service-entrance conductors, the grounding electrode conductor size is required to be determined by the equivalent size of the largest service-entrance conductor required for the load to be served.

6.0.0 INSTALLING THE SERVICE ENTRANCE

In practical applications, the electric service is normally one of the last components of an electrical system to be installed. However, it is one of the first considerations when laying out a residential electrical system. For instance:

- The electrician must know in which direction and to what location to route the circuit homeruns while **roughing in** the electrical wiring.
- Provisions must be made for sleeves through footings and foundations in cases where underground systems (**service laterals**) are used.
- The local power company must be notified as to the approximate size of service required so they may plan the best way to furnish a **service drop** to the property.

6.1.0 SERVICE DROP LOCATIONS

The location of the service drop, electric meter, and load center should be considered first. It is always wise to consult the local power company to obtain their recommendations; where you want the service drop and where they want it may not coincide. A brief meeting with the power company about the location of the service drop can prevent problems later on.

The service drop must be routed so that the service drop conductors have a clearance of not less than three feet from windows, doors, porches, fire escapes, or similar locations. Furthermore, when service drop conductors pass over rooftops, driveways, yards, etc., they must have clearances as specified in **NEC Section 230-24**.

A plot plan (also called a *site plan*) is often available for new construction. The plot plan shows the entire property, with the building or buildings drawn in their proper location on the plot of land. It also shows sidewalks, driveways, streets, and existing utilities—both overhead and underground.

A plot plan of the sample residence is shown in *Figure 14*. In reviewing this drawing, you can see that the closest power pole is located across a public street from the house. By consulting with the local power company, it is learned that the service will be brought to the house from this pole by triplex cable which will hit the residence at a point on its left (west) end. The steel uninsulated conductor of triplex cable acts as both the grounded conductor (neutral) and as a support for the insulated (ungrounded) conductors. It is also suitable for overhead use.

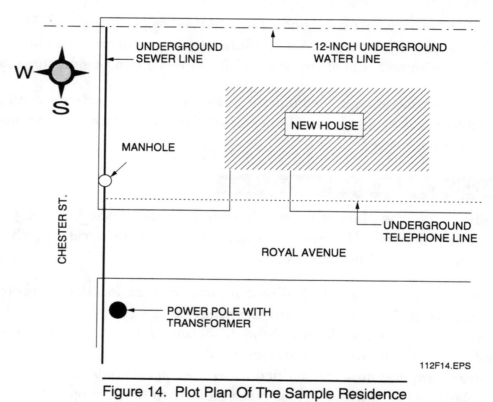

Figure 14. Plot Plan Of The Sample Residence

When Type SE cable is used, it will run directly from the point of attachment and service head to the meter base. However, since the carport is located on the west side of the building, a service mast (*Figure 15*) will have to be installed.

Roof clearances are specified in **NEC Section 230-24(a)**. The NEC requires a clearance of not less than eight feet above the roof surface that is maintained for a distance of not less than three feet in all directions from the roof edge.

6.2.0 VERTICAL CLEARANCES OF SERVICE DROP

NEC Section 225-18 specifies the distances by which service drop conductors must clear the ground. These distances vary according to the surrounding conditions.

ELECTRICAL — TRAINEE TASK MODULE 26112

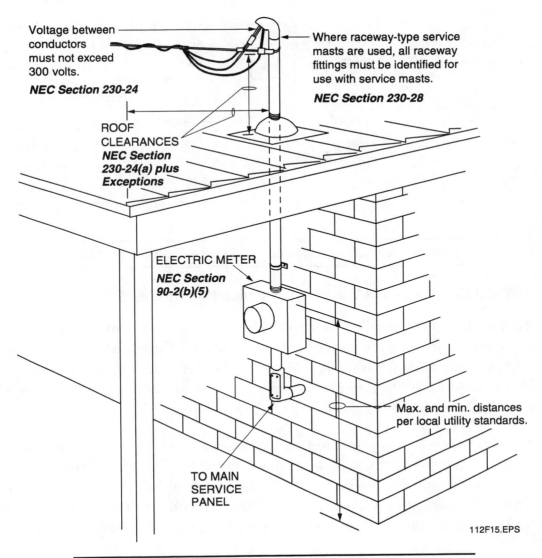

Voltage between conductors must not exceed 300 volts.
NEC Section 230-24

Where raceway-type service masts are used, all raceway fittings must be identified for use with service masts.
NEC Section 230-28

ROOF CLEARANCES
NEC Section 230-24(a) plus Exceptions

ELECTRIC METER
NEC Section 90-2(b)(5)

Max. and min. distances per local utility standards.

TO MAIN SERVICE PANEL

112F15.EPS

Figure 15. NEC Sections Governing Service Mast Installations

In general, the NEC states that the vertical clearances of all service drop conductors that carry 600V or under are based on a conductor temperature of 60°F (15°C) with no wind and with the final unloaded sag in the wire, conductor, or cable. Service drop conductors must be at least 10 feet above the ground or other accessible surfaces at all times. More distance is required under most conditions. For example, if the service conductors pass over residential property and driveways or commercial property that is not subject to truck traffic, the conductors must be at least 15 feet above the ground. However, this distance may be reduced to 12 feet when the voltage is limited to 300V to ground.

In other areas, such as public streets, alleys, roads, parking areas subject to truck traffic, driveways on other-than-residential property, etc., the minimum vertical distance is 18 feet. The conditions of the sample residence are shown in *Figure 16.*

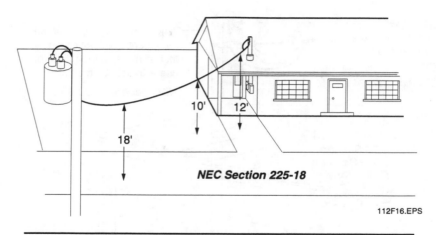

NEC Section 225-18

112F16.EPS

Figure 16. Vertical Clearances For Service Drop Conductors

6.3.0 SERVICE DROP CLEARANCES FOR BUILDING OPENINGS

Service conductors that are installed as open conductors or multiconductor cable without an overall outer jacket must have a clearance of not less than three feet from windows that are designed to be opened, doors, porches, balconies, ladders, stairs, fire escapes, or similar locations (**NEC Section 230-9**). However, conductors run above the top level of a window are permitted to be less than three feet from the window opening.

The three-foot clearance is not applicable to raceways or cable assemblies which have an overall outer jacket approved for use as a service conductor. The intention of this requirement is to protect the conductors from physical damage and/or physical contact with unprotected personnel. The exception allows service conductors, including drip loops and service drop conductors, to be located just above window openings because they are considered to be out of reach.

7.0.0 PANELBOARD LOCATION

The main service disconnect or panelboard is normally located in a portion of an unfinished basement or utility room on an outside wall so that the service cable coming from the electric meter can terminate immediately into the switch or panelboard when the cable enters the building. In the example home, however, there is no basement and the utility room is located in the center of the house with no outside walls. Consequently, a somewhat different arrangement will have to be used. A *load center* is a type of panelboard that is normally located at the service entrance of a residential installation. The load center usually contains a main circuit breaker which is the main disconnect. Circuit breakers are provided for equipment such as electric water heaters, ranges, dryers, air conditioning and heating units, and breakers that feed subpanels such as lighting panels.

NEC Section 230-70 requires that the service disconnecting means be installed in a readily accessible location—either outside or inside the building. If located inside the building, it must be located nearest the point of entrance of the service conductors. In the sample home, there are at least two methods of installing the panelboard in the utility room that will comply with this NEC regulation.

The first method utilizes a weatherproof 100A disconnect (safety switch or circuit breaker enclosure) mounted next to the meter base on the outside of the building. With this method, service conductors are provided with overcurrent protection; the neutral conductor is also grounded at this point, as this becomes the main disconnect switch. Three-wire cable with an additional grounding wire is then routed from this main disconnect to the panelboard in the utility room. All three current-carrying conductors (two ungrounded and one neutral) must be insulated with this arrangement; the equipment ground, however, may be bare. The panelboard containing overcurrent protection devices for the branch circuits, which is located in the utility room, now becomes a subpanel. See *Figure 17*.

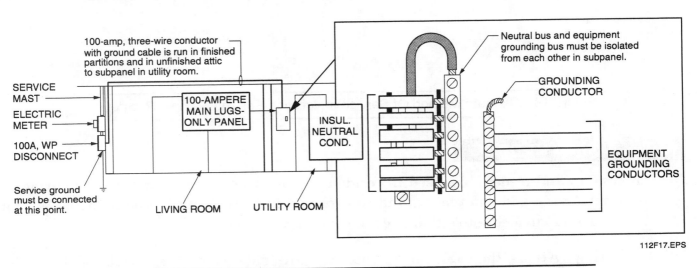

Figure 17. One Method Of Wiring A Panelboard For The Sample Residence

An alternate method utilizes conduit from the meter base that is routed under the concrete slab and then up to a main panelboard located in the utility room. **NEC Section 230-6** considers conductors installed under at least two inches of concrete to be outside the building. The sample residence has a four-inch thick reinforced concrete slab—well within the NEC regulations. Therefore, the service conductors from the meter base that are installed under the concrete slab in conduit are considered to be outside the house, and no disconnect is required at the meter base. When this conduit emerges in the utility room, it will run straight up into the bottom of the panelboard, again meeting the NEC requirement that the panel be located nearest the point of entrance of the service conductors. Details of this service arrangement are shown in *Figure 18*.

Note: Local ordinances in some areas may require a disconnect at the meter base, making the panel in the utility room a subpanel.

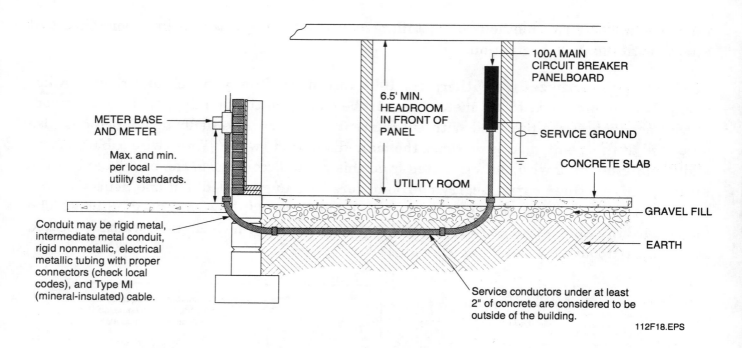

Figure 18. Alternate Method Of Service Installation For The Sample Residence

Labels in figure:
- METER BASE AND METER
- Max. and min. per local utility standards.
- Conduit may be rigid metal, intermediate metal conduit, rigid nonmetallic, electrical metallic tubing with proper connectors (check local codes), and Type MI (mineral-insulated) cable.
- 6.5' MIN. HEADROOM IN FRONT OF PANEL
- UTILITY ROOM
- 100A MAIN CIRCUIT BREAKER PANELBOARD
- SERVICE GROUND
- CONCRETE SLAB
- GRAVEL FILL
- EARTH
- Service conductors under at least 2" of concrete are considered to be outside of the building.
- 112F18.EPS

8.0.0 WIRING METHODS

Branch circuits and feeders are used in residential construction to provide power wiring to operate components and equipment, and control wiring to regulate the equipment. Wiring may be further subdivided into either *open* or *concealed* wiring.

In open wiring systems, the cable and/or raceways are installed on the surface of the walls, ceilings, columns, and other areas where they are in view and are readily accessible. Open wiring is often used in areas where appearance is not important, such as in unfinished basements, attics, and garages.

Concealed wiring systems are installed inside walls, partitions, ceilings, columns, and behind baseboards or moldings where they are out of view and are not readily accessible. This type of wiring is generally used in all new construction with finished interior walls, ceilings, and floors, and it is the preferred type of wiring where appearance is important.

In general, there are two basic wiring methods used in the majority of modern residential electrical systems. They are:

- Sheathed cables of two or more conductors
- Raceway (conduit) systems

The method used on a given job is determined by the requirements of the NEC, the type of building construction, the location of the wiring in the building, and the relative costs of different wiring methods.

In most applications, either of the two methods may be used, and both methods are frequently used in combination.

8.1.0 CABLE SYSTEMS

Several types of cable are used in wiring systems to feed or supply power to equipment. These include nonmetallic-sheathed cable, armored cable, underground feeder cable, and service-entrance cable.

8.1.1 Nonmetallic-Sheathed Cable

Nonmetallic-sheathed (Type NM) cable is manufactured in two- or three-wire configurations with varying sizes of conductors. In both two- and three-wire cables, conductors are color-coded: one conductor is black while the other is white in two-wire cable; in three-wire cable, the additional conductor is red. Both types also have a grounding conductor which is usually bare, but it is sometimes covered with green plastic insulation, depending upon the manufacturer. The jacket or covering consists of rubber, plastic, or fiber. Most also have markings on this jacket giving the manufacturer's name or trademark, wire size, and number of conductors (see *Figure 19*). For example, NM 12-2 W/GRD indicates that the jacket contains two No. 12 AWG conductors along with a grounding wire; NM 12-3 W/GRD indicates three conductors plus a grounding wire. This type of cable may be concealed in the framework of buildings, or in some instances, may be run exposed on the building surfaces. It may not be used in any building exceeding three floors above grade, as a service-entrance cable, in commercial garages having hazardous locations, in theaters and similar locations, in places of assembly, in motion picture studios, in storage battery rooms, in hoistways, embedded in poured concrete or aggregate, or in any hazardous location except as otherwise permitted by the NEC. Type NM cable is frequently referred to as **Romex®**.

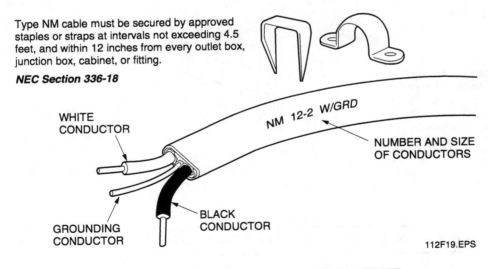

Type NM cable must be secured by approved staples or straps at intervals not exceeding 4.5 feet, and within 12 inches from every outlet box, junction box, cabinet, or fitting.

NEC Section 336-18

WHITE CONDUCTOR

NM 12-2 W/GRD

NUMBER AND SIZE OF CONDUCTORS

GROUNDING CONDUCTOR

BLACK CONDUCTOR

112F19.EPS

Figure 19. Characteristics Of Type NM Cable

Since Type NM cable is the least expensive type for residential use, this wiring method will be the type encountered most often. *Figure 20* shows additional NEC regulations pertaining to the installation of Type NM cable.

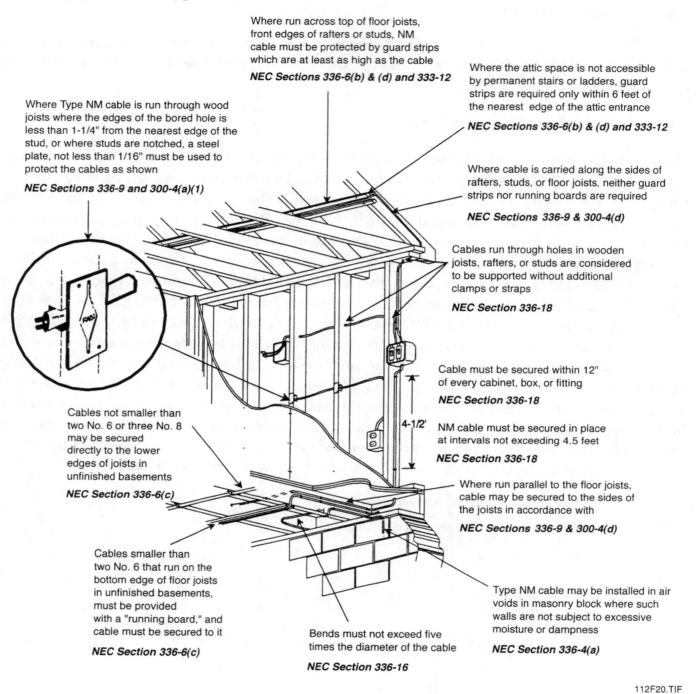

Where run across top of floor joists, front edges of rafters or studs, NM cable must be protected by guard strips which are at least as high as the cable
NEC Sections 336-6(b) & (d) and 333-12

Where the attic space is not accessible by permanent stairs or ladders, guard strips are required only within 6 feet of the nearest edge of the attic entrance
NEC Sections 336-6(b) & (d) and 333-12

Where Type NM cable is run through wood joists where the edges of the bored hole is less than 1-1/4" from the nearest edge of the stud, or where studs are notched, a steel plate, not less than 1/16" must be used to protect the cables as shown
NEC Sections 336-9 and 300-4(a)(1)

Where cable is carried along the sides of rafters, studs, or floor joists, neither guard strips nor running boards are required
NEC Sections 336-9 & 300-4(d)

Cables run through holes in wooden joists, rafters, or studs are considered to be supported without additional clamps or straps
NEC Section 336-18

Cable must be secured within 12" of every cabinet, box, or fitting
NEC Section 336-18

Cables not smaller than two No. 6 or three No. 8 may be secured directly to the lower edges of joists in unfinished basements
NEC Section 336-6(c)

4-1/2'

NM cable must be secured in place at intervals not exceeding 4.5 feet
NEC Section 336-18

Where run parallel to the floor joists, cable may be secured to the sides of the joists in accordance with
NEC Sections 336-9 & 300-4(d)

Cables smaller than two No. 6 that run on the bottom edge of floor joists in unfinished basements, must be provided with a "running board," and cable must be secured to it
NEC Section 336-6(c)

Bends must not exceed five times the diameter of the cable
NEC Section 336-16

Type NM cable may be installed in air voids in masonry block where such walls are not subject to excessive moisture or dampness
NEC Section 336-4(a)

112F20.TIF

Figure 20. NEC Sections Governing The Installation Of Type NM Cable

8.1.2 Armored Cable

Armored (Type AC) cable, commonly called **BX®**, is manufactured in two-, three-, and four-wire assemblies with varying sizes of conductors, and it is used in locations similar to those where Type NM cable is allowed. The metallic spiral covering on Type AC cable offers a greater degree of mechanical protection than with Type NM cable and also provides a continuous grounding bond without the need for additional grounding conductors.

Type AC cable may be used for under-plaster extensions, as provided in the NEC, and embedded in plaster finish, brick, or other masonry, except in damp or wet locations. It may also be run in the air voids of masonry block or tile walls, except where such walls are exposed or subject to excessive moisture or dampness. See *Figures 21* and *22*.

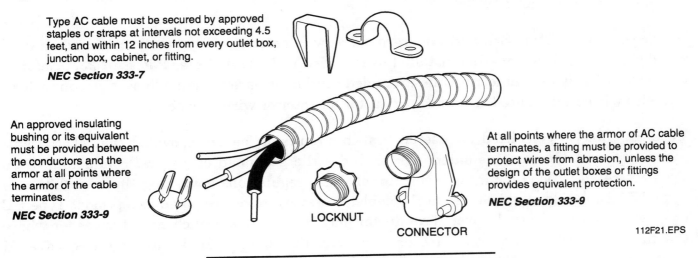

Type AC cable must be secured by approved staples or straps at intervals not exceeding 4.5 feet, and within 12 inches from every outlet box, junction box, cabinet, or fitting.

NEC Section 333-7

An approved insulating bushing or its equivalent must be provided between the conductors and the armor at all points where the armor of the cable terminates.

NEC Section 333-9

LOCKNUT

CONNECTOR

At all points where the armor of AC cable terminates, a fitting must be provided to protect wires from abrasion, unless the design of the outlet boxes or fittings provides equivalent protection.

NEC Section 333-9

112F21.EPS

Figure 21. Characteristics Of Type AC Cable

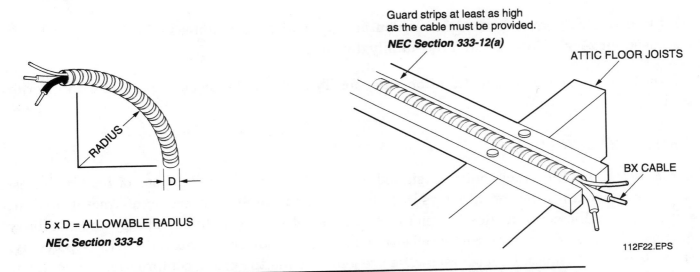

Guard strips at least as high as the cable must be provided.
NEC Section 333-12(a)

ATTIC FLOOR JOISTS

BX CABLE

RADIUS

D

5 x D = ALLOWABLE RADIUS
NEC Section 333-8

112F22.EPS

Figure 22. NEC Sections Governing The Installation Of Type AC Cable

8.1.3 Underground Feeder Cable

Underground feeder (Type UF) cable (**NEC Article 339**) may be used underground, including direct burial in the earth, as a feeder or branch circuit cable when provided with overcurrent protection at the rated ampacity as required by the NEC. When Type UF cable is used above grade where it will come in direct contact with the rays of the sun, its outer covering must be sun-resistant. Furthermore, where Type UF cable emerges from the ground, some means of mechanical protection must be provided. This protection may be in the form of conduit or guard strips. Type UF cable resembles Type NM cable; however, the jacket is constructed of weather-resistant material to provide the required protection for direct-burial wiring installations.

8.1.4 Service-Entrance Cable

Service-entrance (Type SE) and underground service-entrance (Type USE) cable, when used for electrical services, must be installed as specified in **NEC Articles 230 and 338**. Service-entrance cable is available with the grounded conductor bare for outside service conductors, and also with an insulated grounded conductor for interior wiring systems.

Type SE cable is permitted for use on branch circuits or feeders provided that all current-carrying conductors are insulated; this includes the grounded or neutral conductor. When Type SE cable is used for interior wiring, all NEC regulations governing the installation of Type NM cable also apply to Type SE cable; however, there are some exceptions. Type SE cable with an uninsulated, covered conductor may be used as a branch circuit or feeder where the fully-insulated conductors are used for circuit wiring and the uninsulated covered conductor is used only for equipment grounding purposes. Bends of Type SE and USE cable must be made so that the protective covering of the cable is not damaged, and the radius of the bend to the inner edge of the cable is not less than five times the diameter of the cable.

SE Style R (SER) cable is used in residential applications for subfeeds for ranges, and it is also used for service laterals in multi-family dwellings.

Figure 23 summarizes the installation rules for Type SE cable for both exterior and interior wiring.

8.2.0 RACEWAYS

A raceway is any channel that is designed and used solely for the purpose of holding wires, cables, or busbars. Types of raceways include rigid metal conduit, intermediate metal conduit, rigid nonmetallic conduit, flexible metallic conduit, electrical metallic tubing, and auxiliary gutters. Raceways are constructed of either metal or insulating material such as polyvinyl chloride or PVC (plastic). Metal raceways are joined using threaded, compression, or setscrew couplings; nonmetallic raceways are joined using cement-coated couplings. Where a raceway terminates in an outlet box, junction box, or other enclosure, an approved connector must be used.

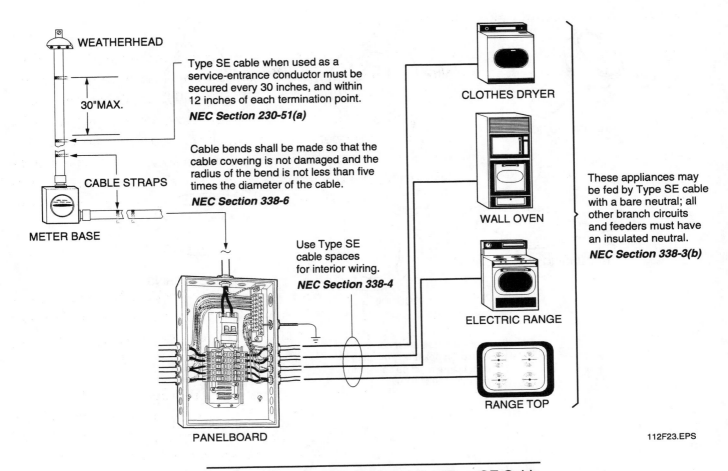

WEATHERHEAD

30"MAX.

CABLE STRAPS

METER BASE

Type SE cable when used as a service-entrance conductor must be secured every 30 inches, and within 12 inches of each termination point.
NEC Section 230-51(a)

Cable bends shall be made so that the cable covering is not damaged and the radius of the bend is not less than five times the diameter of the cable.
NEC Section 338-6

Use Type SE cable spaces for interior wiring.
NEC Section 338-4

PANELBOARD

CLOTHES DRYER

WALL OVEN

ELECTRIC RANGE

RANGE TOP

These appliances may be fed by Type SE cable with a bare neutral; all other branch circuits and feeders must have an insulated neutral.
NEC Section 338-3(b)

112F23.EPS

Figure 23. NEC Sections Governing Type SE Cable

Raceways provide mechanical protection for the conductors that run in them and also prevent accidental damage to insulation and the conducting material. They also protect conductors from corrosive atmospheres and prevent fire hazards to life and property by confining arcs and flames that may occur due to faults in the wiring system. Conduits or raceways are used in residential applications for service masts, underground wiring embedded in concrete, and sometimes in unfinished basements, shops, or garage areas.

Another function of metal raceways is to provide a continuous equipment grounding system throughout the electrical system. To maintain this feature, it is extremely important that all raceway systems be securely bonded together into a continuous conductive path and properly connected to the system ground. The following section explains how this is accomplished.

9.0.0 EQUIPMENT GROUNDING SYSTEM

NEC Article 250 covers grounding requirements. *Figure 24* summarizes the equipment grounding rules for most types of equipment that are required to be grounded. These general guidelines apply to all installations except for specific equipment (special applications) as indicated in the NEC. The NEC also lists specific equipment that is to be grounded regardless of voltage.

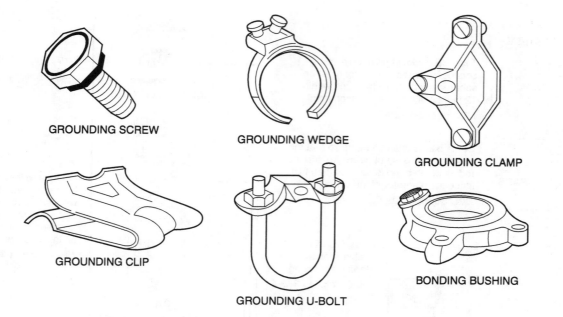

GROUNDING SCREW

GROUNDING WEDGE

GROUNDING CLAMP

GROUNDING CLIP

GROUNDING U-BOLT

BONDING BUSHING

Where splices are made in a junction box, the grounding conductors must be spliced to the metal junction box.

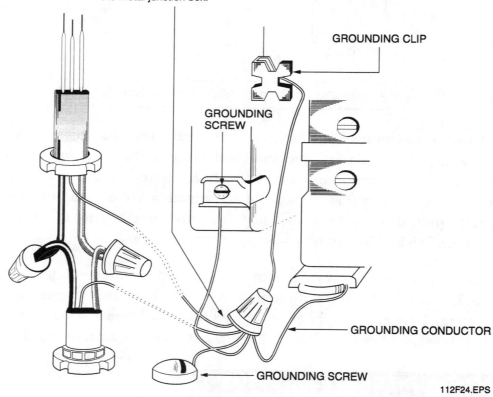

GROUNDING CLIP

GROUNDING SCREW

GROUNDING CONDUCTOR

GROUNDING SCREW

112F24.EPS

Figure 24. Equipment Grounding

In all occupancies, major appliances and many hand-held appliances are required to be grounded. Some of these appliances are refrigerators, freezers, air conditioners, clothes dryers, washing machines, dishwashing machines, sump pumps, and electrical aquarium equipment. Tools that are likely to be used outdoors and in wet or damp locations must also be grounded or have a system of double insulation.

Although most appliance circuits require an equipment grounding conductor, the frames of electric ranges, clothes dryers, and similar appliances that utilize both 120V and 240V may be grounded via the grounded circuit conductor (neutral) under most conditions. In addition, the grounding contacts of any receptacles on the equipment must be bonded to the equipment. If these specified conditions are met, it is not necessary to provide a separate equipment grounding conductor, either for the frames or any outlet or junction boxes that are part of the circuit for these applications.

A bonding jumper is sometimes used to ensure electrical conductivity between metal parts. When the jumper is installed to connect two or more portions of the equipment grounding conductor, the jumper is referred to as an *equipment bonding jumper*. Some specific cases in which a bonding jumper is required are as follows:

- Metal raceways, cable armor, and other metal noncurrent-carrying parts that serve as grounding conductors must be bonded whenever necessary to ensure electrical continuity.
- When flexible metal conduit is used for equipment grounding, an equipment bonding jumper is required if the length of the ground return path exceeds six feet or the circuit enclosed is rated over 20A. When the path exceeds six feet, the circuit must contain an equipment grounding conductor, and the bonding may be accomplished using approved fittings.
- A short length of flexible metal conduit that contains a circuit rated over 20A may not serve as a grounding conductor itself, but a separate bonding jumper can be provided in place of a separate equipment grounding conductor for the circuit. This bonding jumper may be installed inside or outside the conduit, but an outside jumper cannot exceed six feet in length.

10.0.0 BRANCH CIRCUIT LAYOUT FOR POWER

The point at which electrical equipment is connected to the wiring system is commonly called an *outlet*. There are many classifications of outlets: lighting, receptacle, motor, appliance, etc. This section, however, deals with the power outlets normally found in residential electrical wiring systems.

When viewing an electrical drawing, outlets are indicated by symbols (usually a small circle with appropriate markings to indicate the type of outlet). The most common symbols for receptacles are shown in *Figure 25*.

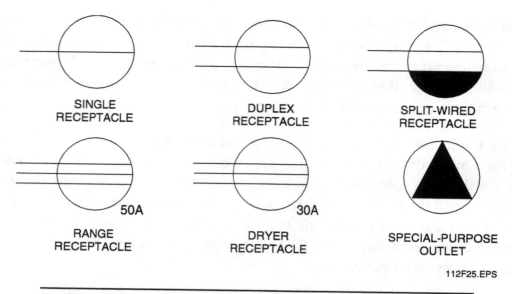

Figure 25. Typical Outlet Symbols Appearing In Electrical Drawings

10.1.0 BRANCH CIRCUITS AND FEEDERS

The conductors that extend from the panelboard to the various outlets are called *branch circuits* and are defined by the NEC as the point of a wiring system that extends beyond the final overcurrent device protecting the circuit. See *Figure 26.*

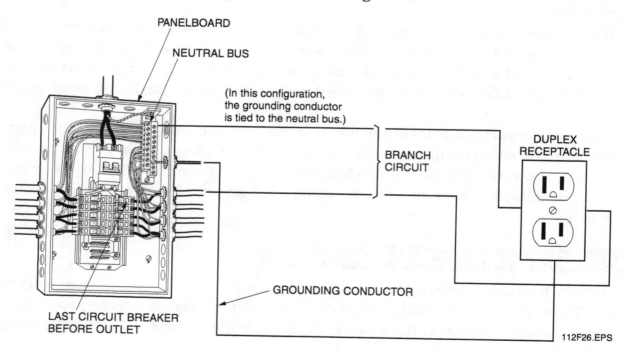

Figure 26. Components Of A Duplex Receptacle Branch Circuit

A feeder consists of all conductors between the service equipment and the final overcurrent device. See *Figure 27.*

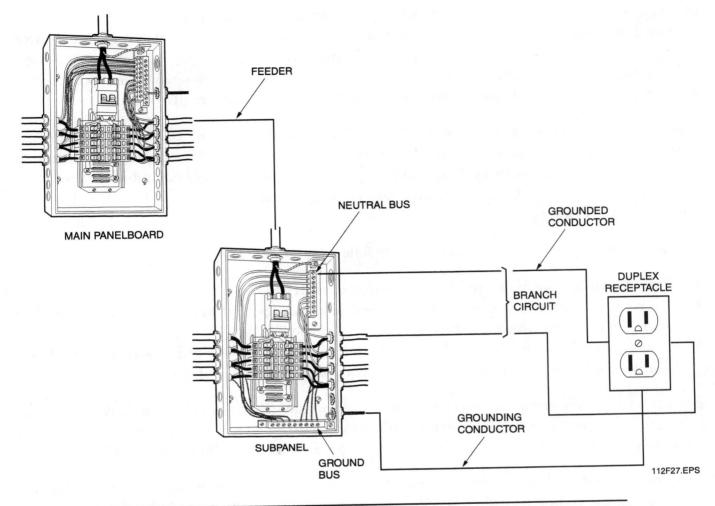

FEEDER

NEUTRAL BUS

GROUNDED
CONDUCTOR

MAIN PANELBOARD

BRANCH
CIRCUIT

DUPLEX
RECEPTACLE

SUBPANEL

GROUND
BUS

GROUNDING
CONDUCTOR

112F27.EPS

Figure 27. A Feeder Being Used To Feed A Subpanel From The Main Panelboard

In general, the size of the branch circuit conductors varies depending upon the load requirements of the electrically-operated equipment connected to the outlet. For residential use, most branch circuits consist of either No. 14 AWG, No. 12 AWG, No. 10 AWG, or No. 8 AWG conductors.

The basic branch circuit requires two wires or conductors to provide a continuous path for the flow of electric current, plus a third wire for equipment grounding. The usual receptacle branch circuit operates at 120V.

Fractional horsepower motors and small electric heaters usually operate at 120V and are connected to 120V branch circuits either by means of a receptacle, junction box, or direct connection.

With the exception of very large residences and tract-development houses, the size of the average residential electrical system of the past has not been large enough to justify the expense of preparing complete electrical working drawings and specifications. Such electrical systems were usually laid out by the architect in the form of a sketchy outlet arrangement or else laid out by the electrician on the job, often only as the work progressed. However, many technical developments in residential electrical use—such as electric heat with sophisticated control wiring, increased use of electrical appliances, various electronic alarm systems, new lighting techniques, and the need for energy conservation techniques—have greatly expanded the demand and extended the complexity of today's residential electrical systems.

Each year, the number of homes with electrical systems designed by consulting engineering firms increases. Such homes are provided with complete electrical working drawings and specifications, similar to those frequently provided for commercial and industrial projects. Still, these are more the exception than the rule. Most residential projects will not have a complete set of drawings.

Circuit layout is provided on the drawings to follow for several reasons:

- They provide a visual layout of house wiring circuitry.
- They provide a sample of electrical residential drawings that are prepared by consulting engineering firms, although the number may still be limited.
- They introduce the method of showing electrical systems on working drawings so that you will have a better foundation for tackling advanced electrical systems.

Branch circuits are shown on electrical drawings by means of a single line drawn from the panelboard (or by homerun arrowheads indicating that the circuit goes to the panelboard) to the outlet or from outlet to outlet where there is more than one outlet on the circuit.

The lines indicating branch circuits can be solid to show that the conductors are to be run concealed in the ceiling or wall, dashed to show that the conductors are to be run in the floor or ceiling below, or dotted to show that the wiring is to be run exposed. *Figure 28* shows examples of these three types of branch circuit lines.

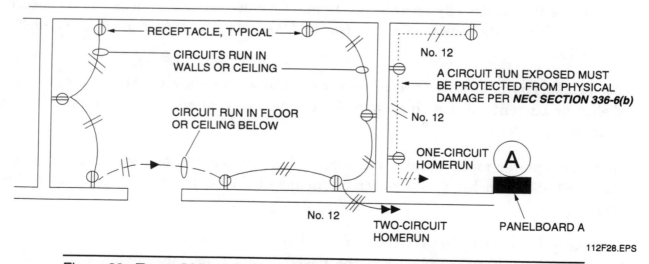

Figure 28. Types Of Branch Circuit Lines Shown On Electrical Working Drawings

In *Figure 28*, No. 12 indicates the wire size. The slash marks shown through the circuits in *Figure 28* indicate the number of current-carrying conductors in the circuit. Although two slash marks are shown, in actual practice, a branch circuit containing only two conductors usually contains no slash marks; that is, any circuit with no slash marks is assumed to have two conductors. However, three or more conductors are always indicated on electrical working drawings—either by slash marks for each conductor, or else by a note.

Never assume that you know the meaning of any electrical symbol. Although great efforts have been made in recent years to standardize drawing symbols, architects, consulting engineers, and electrical drafters still modify existing symbols or devise new ones to meet their own needs. Always consult the symbol list or legend on electrical working drawings for an exact interpretation of the symbols used.

10.2.0   LOCATING RECEPTACLES

NEC Section 210-52 states the minimum requirements for the location of receptacles in dwelling units. It specifies that in each kitchen, family room, and dining room, receptacle outlets shall be installed so that no point along the floor line in any wall space is more than six feet (1.83m), measured horizontally, from an outlet in that space, including any wall space two feet (610mm) or more in width and the wall space occupied by fixed panels in exterior walls, but excluding sliding panels. Receptacle outlets shall, insofar as practicable, be spaced equal distances apart. Receptacle outlets in floors shall not be counted as part of the required number of receptacle outlets unless located within 18 inches (457mm) of the wall.

The NEC defines wall space as a wall that is unbroken along the floor line by doorways, fireplaces, or similar openings. Each wall space that is two feet or more in width must be treated individually and separately from other wall spaces within the room.

The purpose of the above NEC requirement is to minimize the use of cords across doorways, fireplaces, and similar openings.

With this NEC requirement in mind, outlets for our sample residence will be laid out (see *Figure 29*). In laying out these receptacle outlets, the floor line of the wall is measured (also around corners) but not across doorways, fireplaces, passageways, or other spaces where a flexible cord extended across the space would be unsuitable.

In general, duplex receptacle outlets must be no more than 12 feet apart. When spaced in this manner, a six-foot extension cord will reach a receptacle from any point along the wall line.

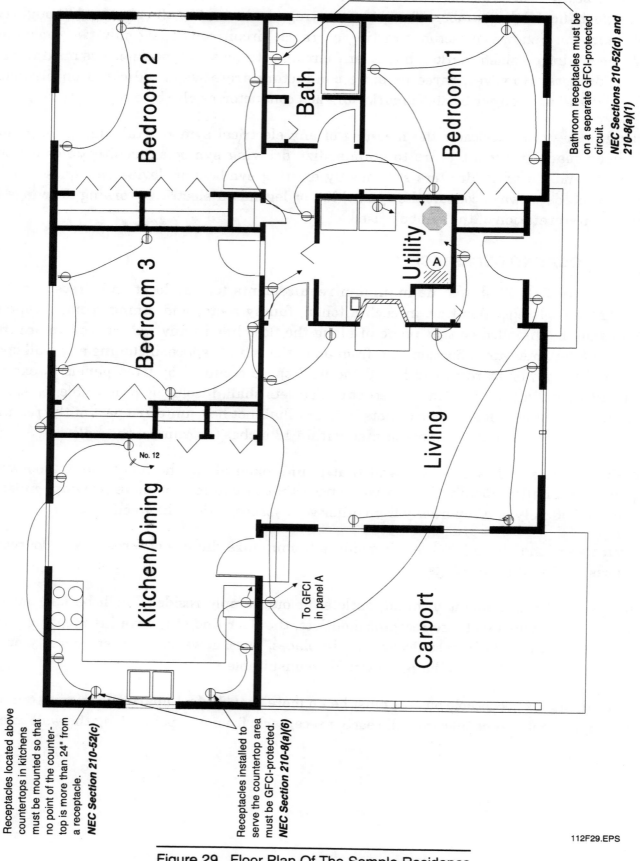

Receptacles located above
countertops in kitchens
must be mounted so that
no point of the counter-
top is more than 24" from
a receptacle.
NEC Section 210-52(c)

Receptacles installed to
serve the countertop area
must be GFCI-protected.
NEC Section 210-8(a)(6)

Bathroom receptacles must be
on a separate GFCI-protected
circuit.
*NEC Sections 210-52(d) and
210-8(a)(1)*

To GFCI
in panel A

No. 12

Bedroom 2

Bath

Bedroom 1

Bedroom 3

Utility

(A)

Living

Kitchen/Dining

Carport

112F29.EPS

Figure 29. Floor Plan Of The Sample Residence

Note that at no point along the wall line are any receptacles more than 12 feet apart or more than six feet from any door or room opening. Where practical, no more than eight receptacles are connected to one circuit. However, *NEC Section 220-3(b)* specifies a demand factor of 180VA per receptacle or group of receptacles. Since a 15A branch circuit is rated at 1,800VA (15A x 120V = 1,800VA), 10 general purpose duplex receptacles may be connected to one circuit.

The utility room has at least one receptacle for the laundry on a separate circuit in order to comply with *NEC Section 210-52(f)*.

One duplex receptacle is located in the vestibule for cleaning purposes, such as feeding a portable vacuum cleaner or similar appliance. It is connected to the living room circuit. An additional duplex receptacle is required per *NEC Section 210-52(h)* in hallways of 10 feet or more.

Although this is not shown in the figure, the living room outlets could be split-wired (the lower half of each duplex receptacle is energized all the time, while the upper half can be switched on or off). The reason for this is that a great deal of the illumination for this area will be provided by portable table lamps, and the split-wired receptacles provide a means to control these lamps from several locations, such as at each entry to the living room, if desired. Split receptacles are discussed in more detail in the next section.

To comply with *NEC Section 210-52(b)*, the kitchen receptacles are laid out as follows. In addition to the number of branch circuits determined previously, two or more 20A small appliance branch circuits must be provided to serve all receptacle outlets (including refrigeration equipment) in the kitchen, pantry, breakfast room, dining room, or similar area of the house. Such circuits, whether two or more are used, must have no other outlets connected to them. All receptacles serving a kitchen countertop require GFCI protection. No small appliance branch circuit shall serve more than one kitchen.

To comply with *NEC Section 210-11(c)(3)*, bathroom receptacle(s) must be on a separate branch circuit supplying only bathroom receptacles or on a circuit supplying a single bathroom with no loads other than that bathroom. All receptacles located within a bathroom require GFCI protection. GFCI protection is also required on garage and exterior receptacles.

Effective January 1, 2002, all bedroom receptacles will be required to have arc-fault circuit interrupter protection to comply with *NEC Section 210-12(b)*.

10.3.0   SPLIT-WIRED DUPLEX RECEPTACLES

In modern residential construction, it is common to have duplex wall receptacles that have one of the outlets wired as a standard duplex outlet (hot all the time) and the other half controlled by a wall switch. This allows table or floor lamps to be controlled by a wall switch and leaves the other outlet available for items that are not to be switched. This wiring method is commonly referred to as a *split receptacle*.

Most duplex 15A and 20A receptacles are provided with a breakoff tab that permits each of the two receptacle outlets to be supplied from a different source or polarity. For example, one outlet would be supplied from the hot leg of a series of outlets and the other outlet supplied from the **switch leg** of a light switch. A diagram of this arrangement is shown in *Figure 30*.

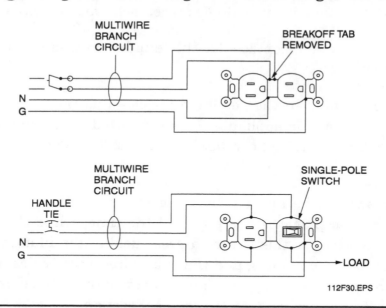

Figure 30. Two 120V Receptacle Outlets Supplied From Different Sources

Another application of split receptacles is shown in *Figure 31*. In this example, one outlet connected from a double-pole circuit breaker supplies 240V for an appliance such as a window air conditioning unit, while the other outlet is connected from one pole of the double-pole circuit breaker and the other side is connected to the neutral or grounded conductor to supply 120V for an appliance such as a lamp. This circuit and the split receptacle mentioned above are both considered multiwire branch circuits.

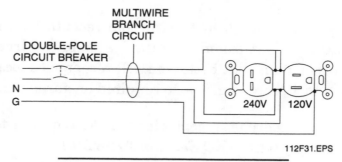

Figure 31. Combination Receptacle

10.4.0   MULTIWIRE BRANCH CIRCUITS

NEC Article 100 defines a multiwire branch circuit as "two or more ungrounded conductors having a potential difference between them, and a grounded conductor having equal potential difference between it and each ungrounded conductor of the circuit and that is connected to the neutral conductor of the system."

Multiwire branch circuits have many advantages, such as three wires doing the work of four (in place of two two-wire branch circuits), less raceway fill, easier balancing and phasing of a system, and less voltage drop.

10.5.0 240-VOLT CIRCUITS

The electric range, clothes dryer, and water heater in the sample residence all operate at 240VAC. Each will be fed by a separate circuit and connected to a two-pole circuit breaker of the appropriate rating in the panelboard. To determine the conductor size and overcurrent protection for the range, proceed as follows:

Step 1   Find the nameplate rating of the electric range. This has previously been determined to be 12kVA.

Step 2   Refer to **NEC Table 220-19**. Since Column A of this table applies to ranges rated at 12kVA (12kW) and under, this will be the column to use in this example.

Step 3   Under the Number of Appliances column, locate the appropriate number of appliances (one in this case), and find the maximum demand given for it in Column A. Column A states that the circuit should be sized for 8kVA (not the nameplate rating of 12kVA).

Step 4   Calculate the required conductor ampacity as follows:

$$\frac{8,000VA}{240V} = 33.33A$$

The minimum branch circuit must be rated at 40A since common residential circuit breakers are rated in steps of 15A, 20A, 30A, 40A, etc. A 30A circuit breaker is too small, so a 40A circuit breaker is selected. The conductors must have a current-carrying capacity that is equal to or greater than the overcurrent protection. Therefore, No. 8 AWG conductors will be used.

If a cooktop and wall oven were used instead of the electric range, the circuit would be sized similarly. The NEC specifies that a branch circuit for a counter-mounted cooking unit and not more than two wall-mounted ovens, all supplied from a single branch circuit and located in the same room, is computed by adding the nameplate ratings of the individual appliances and treating this total as equivalent to one range. Therefore, two appliances of 6kVA each may be treated as a single range with a 12kVA nameplate rating.

Figure 32 shows how the electric range circuit may appear on an electrical drawing. The connection may be made directly to the range junction box, but more often a 50A range receptacle is mounted at the range location and a range cord-and-plug set is used to make the connection. This facilitates moving the appliance later for maintenance or cleaning.

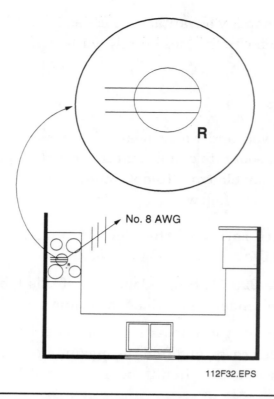

No. 8 AWG

112F32.EPS

Figure 32. Range Circuit Shown On An Electrical Drawing

Figure 33 shows several types of receptacle configurations used in residential wiring applications. You will eventually recognize these configurations at a glance.

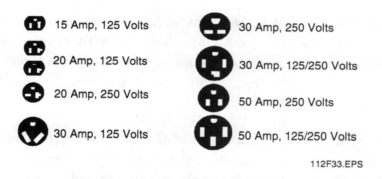

15 Amp, 125 Volts

20 Amp, 125 Volts

20 Amp, 250 Volts

30 Amp, 125 Volts

30 Amp, 250 Volts

30 Amp, 125/250 Volts

50 Amp, 250 Volts

50 Amp, 125/250 Volts

112F33.EPS

Figure 33. Residential Receptacle Configurations

The branch circuit for the water heater in the sample residence must be sized for its full capacity because there is no diversity or demand factor for this appliance. Since the nameplate rating on the water heater indicates two heating elements of 4,500 watts each, the first inclination would be to size the circuit for a total load of 9,000 watts (volt-amperes). However, only one of the two elements operates at a time. See *Figure 34*. Note that each element is controlled by a separate thermostat. The lower element becomes energized when the thermostat calls for heat, and at the same time, the thermostat opens a set of contacts to prevent the upper element from operating. When the lower element's thermostat is satisfied, the lower contacts open, and at the same time, the thermostat closes the contacts for the upper element to become energized to maintain the water temperature.

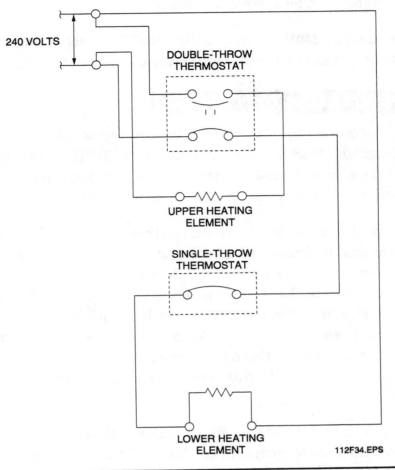

Figure 34. Wiring Diagram Of Water Heater Controls

With this information in hand, the circuit for the water heater may be sized as follows:

$$\frac{4,500VA}{240V} = 18.75A \times 1.25 = 23.44A$$

Since a water heater will more than likely fall under the continuous load category, the conductor and overcurrent protection should be rated at 125%. Therefore, No. 10 AWG wire should be used and protected with a 30A circuit breaker. A direct connection is made to the water heater at the integral junction box on top of the heater.

The NEC specifies that electric clothes dryers must be rated at 5kVA or the nameplate rating, whichever is greater. In this case, the dryer is rated at 5.5kVA, and the conductor current-carrying capacity is calculated as follows:

$$\frac{5,500VA}{240V} = 22.92A$$

A three-wire, 30A circuit will be provided (No. 10 AWG wire). It is protected by a 30A circuit breaker. The dryer may be connected directly, but a 30A dryer receptacle is normally provided for the same reasons as mentioned for the electric range.

Large appliance outlets rated at 240V are frequently shown on electrical drawings using lines and symbols to indicate the outlets and circuits. In some cases, no drawings are provided.

11.0.0 BRANCH CIRCUIT LAYOUT FOR LIGHTING

A simple lighting branch circuit requires two conductors to provide a continuous path for current flow. The usual lighting branch circuit operates at 120V; the white (grounded) circuit conductor is therefore connected to the neutral bus in the panelboard, while the black (ungrounded) circuit conductor is connected to an overcurrent protection device.

Lighting branch circuits and outlets are shown on electrical drawings by means of lines and symbols; that is, a single line is drawn from outlet to outlet and then terminated with an arrowhead to indicate a homerun to the panelboard. Several methods are used to indicate the number and size of conductors, but the most common is to indicate the number of conductors in the circuit by using slash marks through the circuit lines and then indicate the wire size by a notation adjacent to these slash marks. For example, two slash marks indicate two conductors; three slash marks indicate three conductors; etc. Some electrical designers omit slash marks for two-conductor circuits. In this case, the conductor size is usually indicated in the symbol list or legend.

The circuits used to feed residential lighting must conform to standards established by the NEC as well as by local and state ordinances. Most of the lighting circuits should be calculated to include the total load, although at times this is not possible because the electrician cannot be certain of the exact wattage that might be used by the homeowner. For example, an electrician may install four porcelain lampholders for the unfinished basement area, each to contain one 100-watt (100W) incandescent lamp. However, the homeowners may eventually replace the original lamps with others rated at 150W or even 200W. Thus, if the electrician initially loads the lighting circuit to full capacity, the circuit will probably become overloaded in the future.

It is recommended that no residential branch circuit be loaded to more than 80% of its rated capacity. Since most circuits used for lighting are rated at 15A, the total ampacity (in volt-amperes) for the circuit is as follows:

15A x 120V = 1,800VA

Therefore, if the circuit is to be loaded to only 80% of its rated capacity, the maximum initial connected load should be no more than 1,440VA.

Figure 35 shows one possible lighting arrangement for the sample residence. All lighting fixtures are shown in their approximate physical location as they should be installed in the residence.

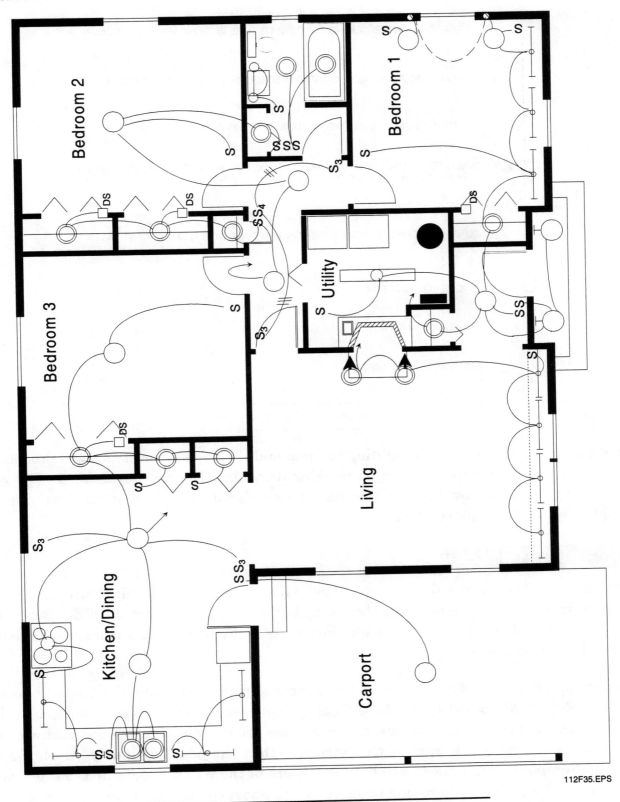

112F35.EPS

Figure 35. Lighting Layout Of The Sample Residence

Electrical symbols are used to show the fixture types. Switches and lighting branch circuits are also shown by appropriate lines and symbols. The meanings of the symbols used on this drawing are explained in the symbol list in *Figure 36*.

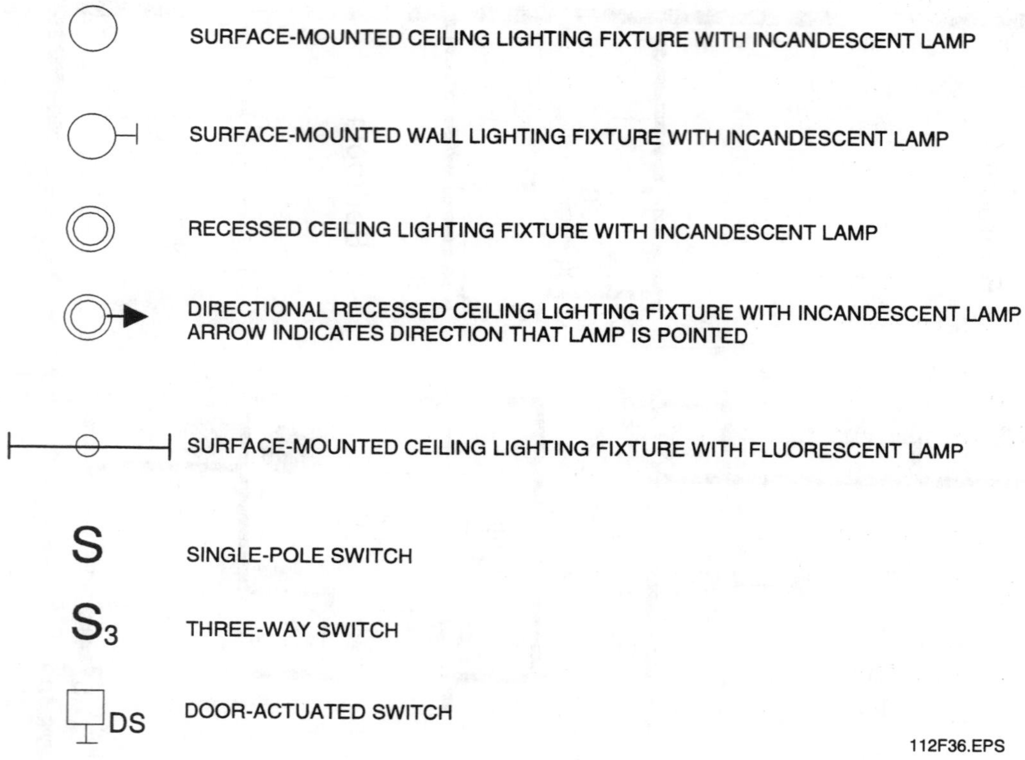

SURFACE-MOUNTED CEILING LIGHTING FIXTURE WITH INCANDESCENT LAMP

SURFACE-MOUNTED WALL LIGHTING FIXTURE WITH INCANDESCENT LAMP

RECESSED CEILING LIGHTING FIXTURE WITH INCANDESCENT LAMP

DIRECTIONAL RECESSED CEILING LIGHTING FIXTURE WITH INCANDESCENT LAMP
ARROW INDICATES DIRECTION THAT LAMP IS POINTED

SURFACE-MOUNTED CEILING LIGHTING FIXTURE WITH FLUORESCENT LAMP

S   SINGLE-POLE SWITCH

S_3 THREE-WAY SWITCH

DS   DOOR-ACTUATED SWITCH

112F36.EPS

Figure 36. Symbols

In actual practice, the location of lighting fixtures and their related switches will probably be the extent of the information shown on working drawings. The circuits shown in *Figure 35* are meant to illustrate how lighting circuits are routed, not to imply that such drawings are typical for residential construction.

12.0.0   OUTLET BOXES

Electricians installing residential electrical systems must be familiar with outlet box capacities, means of supporting outlet boxes, and other requirements of the NEC. Boxes were discussed in detail in an earlier module, but a general review of the rules and necessary calculations is provided here.

The maximum numbers of conductors of the same size permitted in standard outlet boxes are listed in **NEC Table 370-16(a)**. These figures apply where no fittings or devices such as fixture studs, cable clamps, switches, or receptacles are contained in the box and where no grounding conductors are part of the wiring within the box. Obviously, in all modern residential wiring systems there will be one or more of these items contained in every outlet box installed. Therefore, where one or more of the above-mentioned items are present, the total number of conductors will be one less than that shown in the table.

For example, a deduction of two conductors must be made for each strap containing a wiring device entering the box (based on the largest size conductor connected to the device) such as a switch or duplex receptacle; a further deduction of one conductor must be made for one or more equipment grounding conductors entering the box (based on the largest size grounding conductor). For instance, a 3-inch x 2-inch x 2¾-inch box is listed in the table as containing a maximum of six No. 12 wires. If the box contains cable clamps and a duplex receptacle, three wires will have to be deducted from the total of six, providing for only three No. 12 wires. If a ground wire is used—which is always the case in residential wiring—only two No. 12 wires may be used.

For example, to size a metallic outlet box for two No. 12 AWG conductors with a ground wire, cable clamp, and receptacle, proceed as follows:

Step 1 Calculate the total number of conductors and equivalents [*NEC Section 370-16(b)*]. One ground wire plus one cable clamp plus one receptacle (two wires) plus two No. 12 conductors equals a total of six No. 12 conductors.

Step 2 Determine the amount of space required for each conductor. *NEC Table 370-16(b)* gives the box volume required for each conductor. No. 12 AWG equals 2.25 cubic inches.

Step 3 Calculate the outlet box space required by multiplying the number of cubic inches required for each conductor by the total number of conductors:

6 x 2.25 = 13.5 cubic inches

Step 4 Once you have determined the required box capacity, again refer to *NEC Table 370-16(a)* and note that a 3-inch x 2-inch x 2¾-inch box comes closest to our requirements. This box is rated for 14 cubic inches.

Now, size the box for two additional conductors. Where four No. 12 conductors enter the box with two ground wires, only the two additional No. 12 conductors must be added to our previous count for a total of 8 conductors (6 + 2 = 8). Remember, any number of ground wires in a box counts as only one conductor; any number of cable clamps also count as only one conductor. Therefore, the box size required for use with two additional No. 12 conductors may be calculated as follows:

8 x 2.25 = 18 cubic inches

Again, refer to *NEC Table 370-16(a)* and note that a 3-inch x 2-inch x 3½-inch device box with a rated capacity of 18.0 cubic inches is the closest device box that meets NEC requirements. An alternative is to use a 4-inch x 1¼-inch square box with a single-gang plaster ring, as shown in *Figure 37*. This box also has a capacity of 18.0 cubic inches.

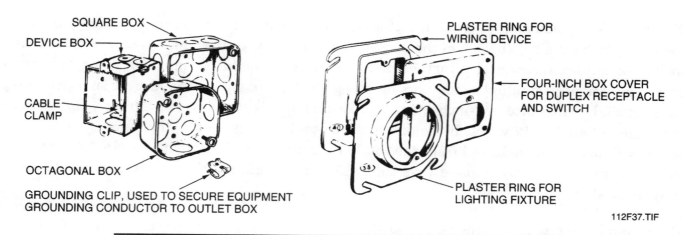

SQUARE BOX

DEVICE BOX

CABLE CLAMP

OCTAGONAL BOX

GROUNDING CLIP, USED TO SECURE EQUIPMENT
GROUNDING CONDUCTOR TO OUTLET BOX

PLASTER RING FOR WIRING DEVICE

FOUR-INCH BOX COVER FOR DUPLEX RECEPTACLE AND SWITCH

PLASTER RING FOR LIGHTING FIXTURE

112F37.TIF

Figure 37. Typical Metallic Outlet Boxes With Extension (Plaster) Rings

Other box sizes are calculated in a similar fashion. When sizing boxes for different size conductors, remember that the box capacity varies as shown in ***NEC Table 370-16(b)***.

12.1.0   MOUNTING OUTLET BOXES

Outlet box configurations are almost endless, and if you research the various methods of mounting these boxes, you will be astonished. In this section, some common outlet boxes and their mounting considerations will be reviewed.

The conventional metallic device box, which is used for residential duplex receptacles and switches for lighting control, may be mounted to wall studs using 16-l (penny) nails placed through the round mounting holes passing through the interior of the box. The nails are then driven into the wall stud. When nails are used for mounting outlet boxes in this manner, the nails must be located within ¼ inch of the back or ends of the enclosure.

Nonmetallic boxes normally have mounting nails fitted to the box for mounting. Other boxes have mounting brackets. When mounting outlet boxes with brackets, use either wide-head roofing nails or box nails about 1¼ inches in length. *Figure 38* shows various methods of mounting outlet boxes.

Before mounting any boxes during the rough wiring process, first find out what type and thickness of finish will be used on the walls. This will dictate the depth to which the boxes must be mounted to comply with NEC regulations. For example, the finish on plastered walls or ceilings is normally ½-inch thick; gypsum board or drywall is either ½-inch or ⅝-inch thick; and wood paneling is normally only ¼-inch thick. (Some tongue-and-groove wood paneling is ½-inch to ⅝-inch thick.)

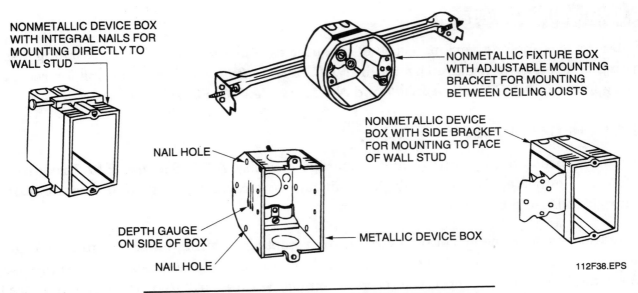

Figure 38. Several Methods Of Mounting Outlet Boxes

The NEC specifies the amount of space permitted from the edge of the outlet box to the finished wall. When a noncombustible wall finish (such as plaster, masonry, or tile) is used, the box may be recessed ¼ inch. However, when combustible finishes are used (such as wood paneling), the box must be flush (even) with the finished wall or ceiling. See *Figure 39* and ***NEC Section 370-20***.

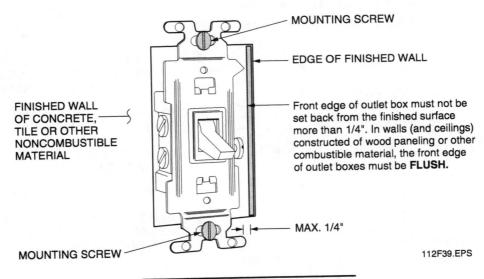

Figure 39. Outlet Box Installation

When Type NM cable is used in either metallic or nonmetallic outlet boxes, the cable assembly, including the sheath, must extend into the box by not less than ¼ inch [***NEC Section 370-17(c)***]. In all instances, all permitted wiring methods must be secured to the boxes by means of either cable clamps or approved connectors. The one exception to this rule is where Type NM cable is used with 2¼-inch x 4-inch (or smaller) nonmetallic boxes where the cable is fastened within eight inches of the box. In this case, the cable does not have to be secured to the box. See ***NEC Section 370-17(c), Exception***.

Wiring devices include various types of receptacles and switches, the latter being used for lighting control. Switches are covered in **NEC Article 380**, while regulations for receptacles may be found in **NEC Section 210-7** and **NEC Article 410, Part L**.

13.1.0 RECEPTACLES

Receptacles are rated by voltage and amperage capacity. **NEC Section 210-7** requires that receptacles connected to a 15A or 20A circuit have the correct voltage and current rating for the application, and be of the grounding type.

Where there is only one outlet on a circuit, the receptacle's rating must be equal to or greater than the capacity of the conductors feeding it. For example, if one receptacle is connected to a 20A residential laundry circuit, the receptacle must be rated at 20A or more. When more than one outlet is on a circuit, the total connected load must be equal to or less than the capacity of the branch circuit conductors feeding the receptacles.

Refer to *Figure 40* for some of the characteristics of a standard 125V, 15A duplex receptacle. Note that the terminals are color coded as follows:

- *Green* – Connection for the equipment grounding conductor
- *Silver* – Connection for the neutral or grounded conductor
- *Brass* – Connection for the ungrounded conductor

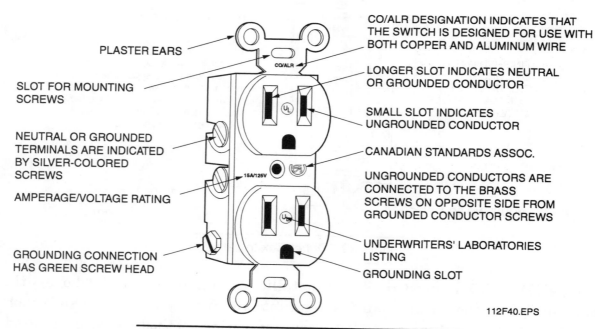

Figure 40. Standard 125V, 15A Duplex Receptacle

A standard 125V, 15A receptacle is also typically imprinted with the following symbols:

- *UL* – Underwriters' Laboratories listing
- *CSA* – Canadian Standards Association
- *CO/ALR* – Designed for use with both copper and aluminum wire
- *15A* – Receptacle rated for a maximum of 15A
- *125V* – Receptacle rated for a maximum of 125V

The UL label means that the receptacle has undergone testing by Underwriters' Laboratories, Inc., and meets minimum safety requirements. Underwriters' Laboratories was created by the National Board of Fire Underwriters to test electrical devices and materials. The UL label is a safety rating only and does not mean that the device or equipment meets any type of quality standard. The CSA label means that the receptacle is approved by the Canadian Standards Association, the Canadian equivalent to Underwriters' Laboratories. The CSA label means that the receptacle is acceptable for use in Canada.

The CO/ALR symbol means that the device is suitable for use with copper, aluminum, or copper-clad aluminum wire. The CO in the symbol stands for *copper* while ALR stands for *aluminum revised*. The CO/ALR symbol replaces the earlier CU/AL mark, which appeared on wiring devices that were later found to be inadequate for use with aluminum wire in the 15A to 20A range. Therefore, any receptacle or wall switch marked with the CU/AL configuration or anything other than CO/ALR should be used only for copper wire.

These same configurations also apply to wall switches used for lighting control. These will be discussed next.

14.0.0   LIGHTING CONTROL

There are many types of lighting control devices. These devices have been designed to make the best use of the lighting equipment provided by the lighting industry. They include:

- Automatic timing devices for outdoor lighting
- Dimmers for residential lighting
- Common single-pole, three-way, and four-way switches

For the purposes of this module, a switch is defined as a device that is used on branch circuits to control lighting. Switches fall into the following basic categories:

- Snap-action switches
- Quiet switches

Snap-action switches – A single-pole snap-action switch consists of a device containing two stationary current-carrying elements, a moving current-carrying element, a toggle handle, a spring, and a housing. When the contacts are open, as shown in *Figure 41*, the circuit is broken and no current flows. When the moving element is closed by manually flipping the toggle handle, the contacts complete the circuit and the lamp is energized. See *Figure 42*.

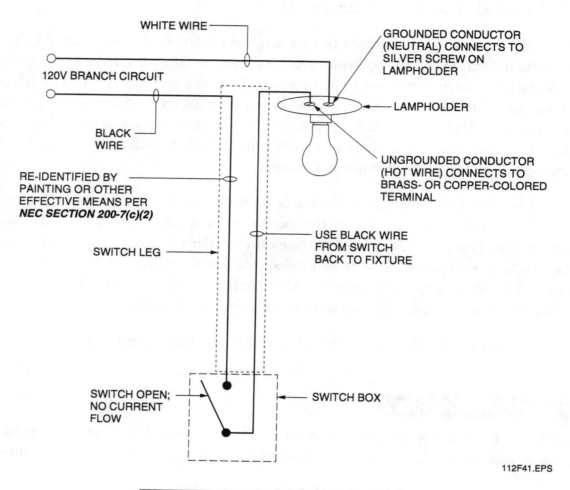

112F41.EPS

Figure 41. Switch Operation, Contacts Open

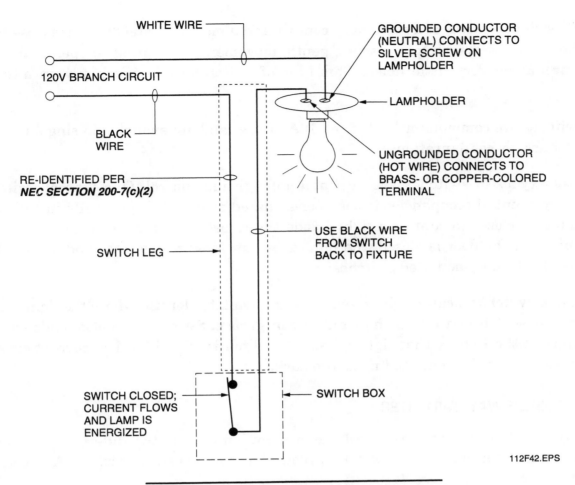

Figure 42. Switch Operation, Contacts Closed

Quiet switches – The quiet switch (*Figure 43*) is the most common switch for use in lighting applications. Its operation is much quieter than the snap-action switch.

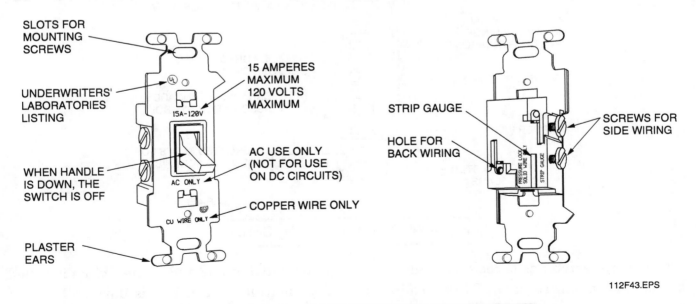

Figure 43. Characteristics Of A Single-Pole Quiet Switch

The quiet switch consists of a stationary contact and a moving contact that are close together when the switch is open. Only a short, gentle movement is required to open and close the switch, producing very little noise. This type of switch may be used only on alternating current.

Quiet switches are common for loads from 10A to 20A, and are available in single-pole, three-way, and four-way configurations.

Many other types of switches are available for lighting control. One type of switch used mainly in residential occupancies is the door-actuated switch. It is generally installed in the door jamb of a closet to control a light inside the closet. When the door is open, the light comes on; when the door is closed, the light goes out. Most refrigerator and oven lights are also controlled by door-actuated switches.

Combination switch/indicator light assemblies are available for use where the light cannot be seen from the switch location, such as an attic or garage. Switches are also made with small neon lamps in the handle that light when the switch is off. These low-current-consuming lamps make the switches easy to find in the dark.

14.1.0 THREE-WAY SWITCHES

Three-way switches are used to control one or more lamps from two different locations, such as at the top and bottom of stairways, in a room that has two entrances, etc. A typical three-way switch is shown in *Figure 44*.

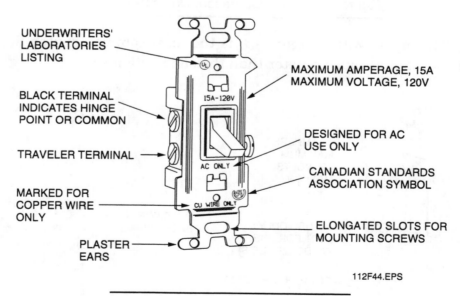

112F44.EPS

Figure 44. Typical Three-Way Switch

A three-way switch has three terminals. The single terminal at one end of the switch is called the *common* or *hinge point*. This terminal is easily identified because it is darker than the other two terminals. The feeder (hot wire) or switch leg is always connected to the common dark or black terminal. The two remaining terminals are called *traveler terminals*. These terminals are used to connect three-way switches together.

The connection of two three-way switches is shown in *Figure 45*. By means of the two switches, it is possible to control the lamp from two locations. By tracing the circuit, it may be seen how these three-way switches operate.

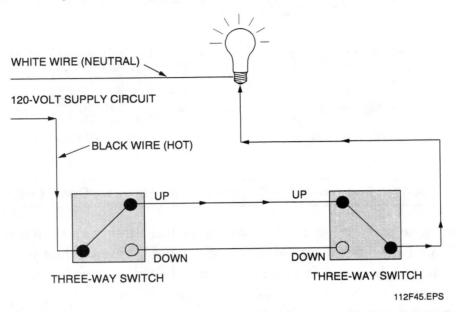

Figure 45. Three-Way Switches In The On Position; Both Handles Are Up

A 120V circuit emerges from the left side of the drawing. The white or neutral wire connects directly to the neutral terminal of the lamp. The hot wire carries current, in the direction of the arrows, to the common terminal of the three-way switch on the left. Since the handle is in the up position, the current continues to the top traveler terminal and is carried by this traveler to the other three-way switch. Note that the handle is also in the up position on this switch; this picks up the current flow and carries it to the common point, which continues to the ungrounded terminal of the lamp to make a complete circuit. The lamp is energized.

Moving the handle to a different position on either three-way switch will break the circuit, which in turn deenergizes the lamp. For example, let's say a person leaves the room at the point of the three-way switch on the left, and the switch handle is flipped down, as shown in *Figure 46*. Note that the current flow is now directed to the bottom traveler terminal, but since the handle of the three-way switch on the right is still in the up position, no current will flow to the lamp.

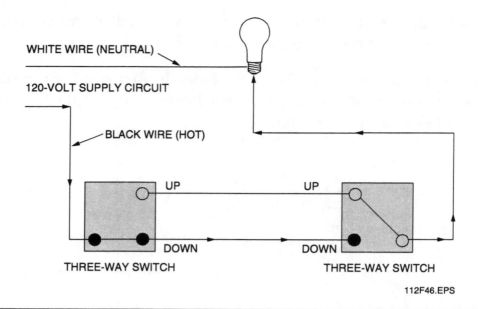

WHITE WIRE (NEUTRAL)

120-VOLT SUPPLY CIRCUIT

BLACK WIRE (HOT)

UP

UP

DOWN

DOWN

THREE-WAY SWITCH

THREE-WAY SWITCH

112F46.EPS

Figure 46. Three-Way Switches In The Off Position; One Handle Is Down, One Handle Is Up

If another person enters the room at the location of the three-way switch on the right, and the handle is flipped downward, as shown in *Figure 47*, this change provides a complete circuit to the lamp, which causes it to be energized. In this example, current flow is on the bottom traveler. Again, changing the position of the switch handle (pivot point) on either three-way switch will deenergize the lamp.

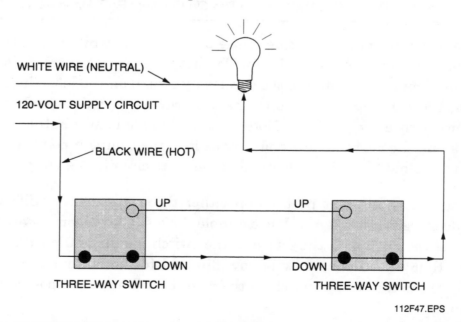

WHITE WIRE (NEUTRAL)

120-VOLT SUPPLY CIRCUIT

BLACK WIRE (HOT)

UP

UP

DOWN

DOWN

THREE-WAY SWITCH

THREE-WAY SWITCH

112F47.EPS

Figure 47. Three-Way Switches With Both Handles Down; The Light Is Energized

In actual practice, the exact wiring of the two three-way switches to control the operation of a lamp will be slightly different than the routing shown in these three diagrams. There are several ways that two three-way switches may be connected. One solution is shown in *Figure 48*. In this case, two-wire, Type NM cable is fed to the three-way switch on the left.

ELECTRICAL — TRAINEE TASK MODULE 26112

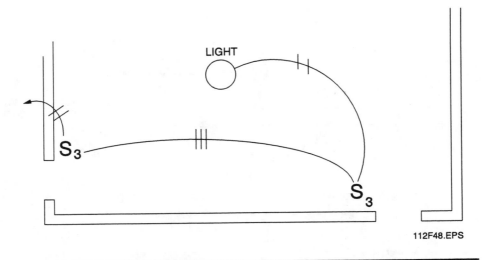

Figure 48. Method Of Showing The Wiring Arrangement On A Floor Plan

The black or hot conductor is connected to the common terminal on the switch, while the white or neutral conductor is spliced to the white conductor of the three-wire, Type NM cable leaving the switch. This three-wire cable is necessary to carry the two travelers plus the neutral to the three-way switch on the right. At this point, the black and red wires connect to the two traveler terminals, respectively. The white or neutral wire is again spliced—this time to the white wire of another two-wire, Type NM cable. The neutral wire is never connected to the switch itself. The black wire of the two-wire, Type NM cable connects to the common terminal on the three-way switch. This cable, carrying the hot and neutral conductors, is routed to the lighting fixture outlet for connection to the fixture.

Another solution is to feed the lighting fixture outlet with two-wire cable. Run another two-wire cable carrying the hot and neutral conductors to one of the three-way switches. A three-wire cable is pulled between the two three-way switches, and then another two-wire cable is routed from the other three-way switch to the lighting fixture outlet.

Some electricians use a shortcut method that eliminates one of the two-wire cables in the preceding method. In this case, a two-wire cable is run from the lighting fixture outlet to one three-way switch. Three-wire cable is pulled between the two three-way switches—two of the wires for travelers and the third for the common point return. This method is shown in *Figure 49*.

14.2.0   FOUR-WAY SWITCHES

Two three-way switches may be used in conjunction with any number of four-way switches to control a lamp, or a series of lamps, from any number of positions. When connected correctly, the actuation of any one of these switches will change the operating condition of the lamp (i.e., either turn the lamp on or off).

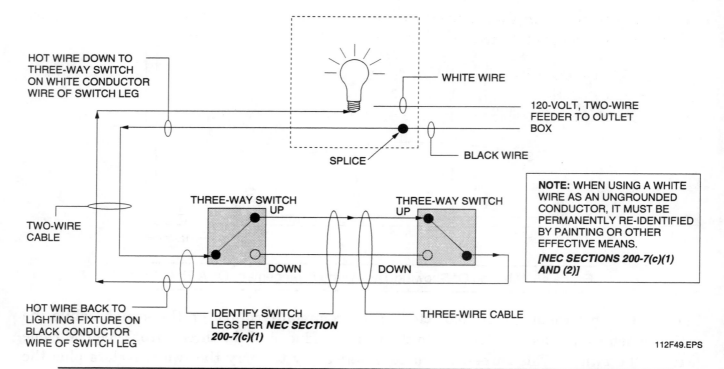

Figure 49. One Way To Connect A Pair Of Three-Way Switches To Control One Lighting Fixture

Figure 50 shows how a four-way switch may be used in combination with two three-way switches to control a device from three locations. In this example, note that the hot wire is connected to the common terminal on the three-way switch on the left. Current then travels to the top traveler terminal and continues on the top traveler conductor to the four-way switch. Since the handle is up on the four-way switch, current flows through the top terminals of the switch and into the traveler conductor going to the other three-way switch.

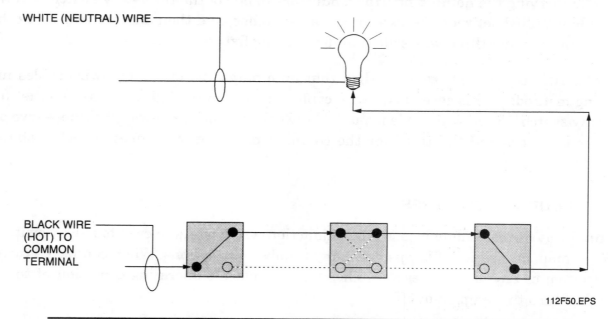

Figure 50. Three- And Four-Way Switches Used In Combination; The Light Is On

Again, the switch is in the up position. Therefore, current is carried from the top traveler terminal to the common terminal and then to the lighting fixture to energize it.

If the position of any one of the three switch handles is changed, the circuit will be broken and no current will flow to the lamp. For example, assume that the four-way switch handle is flipped downward. The circuit will now appear as shown in *Figure 51*, and the light will be out.

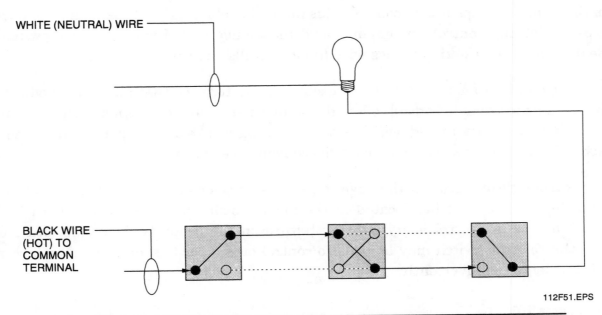

WHITE (NEUTRAL) WIRE

BLACK WIRE
(HOT) TO
COMMON
TERMINAL

112F51.EPS

Figure 51. Three- And Four-Way Switches Used In Combination; The Light Is Off

Remember, any number of four-way switches may be used in combination with two three-way switches, but two three-way switches are always necessary for the correct operation of one or more four-way switches.

14.3.0 PHOTOELECTRIC SWITCH

The chief application of the photoelectric switch is to control outdoor lighting, especially the dusk-to-dawn lights found in suburban areas. This switch has an endless number of possible uses and is a great tool for electricians dealing with outdoor lighting situations.

14.4.0 RELAYS

Next to switches, relays play the most important part in the control of light. However, the design and application of relays is a study in itself, and they are far beyond the scope of this module. Still, a brief mention of relays is necessary to round out your knowledge of lighting controls.

An electric relay is a device whereby an electric current causes the opening or closing of one or more pairs of contacts. These contacts are usually capable of controlling much more power than is necessary to operate the relay itself. This is one of the main advantages of relays.

One popular use of the relay in residential lighting systems is that of remote control lighting. In this type of system, all relays are designed to operate on a 24V circuit and are used to control 120V lighting circuits. They are rated at 20A, which is sufficient to control the full load of a normal lighting branch circuit, if desired.

Remote control switching makes it possible to install a switch wherever it is convenient and practical to do so or wherever there is an obvious need for a switch, no matter how remote it is from the lamp or lamps it is to control. This method enables lighting designs to achieve new advances in lighting control convenience at a reasonable cost. Remote control switching is also ideal for rewiring existing homes with finished walls and ceilings.

One relay is required for each fixture or each group of fixtures that are controlled together. Switch locations for remote control follow the same rules as for conventional direct switching. However, since it is easy to add switches to control a given relay, no opportunities should be overlooked for adding a switch to improve the convenience of control.

Remote control lighting also has the advantage of using selector switches at central locations. For example, selector switches located in the master bedroom or in the kitchen of a home enable the owner to control every lighting fixture on the property from this location. For example, the selector switch may be used to control outside or basement lights which might otherwise be left on inadvertently.

14.5.0   DIMMERS

Dimming a lighting system provides control of the quantity of illumination. It may be done to create certain moods or to blend the lighting from different sources for various lighting effects.

For example, in homes with formal dining rooms, a chandelier mounted directly above the dining table and controlled by a dimmer switch becomes the centerpiece of the room while providing general illumination. The dimmer adds versatility since it can set the mood for the activity—low brilliance (candlelight effect) for formal dining or bright for an evening of playing cards. When chandeliers with exposed lamps are used, the dimmer is essential to avoid a garish and uncomfortable atmosphere. The chandelier should be sized in proportion to the dining area.

Note: It is very important that dimmers be matched to the wattage of the application. Check the manufacturer's data.

14.6.0   SWITCH LOCATIONS

Although the location of wall switches is usually provided for convenience, the NEC also stipulates certain mandatory locations for lighting fixtures and wall switches. These locations are deemed necessary for added safety in the home for both the occupants and service personnel.

For example, the NEC requires adequate light in areas where HVAC equipment is placed. Furthermore, these lights must be conveniently controlled so that homeowners and service personnel do not have to enter a dark area where they might come in contact with dangerous equipment. Three-way switches are required under certain conditions. The NEC also specifies regulations governing lighting fixtures in clothes closets, along with those governing lighting fixtures that may be mounted directly to the outlet box without further support. *Figure 52* summarizes some of the NEC requirements for light and switch placement in the home. For further details, refer to the appropriate sections in the NEC.

14.7.0 LOW-VOLTAGE ELECTRICAL SYSTEMS

Conventional lighting systems operate and are controlled by the same system voltage, generally 120V in residential lighting circuits. The NEC permits the use of low-voltage systems to control lighting circuits. There are some advantages to low-voltage systems. One advantage is that the control of lighting from several different locations is more easily accomplished, such as with the remote control system discussed earlier. For example, outside flood lighting can be controlled from several different rooms in a house. The cost of the control wiring is less in that it is rated for a lower voltage and only carries a minimum amount of current compared to a standard lighting system. When extensive or complex lighting control is required, low-voltage systems are preferred. Also, since these circuits are low-energy circuits, circuit protection is not required.

14.7.1 NEC Requirements For Low-Voltage Systems

NEC Article 725 governs the installation of low-voltage system wiring. These provisions apply to remote control circuits, low-voltage relay switching, low-energy power circuits, and low-voltage circuits. The NEC divides these circuits into three categories:

- Remote control
- Signaling
- Power-limited circuits

As mentioned earlier, circuit protection of the low-voltage circuit is not required; however, the high-voltage side of the transformer that supplies the low-voltage system must be protected. *NEC Chapter 9, Tables 11(a) and 11(b)* cover circuits that are inherently limited in power output and therefore require no overcurrent protection or are limited by a combination of power source and overcurrent protection.

There are a number of requirements of the power systems described in *NEC Chapter 9, Tables 11(a) and 11(b)* and the notes preceding the tables. You should read and study all applicable portions of the NEC before installing low-voltage power systems.

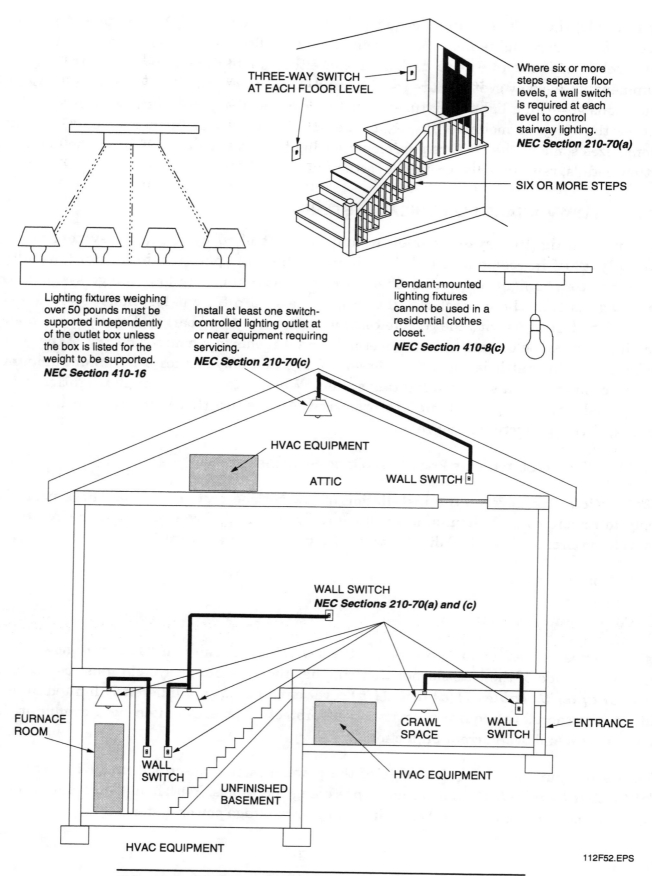

THREE-WAY SWITCH AT EACH FLOOR LEVEL

Where six or more steps separate floor levels, a wall switch is required at each level to control stairway lighting. **NEC Section 210-70(a)**

SIX OR MORE STEPS

Lighting fixtures weighing over 50 pounds must be supported independently of the outlet box unless the box is listed for the weight to be supported. **NEC Section 410-16**

Install at least one switch-controlled lighting outlet at or near equipment requiring servicing. **NEC Section 210-70(c)**

Pendant-mounted lighting fixtures cannot be used in a residential clothes closet. **NEC Section 410-8(c)**

HVAC EQUIPMENT

ATTIC

WALL SWITCH

WALL SWITCH
NEC Sections 210-70(a) and (c)

FURNACE ROOM

WALL SWITCH

CRAWL SPACE

WALL SWITCH

ENTRANCE

HVAC EQUIPMENT

UNFINISHED BASEMENT

HVAC EQUIPMENT

112F52.EPS

Figure 52. NEC Requirements For Light And Switch Placement

14.7.2 Relays

A low-voltage remote control wiring system is relay operated. The relay is controlled by a low-voltage switching system and in turn it controls the power circuit that is connected to it (*Figure 53*). The low-voltage, split coil relay is the heart of the low-voltage remote control system (*Figure 54*). When the ON coil of the relay is energized, the solenoid mechanism causes the contacts to move into the ON position to complete the power circuit. The contacts stay in this position until the OFF coil is energized. When this occurs, the contacts are withdrawn and the power circuit is opened. The red wire is the ON wire; the black wire is the OFF wire, and the blue wire is common to the transformer.

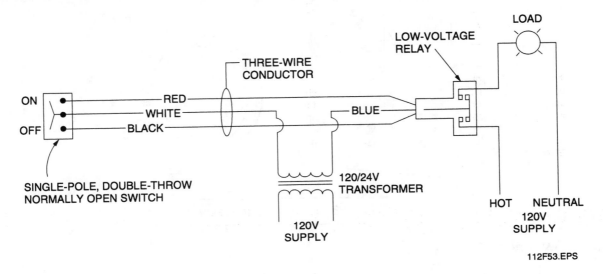

112F53.EPS

Figure 53. Low-Voltage Remote Control System

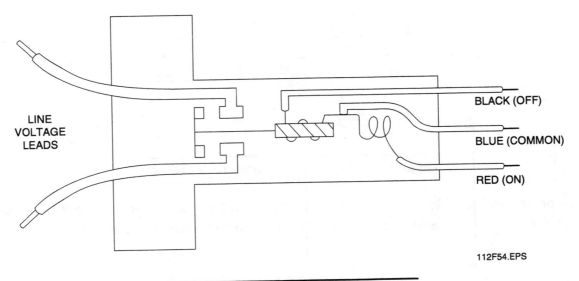

112F54.EPS

Figure 54. Low-Voltage Relay Diagram

The low-voltage relay is available in two mounting styles. One type is designed to mount through a ½-inch knockout opening (*Figure 55*). For a ¾-inch knockout, a rubber grommet is inserted to isolate the relay from the metal. This practice should ensure quieter relay operation. The second relay mounting style is the plug-in relay. This type of relay is used in an installation where several relays are mounted in one enclosure. The advantage of the plug-in relay is that it plugs directly into a busbar. As a result, it is not necessary to splice the line voltage leads.

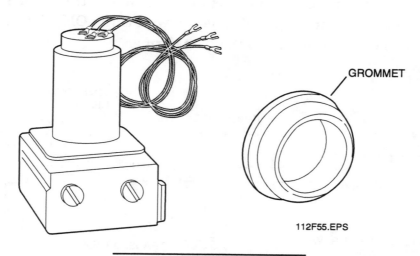

112F55.EPS

Figure 55. Low-Voltage Relay

14.7.3   Switches

The switch used in the low-voltage remote control system is a normally open, single-pole, double-throw, momentary contact switch, as shown in *Figure 56*. This switch is approximately one-third the size of a standard single-pole switch.

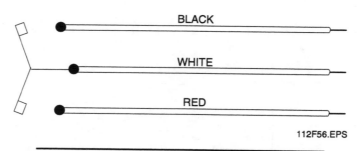

112F56.EPS

Figure 56. Single-Pole Double-Throw Switch

In a low-voltage switch, the white wire is common and is connected to the 24V transformer source. The red wire connects to the ON circuit, and the black wire connects to the OFF circuit.

14.7.4 Wiring Methods

NEC Article 725 governs the installation of low-voltage systems. This article covers remote control circuits, low-voltage relay switching, low-energy power circuits, and low-voltage circuits.

15.0.0 ELECTRIC HEATING

The use of electric heating in residential occupancies has risen tremendously over the past decade or so, and the practice will no doubt continue. This is due to the following advantages of electric heat over most other heating systems:

- Electric heat is noncombustible and is therefore safer than combustible fuels.
- It requires no storage space, fuel tanks, or chimneys.
- It requires little maintenance.
- The initial installation cost is relatively inexpensive when compared to other types of heating systems.
- The comfort level may be improved since each room may be controlled separately by its own thermostat.

There are also some disadvantages to using electric baseboard heat, especially in northern climates. Some of these disadvantages include:

- Electric heat is often more expensive to operate than other types of fuels.
- Receptacles must not be installed above electric baseboard heaters.
- Electric baseboard heaters tend to discolor the wall area immediately above the heater, especially if there are smokers in the home.

Note: It is very important to calculate the extra electric load of an electric heater installation (especially in an add-on situation). Ensure that the extra load does not exceed the maximum amperage draw of either the circuit or the panel.

The type of electric heating system used for a given residence will usually depend on the structural conditions, the kind of room, and the activities for which the room will be used. The homeowner's preference will also enter into the final decision.

Electric heating equipment is available in baseboard, wall, ceiling, kick space, and floor units; in resistance cable embedded in the ceiling or concrete floor; in forced-air duct systems similar to conventional oil- or gas-fired hot air systems; and in electric boilers for hot water baseboard heat.

Electric heat pumps have also become popular for residential heating, ventilating, and air conditioning (HVAC) systems in certain parts of the country. The term *heat pump*, as applied to a year-round air conditioning system, commonly denotes a system in which refrigeration equipment is used in such a manner that heat is taken from a heat source and transferred to the conditioned space when heating is desired; heat is removed from the space and discharged to a heat sink when cooling and dehumidification are desired.

A heat pump has the unique ability to furnish more energy than it consumes. This is due to the fact that under certain outdoor conditions, electrical energy is required only to move the refrigerant and run the fan; thus, a heat pump can attain a heating efficiency of two or more to one; that is, it will put out an equivalent of two or three watts of heat for every watt consumed. For this reason, its use is highly desirable for the conservation of energy.

In general, electric baseboard heating equipment should be located on the outside wall near the areas where the greatest heat loss will occur, such as under windows, etc. The controls for wall-mounted thermostats should be located on an interior wall, about 50 inches above the floor to sense the average room temperature. *Figure 57* shows an electric heating arrangement for the sample residence. NEC regulations governing the installation of these units are also noted.

16.0.0 RESIDENTIAL SWIMMING POOLS, SPAS, AND HOT TUBS

The NEC recognizes the potential danger of electric shock to persons in swimming pools, wading pools, and therapeutic pools, or near decorative pools or fountains. This shock could occur from electric potential in the water itself or as a result of a person in the water or a wet area touching an enclosure that is not at ground potential. Accordingly, the NEC provides rules for the safe installation of electrical equipment and wiring in or adjacent to swimming pools and similar locations. **NEC Article 680** covers the specific rules governing the installation and maintenance of swimming pools, spas, and hot tubs.

The electrical installation procedures for hot tubs and swimming pools are too vast to be covered in detail in this module. However, the general requirements for the installation of outlets, overhead fans and lighting fixtures, and other items are summarized in *Figure 58*.

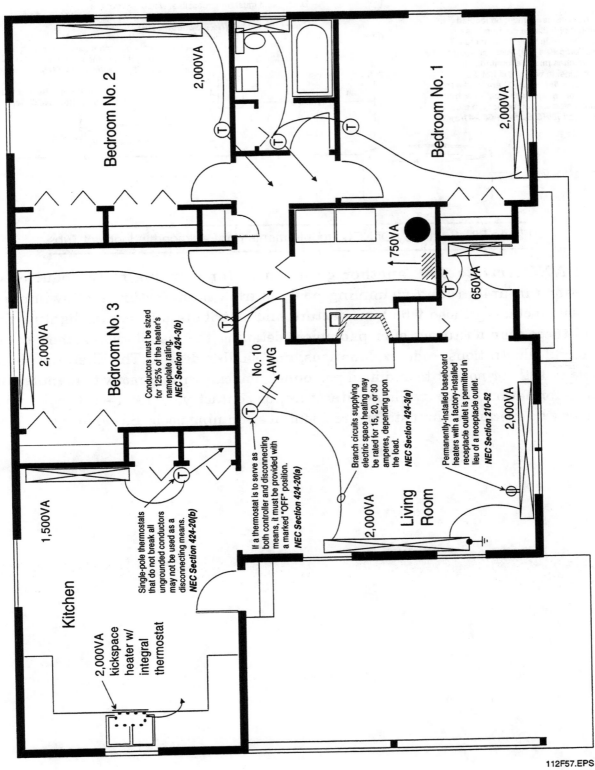

750VA

2,000VA

Bedroom No. 2

Bedroom No. 1

2,000VA

750VA

650VA

2,000VA

Bedroom No. 3

Conductors must be sized for 125% of the heater's nameplate rating. *NEC Section 424-3(b)*

No. 10 AWG

Branch circuits supplying electric space heating may be rated for 15, 20, or 30 amperes, depending upon the load. *NEC Section 424-3(a)*

Permanently-installed baseboard heaters with a factory-installed receptacle outlet is permitted in lieu of a receptacle outlet. *NEC Section 210-52*

2,000VA

1,500VA

Single-pole thermostats that do not break all ungrounded conductors may not be used as a disconnecting means. *NEC Section 424-20(b)*

If a thermostat is to serve as both controller and disconnecting means, it must be provided with a marked "OFF" position. *NEC Section 424-20(a)*

2,000VA

Living Room

Kitchen

2,000VA kickspace heater w/ integral thermostat

112F57.EPS

Figure 57. Electric Heating Arrangement For The Sample Residence

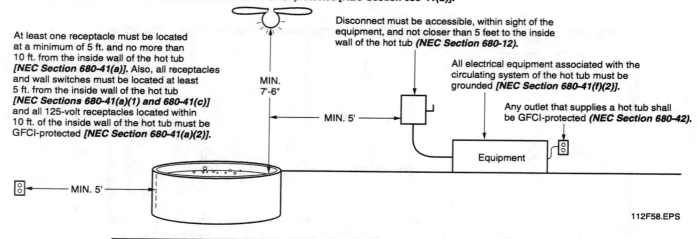

Lighting fixtures, lighting outlets, and ceiling fans located over the hot tub or within 5 ft. from its inside walls shall be a minimum of 7 ft. 6 in. above the maximum water level and shall be GFCI-protected *[NEC Section 680-41(b)]*.

At least one receptacle must be located at a minimum of 5 ft. and no more than 10 ft. from the inside wall of the hot tub *[NEC Section 680-41(a)]*. Also, all receptacles and wall switches must be located at least 5 ft. from the inside wall of the hot tub *[NEC Sections 680-41(a)(1) and 680-41(c)]* and all 125-volt receptacles located within 10 ft. of the inside wall of the hot tub must be GFCI-protected *[NEC Section 680-41(a)(2)]*.

Disconnect must be accessible, within sight of the equipment, and not closer than 5 feet to the inside wall of the hot tub *(NEC Section 680-12)*.

All electrical equipment associated with the circulating system of the hot tub must be grounded *[NEC Section 680-41(f)(2)]*.

Any outlet that supplies a hot tub shall be GFCI-protected *(NEC Section 680-42)*.

MIN. 7'-6"

MIN. 5'

MIN. 5'

Equipment

112F58.EPS

Figure 58. Summary Of NEC Requirements For Packaged Indoor Hot Tubs

Besides ***NEC Article 680***, another good source for learning more about electrical installations in and around swimming pools is from manufacturers of swimming pool equipment, including those who manufacture and distribute underwater lighting fixtures. Many of these manufacturers offer pamphlets detailing the installation of their equipment with helpful illustrations, code explanations, and similar details. This literature is usually available at little or no cost to qualified personnel. You can write directly to manufacturers to request information about available literature, or contact your local electrical supplier or contractor who specializes in installing residential swimming pools. See *Figure 59*.

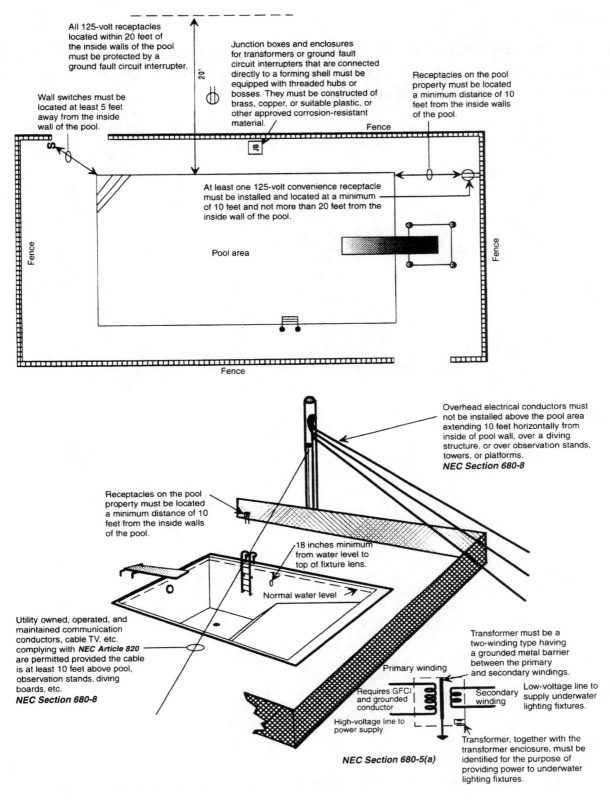

All 125-volt receptacles located within 20 feet of the inside walls of the pool must be protected by a ground fault circuit interrupter.

Junction boxes and enclosures for transformers or ground fault circuit interrupters that are connected directly to a forming shell must be equipped with threaded hubs or bosses. They must be constructed of brass, copper, or suitable plastic, or other approved corrosion-resistant material.

Receptacles on the pool property must be located a minimum distance of 10 feet from the inside walls of the pool.

Wall switches must be located at least 5 feet away from the inside wall of the pool.

20'

Fence

At least one 125-volt convenience receptacle must be installed and located at a minimum of 10 feet and not more than 20 feet from the inside wall of the pool.

Fence

Pool area

Fence

Fence

Overhead electrical conductors must not be installed above the pool area extending 10 feet horizontally from inside of pool wall, over a diving structure, or over observation stands, towers, or platforms. *NEC Section 680-8*

Receptacles on the pool property must be located a minimum distance of 10 feet from the inside walls of the pool.

18 inches minimum from water level to top of fixture lens.

Normal water level

Utility owned, operated, and maintained communication conductors, cable TV, etc. complying with *NEC Article 820* are permitted provided the cable is at least 10 feet above pool, observation stands, diving boards, etc. *NEC Section 680-8*

Transformer must be a two-winding type having a grounded metal barrier between the primary and secondary windings.

Primary winding

Requires GFCI and grounded conductor

Secondary winding

Low-voltage line to supply underwater lighting fixtures.

High-voltage line to power supply

Transformer, together with the transformer enclosure, must be identified for the purpose of providing power to underwater lighting fixtures.

NEC Section 680-5(a)

112F59.TIF

Figure 59. Summary Of NEC Regulations For Typical Swimming Pool Installations

SUMMARY

This module covered the basics of residential wiring, including switches, outlets, conductor sizes, and service sizes. A thorough knowledge of the NEC is essential to the successful installation of residential wiring systems.

References

For advanced study of topics covered in this task module, the following books are suggested:

Illustrated Dictionary for Electrical Workers, Latest Edition, Delmar Publishers, Albany, NY.
Illustrated Guide to the NEC, Latest Edition, Craftsman Book Co., Carlsbad, CA.
National Electrical Code Handbook, Latest Edition, National Fire Protection Association, Quincy, MA.

REVIEW QUESTIONS

1. When sizing electrical services, at what percentage is the first 3,000VA rated?
   a. 20%
   b. 45%
   c. 80%
   d. 100%

2. What section of the NEC requires that fittings be identified for use with service masts?
   a. *NEC Section 230-28*
   b. *NEC Section 230-40*
   c. *NEC Section 250-46*
   d. *NEC Section 250-83*

3. A service conductor without an outer jacket must be _____ feet above a window that can be opened.
   a. two feet
   b. three feet
   c. eight feet
   d. ten feet

4. *NEC Section 230-6* considers conductors installed under at least _____ inch(es) of concrete to be outside the building.
   a. one
   b. two
   c. four
   d. five

5. Type NM cable may not be used in _____ .
   a. buildings over three stories above grade
   b. the framework of a building
   c. protective strips
   d. attic spaces

6. Type AC cable may not be used in _____ .
   a. concrete or plaster where dry
   b. dry masonry
   c. attic spaces
   d. wet locations

7. Type SE cable is available with _____ for interior wiring systems.
   a. a non-insulated ground or neutral conductor
   b. an insulated grounded conductor
   c. no ground conductor
   d. none of the above

8. Type SER cable may be used _____ .
   a. in overhead applications
   b. underground
   c. as a subfeed under certain conditions
   d. none of the above

9. Identify the following receptacles by number.
   a. _____ single receptacle
   b. _____ split-wired receptacle
   c. _____ dryer receptacle
   d. _____ range receptacle
   e. _____ duplex receptacle
   f. _____ special-purpose outlet

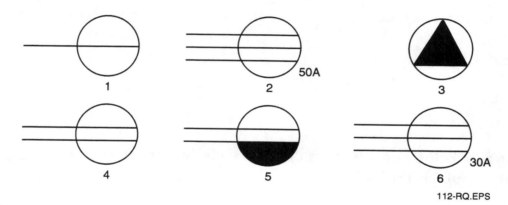

112-RQ.EPS

10. Calculate the cubic inches required for the receptacle outlets shown in the table below.

Number and Size of Conductors in Box	Free Space Within Box for Each Conductor	Total Cubic Inches of Box Space Required	What Size Metallic Box May be Used?
Six No. 12 conductors and three ground wires	2.25		
Seven No. 12 conductors and three ground wires with one receptacle	2.25		
Two No. 14 conductors and one ground wire	2.0		
Four No. 14 conductors and two ground wires	2.0		
Six No. 14 conductors and three ground wires with one receptacle	2.0		

notes

ANSWERS TO REVIEW QUESTIONS

Answer	Section
1. d	2.3.3
2. a	6.1.0/Figure 16
3. b	6.3.0
4. b	7.0.0
5. a	8.1.1
6. d	8.1.2
7. b	8.1.4
8. c	8.1.4
9. a. 1	10.0.0
b. 5	
c. 6	
d. 2	
e. 4	
f. 3	
10. see table below	12.0.0

Number and Size of Conductors in Box	Free Space Within Box for Each Conductor	Total Cubic Inches of Box Space Required	What Size Metallic Box May be Used?
Six No. 12 conductors and three ground wires	2.25	15.75	4 x 1¼ square box
Seven No. 12 conductors and three ground wires with one receptacle	2.25	22.5	4 ¹¹⁄₁₆ x 1¼ square box
Two No. 14 conductors and one ground wire	2.0	6	3 x 2 x 1½ device box
Four No. 14 conductors and two ground wires	2.0	10	3 x 2 x 2 device box
Six No. 14 conductors and three ground wires with one receptacle	2.0	18	3 x 2 x 3½ device box

REFERENCE INFORMATION — OTHER CODES AND ELECTRICAL STANDARDS THAT APPLY TO ELECTRICAL INSTALLATIONS

Note: This information is provided as a general reference only. Refer to operative local codes and the latest editions of codes in effect in your area.

For your reference, information is included here on several other codes and standards that apply to electrical installations in residential structures. It is suggested that to be thoroughly competent in the electrical trade, you should become familiar with the contents of these codes and the terminology used in them. Listed below are the building codes referred to, as well as their addresses. Also included is a list of definitions and phrases that are of particular interest in residential electrical construction.

International Conference of Building Officials (ICBO)

>   5360 South Workman Mill Road

>   Whittier, California 90601

>   (213) 699-0541

Building Official & Code Administrators International, Inc. (BOCA)

>   4051 West Flossmoor Road

>   Country Club Hills, Illinois 60478-5795

>   (708) 799-2300

Southern Building Code Congress International, Inc. (SBCCI)

>   900 Montclair Road

>   Birmingham, Alabama 35213-1206

>   (205) 591-1853

International Association of Plumbing and Mechanical Officials (IAPMO)

>   2001 Walnut Drive South

>   Walnut, California 91789-2825

>   (714) 595-8449

UMC–UNIFORM MECHANICAL CODE

Installation of Appliances

Appliances installed in garages, warehouses, or other areas where they may be subjected to mechanical damage, are required to be suitably guarded against such damage by being installed behind protective barriers or by being elevated or located out of the normal path of vehicles. Heating and cooling equipment that generates a glow, spark, or flame that is capable of igniting flammable vapors located in a garage are required to be installed with the pilots and burners or heating elements and switches at least 18 inches above the floor level.

Where such appliances are installed within a garage and are enclosed in a separate approved compartment, having access only from outside the garage, they are permitted to be installed at floor level only where the required combustion air is taken from and discharged to the exterior of the garage. See UMC Section 508.

Disconnecting Means, Receptacle and Lighting Outlet

Equipment requiring electrical connections of more than 50 volts must have a positive means of disconnect. This disconnect must be adjacent to, and in sight from, the equipment served. A 120-volt receptacle is required to be located within 25 feet of the equipment for service and maintenance purposes. The receptacle is not required to be located on the same level as the equipment.

Where installed within a structure, low-voltage wiring of 50 volts or less is required to be installed in a manner to prevent physical damage. See UMC Section 509.

Where a furnace is installed in an attic, a permanent electric outlet and switch-controlled lighting fixture is required to be located at the required passageway opening. This lighting fixture is required to be installed at or near the furnace. See UMC Section 708.

Where a warm air furnace is located in an under-floor space, a permanent electric outlet and switch-controlled lighting fixture is required to be located at the required passageway opening. This lighting fixture is required to be installed at or near the furnace. See UMC Section 709(5).

Cooling Systems Lighting In Concealed Spaces

Where access is required to equipment located in an under-floor space, attic, or furred space, a permanent electric light outlet and lighting fixture must be installed at or near the equipment. The light must be controlled by a switch located at the required passageway opening.

Exception: Lighting fixtures are not required to be installed where the fixed lighting in the building will provide sufficient light for safe servicing of the equipment. See UMC Section 1203.

Where cooling system equipment requires access for service or maintenance, it must be provided with an unobstructed working space on the control or servicing side of the equipment. This space cannot be less than 30 inches in depth and not less than 6 feet 6 inches in height. Note: Also see the Electrical Code for working space requirements about electrical equipment. See UMC Sections 1202 and Section 1204, Exception.

A cooling system includes all equipment intended or installed for the purpose of cooling air by mechanical means and discharging such air into any room or space. This definition does not include any evaporative cooler. See UMC – 91.

BOCA–NATIONAL MECHANICAL CODE

Convenience Outlet

A 125-volt AC grounding-type outlet must be available for all appliances (see definition later in this chapter.) The convenience outlet must be located on the same level within 75 feet of the appliance. The outlet must not be connected to the load side of the appliance disconnect switch. See BOCA Section M-407.2.

Lighting

Permanent lighting must be provided to illuminate the area in which an appliance is located. For remote locations, the light switch must be located near the access opening leading to the appliance. See BOCA M-407.3.

SBCCI–STANDARD MECHANICAL CODE

Air Conditioning, Heating, And Ventilation Equipment

Every appliance must be located with respect to building construction and other equipment so as to permit access for service of the appliance. Sufficient clearance must be maintained to permit cleaning of heating surfaces, the replacement of filters, blowers, motors, burners, controls, and vent connections; the lubrication of moving parts where required; and the adjustment and cleaning of burners and pilots.

SBCCI Mechanical Code Section 303.2.1. – Outdoor Installations

Appliances listed for outdoor installations are permitted to be installed without protection in accordance with the provisions of their listing and must be accessible for servicing. Appliances listed for outdoor installation only must not be installed inside a building. See SBCCI Mechanical Code Section 303.2.2.

Attic Installations

Every attic or furred space in which mechanical equipment is installed must be accessible by an opening and passageway as large as the largest piece of the equipment and in no case less than 22 x 36 inches continuous from the opening to the equipment and its controls. The opening to the passageway must be located not more than 20 ft. from the equipment measured along the center line of such passageway. Every passageway must be unobstructed and shall have solid continuous flooring not less than 24 inches wide from the entrance opening to the equipment. On the control side and other sides where access is necessary for servicing the equipment, a level working platform extending a minimum of 30 inches from the edge of the equipment with a 36-inch high clear working space must be provided. Top or bottom service equipment must have a full clearance above or below the unit for component removal. See SBCCI Mechanical Code Section 303.4.1.

A permanent electric light outlet and lighting fixture must be provided at, or near, the mechanical equipment and controlled by a switch located at the required passageway opening. See SBCCI Mechanical Code Section 303.4.5.

Roof or Exterior Wall Installation

Mechanical equipment installations on roofs or exterior walls of buildings must comply with the requirements for roof and wall structures as specified in the Standard Building Code, and must be listed and approved for such use. See SBCCI Mechanical Code Section 303.6.1. Each appliance must have an accessible weatherproof disconnect switch and a 110- to 120-volt AC grounding-type convenience outlet on the roof adjacent to the appliance. The convenience outlet must be on the supply side of the disconnect switch. See SBCCI Mechanical Code Section 303.6.2.

Nameplate Marking

Every electric comfort heating appliance must bear a permanent and legible factory-applied nameplate that must include the following:

- Name and trademark of the manufacturer
- The catalog (model) number or equivalent
- The electric rating in volts, ampacity, and phase
- BTU output rating
- Individual marking for each electrical component in amperes or watts, volts and phase
- Required clearances from combustibles
- A seal indicating approval of the appliance by an approved testing agency (see SBCCI Mechanical Code Section 303.11.2)

Panel Heating Systems

Panel heating is a method of radiant space heating in which heat is supplied by large heated areas of room surfaces. The heating element usually consist of warm water piping, warm air ducts, or electrical resistance elements embedded in or located behind ceiling, wall, or floor surfaces. See SBCCI Mechanical Code Section 303.12.1.

The installation of panel heating systems must be designed and installed in strict accordance with the accepted engineering practices and the requirements of the SBCCI Mechanical Code Section 303.12.2.

SBCCI–STANDARD GAS CODE

Electrical Connections for Electrical Ignition and Control Devices

Devices employing or depending upon an electrical current must not be used to control or ignite a gas supply if of such character that failure of the electrical current could result in the escape of unburned gas or in failure to reduce the supply of gas under conditions which would normally result in its reduction unless other means are provided to prevent the development of dangerous temperatures, pressures, or the escape of gas. See SBCCI Standard Gas Code Section 404.2.

Switches in Air Conditioning Electrical Supply Line

Means for interrupting the electrical supply to the air conditioning appliance and to its associated cooling tower (if supplied and installed in a location remote from the air conditioner) must be provided within sight of and not over 50 ft. from the air conditioner and cooling tower. See Section 508.5 SBCCI Standard Gas Code.

Electric Ignition Systems

Electric ignition systems must ignite only a pilot. The input to the pilot shall not exceed 3% of the maximum input to the main burner as fired. If ignition of the pilot is not obtained within 15 seconds, the gas must shut off automatically. See Section 807.3 SBCCI Standard Gas Code.

SBCCI–STANDARD PLUMBING CODE

Existing Metallic Water Service Piping

Existing metallic water service piping used for electrical grounding must not be replaced with nonmetallic pipe or tubing until other grounding means are provided which are satisfactory to the proper administrative authority having jurisdiction. See Section 504.3.3 SBCCI Standard Plumbing Code.

IAPMO–UNIFORM PLUMBING CODE

Water Heater Locations

Attic and under-floor water heater locations must be provided with an electric light at or near the water heater. Such light must be controlled by a switch located adjacent to the opening or trap door. See Section 1311(e) IAPMO – UPC – 91.

The International Association of Plumbing and Mechanical Officials include their Installation Standards in IAPMO – UPC – 91. These standards are not considered as part of the Uniform Plumbing Code (UPC) unless they are formally adopted as such. Some sections include the following.

Recessed light fixtures: Polybutylene (PB) tubing shall not be installed within 12 inches of a recessed light fixture.

Identification: A permanent sign with the legible words THIS BUILDING HAS NONMETALLIC INTERIOR WATER PIPING must be fastened on or inside the main electric service panel.

UBC–UNIFORM BUILDING CODE

Requirements for Group R Occupancies

A Group R occupancy includes Division 3 – Dwellings and Lodging Houses. A complete code for construction of detached one- and two-family dwellings is located in Appendix Chapter 12 of the UBC. See UBC Section 1201.

Smoke Detectors

Smoke detectors are required in dwelling units that are used for sleeping purposes. They must be installed in accordance with the manufacturer's instructions.

The power source used for a smoke detector must be by a permanently-connected wiring method and must be provided with a battery backup where installed in new construction. This supply cannot be operated by any type of switching device or other disconnecting means. This restriction does not include branch-circuit overcurrent protection.

A smoke detector must be installed in all sleeping rooms. Additional smoke detectors must be installed so they are located centrally in the corridor or area giving access to each separate sleeping area. See UBC Section 1210.

Heating

Dwelling units must be provided with heating facilities capable of maintaining a room temperature of 70°F at a point three feet above the floor in all habitable rooms. See UBC Section 1212.

Exterior openings exposed to the weather must be protected and must be flashed in such a manner as to make them weatherproof. See UBC Section 1708(b).

Cutting and Notching Wood

In exterior walls and bearing partitions, any wood stud may be cut or notched to a depth not exceeding 25 percent of its width. Cutting or notching of studs to a depth not greater than 40 percent of the width of the stud is permitted in nonbearing partitions supporting no loads other than the weight of the partition. See UBC Section 2517(g)(8).

Bored Holes in Wooden Studs

A hole not greater in diameter than 40 percent of the stud width may be bored in any wood stud. Bored holes not greater than 60 percent of the width of the stud are permitted in nonbearing partitions or in any wall where each bored stud is doubled, provided not more than two such successive doubled studs are so bored.

In no case shall the edge of the bored hole be nearer than ⅝ inch to the edge of the stud. Bored holes must not be located at the same section of stud as a cut or notch. See UBC Section 2517(g)(9).

Wall and Ceiling Coverings

Where electrical radiant heat cables are installed on ceilings, the stripping, if conductive, may be omitted a distance not to exceed 12 inches from the wall. See UBC Section 4705(a).

BOCA–NATIONAL BUILDING CODE

Bathroom and Toilet Room Lighting

Every bathroom and toilet room must be provided with artificial light. The illumination must have an average intensity of three footcandles measured at a level of 30 inches above the floor. See BOCA Section 703.1.1.

Grounding Metal Veneers

Grounding of metal veneers on all buildings must comply with the requirements of BOCA Article 27 and NFPA 70. See BOCA Section 2103.4.3. See *NEC Section 250-116 (FPN)*.

SBCCI Standard Building Code

Underground inspection: To be made after trenches or ditches are excavated, conduit or cable installed, and before any backfill is put in place.

Rough-in inspection: To be made after the roof, framing, fireblocking, and bracing is in place and prior to the installation of wall or ceiling membranes.

Final inspection: To be made after the building is complete, all required electrical fixtures are in place and properly connected or protected, and the structure is ready for occupancy. See SBCCI Section 103.8.6.

Separation Between Townhouses

Where not more than three stories in height, townhouses (which are defined as a single family dwelling constructed in a series or group of attached units with property lines separating each unit) may be separated by a single wall meeting the following requirements.

The wall is required to provide not less than a two-hour fire resistance rating. Plumbing, piping, ducts, and electrical or other building services are not permitted to be installed within or through the two-hour wall, unless such materials and methods of penetration have been tested in accordance with SBCCI Section 1001.1. See SBCCI Section 403.5.2.

Metal Veneers

Metal veneers fastened to supporting elements which are not a part of the grounded metal framing of a building must be made electrically continuous by contact or interconnection of individual units and must be effectively grounded. The conductor used to ground the veneer must have no greater resistance than the conductor used to ground the electrical system within the building. Where a metal veneer is applied to a building with no electrical wiring system, grounding is required only if determined to be necessary by the Building Official. See SBCCI Section 811.4.6.

Note: This requirement must be coordinated with the electrical inspector as well. See *NEC Section 250-116 (FPN)*.

Walls, Floors, and Partitions

When walls, floors, and partitions are required to have a minimum of one hour or greater fire resistance rating, cabinets, bathroom components, lighting, and other fixtures must be so installed such that the required fire resistance will not be reduced.

Exception: Fixtures which are listed for such installation are permitted. See SBCCI Section 1001.3.7.

SUMMARY OF APPLICABLE DEFINITIONS FROM BUILDING CODES ABOVE

The following definitions are selected from the codes referenced above and that relate to electrical installations in residential structures. Where the code definitions differ, each definition is included.

ACCESSIBLE – Having access to, but which first may require the removal of, a panel, door, or similar obstruction. BOCA – MECH – 90; SBCCI – MECH – 91; SBCCI – 91; SBCCI – 91 – GAS; UBC/UMC – 91

AIR CONDITIONING – The treatment of air so as to control simultaneously its temperature, humidity, cleanliness, and distribution to meet the requirements of a conditioned space. BOCA – MECH – 90; SBCCI – MECH – 91; SBCCI – 91 – GAS

AIR CONDITIONING SYSTEM – An air conditioning system consists of heat exchangers, blowers, filters, supply, exhaust, and return ducts, and shall include any apparatus installed in connection therewith. BOCA – MECH – 90; SBCCI – CI – MECH – 91

AIR-HANDLING UNIT – A blower or fan used for the purpose of distributing supply air to a room, space, or area. UBC/UMC – 91

ALLEY – Any public way or thoroughfare less than 16 feet but not less than 10 feet in width which has been dedicated or deeded to the public for public use. UBC – 91

ALTERATION – A change in an air conditioning, heating, ventilating, or refrigeration system that involves an extension, addition, or change to the arrangement, type, or purpose of the original installation. SBCCI – MECH – 91

AND/OR – In a choice of two code provisions, signifies that use of both provisions will satisfy the code requirement and use of either provision is acceptable also. SBCCI 91 – GAS

APPLIANCE – A device which utilizes fuel or other forms of energy to produce light, heat, power, refrigeration, or air conditioning. This definition also shall include a vented decorative appliance. UBC/UMC – 91

APPLIANCE – Utilization equipment normally built in standardized sizes which is installed or connected as a unit to perform one or more functions. SBCCI – MECH – GAS – 91

APPLIANCE, AUTOMATICALLY CONTROLLED – Appliances equipped with an automatic burner which accomplish complete turn-on and shut-off of the gas to the main burner or burners, and graduate the gas supply to the burner or burners (it does not affect complete shut-off of the gas). SBCCI – 91 – GAS

APPLIANCE (MECHANICAL) – A device or apparatus—including any attachments or apparatus designed to be attached—which is manufactured and designed to use electricity, natural gas, manufactured gas, mixed gas, liquefied petroleum products, solid fuel, oil, or any gas as a fuel for heating, cooling, or developing light or power. BOCA – MECH – 90

ATTIC – The space between the ceiling beams of the top story and the roof rafters. BOCA – 90

ATTIC, HABITABLE – A habitable attic is an attic which has a stairway as a means of access and egress and in which the ceiling area at a height of $7\frac{1}{3}$ feet above the attic floor is not more than one third of the area of the next floor below. BOCA – 90

BASEMENT – Any building story having a floor below grade. SBCCI – 91

BASEMENT – Any floor level below the first story in a building, except that a floor level in a building having only one floor level must be classified as a basement unless such floor level qualifies as a first story as defined herein. UBC – 91

BASEMENT – That portion of a building which is partly or completely below grade (see Story above grade). BOCA – 90

BATHROOM – A room containing a bathtub or shower for use by a person to bathe or cleanse one's self, located in or adjacent to a residence, apartment, hotel, motel, or similar type building. SBCCI – 91 – GAS

BATHROOM – A room equipped with a shower or bathtub. IAPMO – UPC – 91

BUILDING – A building is a structure built, erected, and framed of component structural parts designed for the housing, shelter, enclosure, or support of persons, animals, or property of any kind. IAPMO – UPC – 91

BUILDING – Any structure that encloses a space used for sheltering any occupancy. Each portion of a building separated from other portions by a fire wall must be considered as a separate building. SBCCI – 91

BUILDING – Any structure used or intended for supporting or sheltering any use or occupancy. UBC – 91

BUILDING – Any structure used or intended for supporting or sheltering any use or occupancy. For application of this code, each portion of a building which is completely separated from other portions by fire wall complying with Section 907.0 must be considered as a separate building. BOCA – 90

BUILDING CODE – The Uniform Building Code promulgated by the International Conference of Building Officials, as adopted by this jurisdiction. ICBO ELECTRICAL CODE

BUILDING, EXISTING – Any structure erected prior to the adoption of the appropriate code, or one for which a legal building permit has been issued. BOCA – 90; SBCCI – 91; UBC – 91

BUILDING SERVICE EQUIPMENT – The mechanical, electrical, and elevator equipment including piping, wiring, fixtures, and other accessories, which provides sanitation, lighting, heating, ventilation, fire-fighting, and transportation facilities essential for the habitable occupancy of the building or structure for its designated use and occupancy. BOCA – 90

COMBUSTION – The rapid oxidation of fuel gases accompanied by the production of heat, or heat and light. SBCCI – 91 – GAS

COMBUSTIBLE CONSTRUCTION – A wall or surface constructed of wood, composition, or of wooden studding and lath and plaster. SBCCI – 91 – GAS

DWELLING – A building occupied exclusively for residential purposes by not more than two families, unless qualified otherwise in code text. SBCCI – 91

DWELLING – Any building or portion thereof which contains not more than two dwelling units. UBC – 91

DWELLING UNIT – Any building or portion thereof which contains living facilities, including provisions for sleeping, eating, cooking, and sanitation, as required by this code, for not more than one family, or a congregate residence for 10 or fewer persons. UBC – 91

DWELLING UNIT – A single unit providing complete, independent living facilities for one or more persons, including permanent provisions for living, sleeping, eating, cooking, and sanitation. BOCA – 90; SBCCI – 91

DWELLING, ONE-FAMILY – A building containing one dwelling unit with not more than five lodgers or boarders. BOCA – 90

DWELLING, TWO-FAMILY – A building containing two dwelling units with not more than five lodgers or boarders per family. BOCA – 90

EFFICIENCY DWELLING UNIT – A dwelling unit containing only one habitable room. UBC – 91

ELECTRIC HEATING APPLIANCE – A device which produces heat energy to create a warm environment by the application of electric power to resistance elements, refrigerant compressors, or dissimilar material junctions. Section 407 – UBC/UMC – 91

ELECTRIC SPACE HEATERS, PORTABLE – Heaters not intended for permanent connection to a structure or to electric wiring. SBCCI – MECH – 91

ELECTRIC SPACE HEATERS, STATIONARY – Heaters permanently mounted in a structure (air duct system) and which are permanently connected to electric wiring. SBCCI – MECH – 91

EQUIPMENT, EXISTING – Any equipment covered by this code which was installed prior to the effective date of this code, or for which an application for permit to install was filed with the code official prior thereto. BOCA – 90

EXISTING WORK – A mechanical system, or any part thereof, installed prior to the effective date of this code. BOCA – MECH – 90

EXIT – See Section 3301(b) Definitions of: Balcony, Exterior Exit; Exit; Exit Court; Exit Passageway; and Horizontal Exit. UBC – 91

FIRE PROTECTION – The provision of construction safeguards and exit facilities; and the installation of fire alarm, fire detecting, and fire extinguishing service equipment, to reduce the fire risk and the conflagration hazard. BOCA – 90

FLOOR AREA – The area included within the surrounding exterior walls of a building or portion thereof, exclusive of vent shafts and courts. The floor area of a building, or portion thereof, not provided with surrounding exterior walls must be the usable area under the horizontal projection of the roof or floor above. UBC – 91

FURNACE – A completely self-contained unit that produces heat by utilizing electric energy or by burning fuel. SBCCI – MECH – 91

FURNACE, CENTRAL – A self-contained indirect fired or electrically heated appliance designed to supply heated air through ducts to spaces remote from or adjacent to the appliance. SBCCI – MECH – 91

GARAGE, PRIVATE – A building or a portion of a building, not more than 1,000 square feet in area, in which only motor vehicles used by the tenants of the building or buildings on the premises are stored or kept (see UBC Chapter 11). UBC – 91

GARAGE, PRIVATE – A garage for four or less passenger motor vehicles without provision for repairing or servicing such vehicles for profit (see Section 608.0). BOCA – 90

GRADE – A reference plane representing the average of finished ground level adjoining the building at all exterior walls. When the finished ground level slopes away from the exterior walls, the reference plane must be established by the lowest points within the area between the building and the lot line or between the building and a point six feet from the building, whichever is closer to the building. SBCCI – 91

HABITABLE SPACE – A space in a structure for living, sleeping, eating, or cooking. Bathrooms, toilet compartments, closets, halls, storage or utility spaces, and similar areas are not considered habitable spaces. BOCA – 90; SBCCI – 91

HABITABLE SPACE (ROOM) – Space in a structure for living, sleeping, eating, or cooking. Bathrooms, toilet compartments, closets, halls, storage or utility spaces, and similar areas, are not considered habitable spaces. UBC – 91; BOCA – 90

HEAT PUMP – An appliance having heating or heating/cooling capability and which uses refrigerants to extract heat from air, liquid, or earth sources. UBC/UMC – 91

HEATING EQUIPMENT – Includes all warm-air furnaces, warm-air heaters, combustion products vents, heating air distribution ducts and fans, all steam and hot water piping, together with all control devices and accessories installed as part of, or in connection with, any environmental heating system or appliance regulated by this code. See UBC/UMC Section 410.

HEATING SYSTEM – A warm-air heating plant consisting of a heat exchanger enclosed in a casing, from which the heated air is distributed through ducts to various rooms and areas. A heating system includes the outside air, return air, and supply air system, and all accessory apparatus and equipment installed in connection therewith. UBC/UMC – 91

HEATING SYSTEM, CENTRAL WARM AIR – A heating system consisting of an air heating appliance from which the heated air is distributed by means of ducts, pipes, or plenums including any accessory apparatus and equipment installed in connection therewith. SBCCI – MECH – 91

LISTED and LISTING – Terms referring to equipment and materials which are shown in a list published by an approved testing agency, qualified and equipped for experimental testing and maintaining an adequate periodic inspection of current productions and which listing states that the material or equipment complies with accepted national standards which are approved or standards which have been evaluated for conformity with approved standards.

MASONRY – That form of construction composed of stone, brick, concrete, gypsum, hollow clay tile, concrete block or tile, glass block or other similar building units or materials, or a combination of these materials laid up by unit and set in mortar. UBC – 91

READILY ACCESSIBLE – Having direct access without the need of removing or moving any panel, door or similar obstruction. BOCA – MECH – 90

READILY ACCESSIBLE – Having direct access without the need of removing any panel, door, or similar covering of the item described, and without requiring the use of portable ladders, chairs, etc. SBCCI – MECH – GAS – 91

SINGLE-FAMILY DWELLING – A single-family dwelling shall mean a building designed to be used as a home by the owner of such building, and must be the only dwelling located on a parcel of ground with the usual accessory buildings. IAPMO – UPC – 91

SMOKE DETECTOR – An approved device that senses visible or invisible particles of combustion. UBC – 91

SMOKE DETECTOR – An approved listed detector sensing either visible or invisible particles of combustion. SBCCI – 91

STAIRWAY – One or more flights of stairs, either exterior or interior, with the necessary landings and platforms connecting them, to form a continuous and uninterrupted passage from one level to another in a building or structure. SBCCI – 91

STRUCTURE – That which is built or constructed, an edifice or building of any kind, or any piece of work artificially built up or composed of parts joined together in some definite manner. UBC – 91

TOWNHOUSE – A single-family dwelling constructed in a series or group of attached units with property lines separating each unit. SBCCI – 91

USABLE CRAWL SPACE – A crawl space designed to be used for equipment or storage. SBCCI – 91

WATER HEATER – An appliance designed primarily to supply hot water and is equipped with automatic controls limiting water temperature to a maximum of 210°F. UBC – 91

The NCCER makes every effort to keep these manuals up-to-date and free of technical errors. We appreciate your help in this process. If you have an idea for improving this manual, or if you find an error, a typographical mistake, or an inaccuracy in the NCCER's Craft Training Manuals, please write us, using this form or a photocopy. Be sure to include the exact module number, page number, a description of the problem, and the correction, if possible. Your input will be brought to the attention of the Technical Review Committee. Thank you for your assistance.

Instructors – If you found that additional materials were necessary in order to teach this module effectively, please let us know so that we may include them in the Equipment/Materials list in the Instructor's Guide.

Write: Curriculum Development and Revision Department
National Center for Construction Education and Research
P.O. Box 141104
Gainesville, FL 32614-1104

Fax: 352-334-0932

Craft _____   Module Name _____

Copyright Date _____   Module Number _____   Page Number(s) _____

Description of Problem

_____

_____

_____

_____

(Optional) Correction of Problem

_____

_____

_____

(Optional) Your Name and Address

_____

_____

_____